2000

INVERTEBRATES AS WEBMASTERS IN ECOSYSTEMS

This is an edited volume honouring the contributions of
Professor D.A. (Dac) Crossley, Jr to the field of ecosystem science.

Invertebrates as Webmasters in Ecosystems

Edited by

D.C. Coleman

and

P.F. Hendrix
Institute of Ecology
University of Georgia
Athens, USA

CABI *Publishing*

CABI *Publishing* is a division of CAB *International*

CABI Publishing
CAB International
Wallingford
Oxon OX10 8DE
UK

CABI Publishing
10 E 40th Street,
Suite 3203
New York, NY 10016
USA

Tel: +44 (0)1491 832111
Fax: +44 (0)1491 833508
Email: cabi@cabi.org

Tel: +1 212 481 7018
Fax: +1 212 686 7993
Email: cabi-nao@cabi.org

A catalogue record for this book is available from the British Library, London, UK

Library of Congress Cataloging-in-Publication Data
Invertebrates as webmasters in ecosystems / edited by D.C.
   Coleman and P.F. Hendrix.
      p.   cm.
   Includes bibliographical references and index.
   ISBN 0–85199–394–X (alk. paper)
      1. Invertebrates––Ecology.   I. Coleman, David C.,
   1938–   .   II. Hendrix, Paul F.
   QL364.4.I58   2000
   592.17––dc21
                                                                    99–41246
                                                                          CIP

ISBN 0 85199 394 X

Typeset in 10/12pt Photina by Columns Design Ltd, Reading
Printed and bound in the UK by Biddles Ltd, Guildford and King's Lynn.

Contents

Contributors

J.M. Anderson, School of Biological Sciences, University of Exeter, Hatherly Laboratories, Prince of Wales Road, Exeter EX4 4PS, UK

V. Behan-Pelletier, Biodiversity Program, Research Branch, Agriculture and Agri-Food Canada, K.W. Neatby Building, Ottawa, Canada, K1A 0C6

J.M. Blair, Division of Biology, Kansas State University, 232 Ackert Hall, Manhattan, KS 66506, USA

T.M. Bolger, Department of Zoology, University College Dublin, Belfield, Dublin 4, Ireland

M.A. Callaham, Jr, Division of Biology, Kansas State University, 201 Bushnell Hall, Manhattan, KS 66506, USA

C.R. Carroll, Institute of Ecology, University of Georgia, Ecology Building, Athens, GA 30602–2202, USA

D.C. Coleman, Institute of Ecology, University of Georgia, Ecology Annex, Athens, GA 30602–2360, USA

R.N. Coulson, Knowledge Engineering Laboratory, Department of Entomology, Texas A&M University, Campus MS 2475, College Station, TX 77843, USA

M.I. Dyer, Institute of Ecology, University of Georgia, Ecology Building, Athens, GA 30602–2202, USA

C.A. Edwards, Soil Ecology Program, Botany and Zoology Building, The Ohio State University, 1735 Neil Avenue, Columbus, OH 43210, USA

P.M. Groffman, Institute of Ecosystem Studies, Box AB, Millbrook, NY 12545, USA

R.A. Hansen, Department of Biological Sciences, University of South Carolina, Coker Life Sciences 402, Columbia, SC 29208, USA

P.F. Hendrix, Institute of Ecology, University of Georgia, Ecology Building, Athens, GA 30602–2202, USA

L.J. Heneghan, Environmental Sciences Program, DePaul University, 2320 North Kenmore Avenue, Chicago, IL 60614-3298, USA

C.A. Hoffman, Institute of Ecology, University of Georgia, Athens, GA 30602–2202, USA

J.J. Hutchens, Jr, Institute of Ecology, University of Georgia, 413 Biological Sciences Building, Athens, GA 30602, USA

C.G. Jones, Institute of Ecosystem Studies, New York Botanical Gardens, Cary Arboretum, Millbrook, NY 12545, USA

J.C. Moore, Department of Biological Sciences, University of Northern Colorado, Greeley, CO 80639, USA

P. Neville, Department of Zoology, University College Dublin, Belfield, Dublin 4, Ireland

P.C. de Ruiter, Department of Environmental Studies, University of Utrecht, PO Box 80115, 3508 TC Utrecht, The Netherlands

J. Rusek, Institute of Soil Biology, Academy of Sciences of Czech Republic, Nasádkách 7, 370 05 České Budějovice, Czech Republic

T.D. Schowalter, Department of Entomology, Oregon State University, 2046 Cordley Hall, Corvallis, OR 97331–2907, USA

T.R. Seastedt, Institute of Arctic and Alpine Research, University of Colorado, Boulder, CO 80309-0450, USA

T.C. Todd, Department of Plant Pathology, Kansas State University, Manhattan, KS 66506, USA

J.B. Wallace, Professor of Entomology and Ecology, 711 Biological Science Building, University of Georgia, Athens, GA 30602, USA

D.E. Walter, Department of Zoology and Entomology, University of Queensland, St Lucia, Queensland 4072, Australia

W.G. Whitford, USDA/ARS Jornada Experimental Range, New Mexico State University, PO Box 30003, MSC 3JER, Las Cruces, NM 88003–8003, USA

D.F. Wunneburger, College of Architecture, Texas A&M University, College Station, TX 77843, USA

Preface

The last frontiers on Earth for the study of biocomplexity and ecosystem function are in the depths of the oceans, the canopies of forests and in soils (Haagvar, 1998). Remoteness and inaccessibility contribute to our lack of knowledge of oceanic and high canopy systems, but it is remarkable that life processes in the very underpinnings of our otherwise familiar terrestrial environment remain veiled. Obstacles to studying organisms and biological processes in soil include the opacity of the medium in which they exist and our inability to observe, collect and measure them without altering their characteristics (Coleman, 1985; Wall and Moore, 1999). Despite scientific limitations (or perhaps because of them), in all three situations there has been no shortage of innovative methods for studying biology along the frontiers or of creative ideas to explain phenomena observed there. The present state of knowledge from this work, as limited as it may be, clearly is that invertebrates are conspicuous, ecologically influential components in all of these systems. This fact and the exciting questions that it raises are the bases for this book, which aims to review and assess our current understanding of invertebrates in terrestrial and terrestrially dominated (i.e. lower-order stream) ecosystems.

In computing terminology, 'A webmaster is one who designs, organizes, and maintains a webpage. The webmaster has a global, not a local perspective. No matter what language is used, the webmaster facilitates access to the web' (Aram Rouhani, personal communication). While not imputing purpose in a human-oriented sense to actions of individual species in ecosystems, we suggest that entire assemblages of invertebrates occupying many hot-spots in soils, such as the rhizosphere and drilosphere, and other portions of both terrestrial and aquatic ecosystems, assume an organizing function and, hence, may be considered as 'webmasters'. This theme emphasizes the centrality of the activities of invertebrates, which influence ecosystem function far out of proportion to their physical mass in a wide range of ecosystems, particularly at the interfaces between land and air (litter/soil), water and land (sediments), and in tree canopies and root/soil systems. The webmasters concept reflects both direct and indirect influences of organismal activities, for example on nutrient dynamics in entire watersheds, and is thus qualitatively different from the keystone species concept, which relates to impacts of a particular species on other species and communities in a given habitat. The webmasters concept spans scales ranging from microsites to aspects of global climate change.

We have invited 17 internationally renowned researchers from five countries to participate in this focused, edited volume. Their papers cover a variety of current topics on invertebrate ecology, including spatial scale and nature of invertebrate controls on ecosystem processes; linkages between population/community processes and ecosystem function; food web organization and energetics; feeding strategies, trophic interactions and feedbacks; adaptations to specialized habitats; explicit and implicit models of invertebrate functions; responses and adaptations to disturbance; exotic species invasion; biodiversity; and climate change. As varied as the subject matter is, we have organized the papers into groupings on the basis of: (i) ecosystem function; (ii) feedback interactions; (iii) ecosystem diversity; and (iv) regional and global scales, in all cases noting the common thread of reverberating effects of the array of invertebrates studied.

Finally, we note that Professor Crossley has been among the leaders, worldwide, in recognizing and advocating the importance of invertebrates in ecosystem function. He has taken a broad viewpoint, noting that it is instructive to look for emergent properties in virtually any ecosystems being studied. Dac pioneered the use of isotopic tracers, particularly radioisotopes, in ecosystem studies, writing some veritable citation classics with colleagues at Oak Ridge National Laboratory in the early 1960s. The work by Dac and his colleagues, including two generations of scores of graduate students and postdocs at the Institute of Ecology, University of Georgia, has helped to elucidate the weblike nature of ecological interactions in ecosystems. We have been privileged to work with him on several research projects, mostly funded by the National Science Foundation. This work includes two long-term studies: the LTER project at the Coweeta Hydrologic Laboratory, of which Dac was lead Principal Investigator (PI) for 10 years; and the Horseshoe Bend Agroecosystem Project, on which Dac has been PI or Co-PI for 22 years of continuous funding. It has become clear that many of the responses to change of ecosystem function take years to decades to develop, thus long-term research is of importance in ecosystem studies. Dac's influence on this research and those who conduct it will be apparent for many decades to come.

We are confident that *Invertebrates as Webmasters in Ecosystems* will appeal to a wide range of ecologists in basic and applied aspects of ecology internationally. We hope that you, the reader, will find the chapters as enjoyable to read as we have found to invite, review, edit and publish in a volume devoted to this topic.

Support for the symposium and this volume which resulted from it, came from the office of the Vice President for Research, the Franklin College of Arts and Sciences, and the Institute of Ecology, all at the University of Georgia, and we thank them all for their generosity.

Paul F. Hendrix and David C. Coleman
Institute of Ecology,
University of Georgia,
Athens, USA

References

Coleman, D.C. (1985) Through a ped darkly – an ecological assessment of root– soil– microbial–faunal interactions. In: Fitter, A.H., Atkinson, D., Read, D.J. and Usher, M.B. (eds) *Ecological Interactions in the Soil: Plants, Microbes and Animals*. British Ecological Society Special Publication 4. Blackwells, Oxford, pp. 1–21.

Haagvar, S. (1998) The relevance of the Rio-Convention on biodiversity to conserving the biodiversity of soils. *Applied Soil Ecology* 9, 1–7.

Wall, D.H. and Moore, J.C. (1999) Interactions underground. Soil biodiversity, mutualism, and ecosystem processes. *BioScience* 49, 109–117.

Webmaster Functions in Ecosystems

Food Web Functioning and Ecosystem Processes: Problems and Perceptions of Scaling

J.M. Anderson

School of Biological Sciences, University of Exeter, Exeter EX4 4PS, UK

Introduction

The assumption that all species and their interactions in communities are important for ecosystem functioning (e.g. Ehrlich, 1993) may owe more to concern over rates of extinction and imperatives for conserving biodiversity than to empirical evidence for the functional importance of food web interactions at system level. Despite considerable research into the relationships between community and systems ecology (see Verhoef and Brussaard, 1990; Anderson, 1993; Coleman *et al.*, 1994) there is still poor understanding of how to link organism activities, population processes and community dynamics to ecosystem processes, and little general theory which relates ecosystem properties to the activities, dynamics and assemblages of species (Lawton and Brown, 1993; Johnson *et al.*, 1996). In trying to link food webs and system functioning we need to address whether this apparent hiatus in understanding is a question of the experimental approaches to investigating process controls and system functioning, or is there a hierarchy of process controls which override the specifics of food web interactions at higher levels of organization? Both issues involve questions of the scales at which processes are measured and systems are defined.

Studies of plant biodiversity in grasslands, for example, suggest that primary production is less variable in species-rich than species-poor swards, especially after drought stress (Dodd *et al.*, 1994; Tilman and Downing, 1994). Similarly, the variability of community respiration in aquatic communities has been shown to decrease with increasing complexity of species assemblages (McGrady-Steed *et al.*, 1997). Hence, the contribution of species to fluxes is progressively masked as measurements are made over large areas containing more species, or by adding more species per unit area. Conversely, the smaller the scale at which measurements are made, the fewer species will be found in a functional group and the more their particular functional characteristics are apparent. The scale at which measurements are made in relation to the size of the organisms therefore predetermines perceived relationships between biodiversity and ecosystem function. Kareiva and Anderson (1988) showed that for nearly 100 experiments in community ecology, half were carried out in plots of a metre or less despite considerable differences in the sizes of the organisms concerned.

At the other extreme, catchment studies, involving measurements of mass fluxes over large areas and long time-periods, show that similar types of ecosystems such as mixed temperate forests, monospecific stands and plantations, exhibit similar functional properties in terms of input–output balances of water, nutrients and carbon (Bormann and Likens, 1979; Swank, 1986). At this scale, system type and functioning is primarily related to climate and the composition of parent material. Few functional attributes reflect the composition of animal and plant communities except where genera, such as *Rhizobium*, certain termites and outbreaks of insect herbivore species dominate these functional groups (Anderson, 1993). Within this envelope of physical controls, carbon, nutrient, water and sediment retention and losses are determined by the integrity of rooting systems coupling plants and soil, structural properties of vegetation cover protecting the soil, the size of soil organic matter (SOM) pools and geochemical processes buffering perturbations to the system. Detecting attributes of plant community structure requires scaling down to tree gaps, smaller patch scales in grassland communities and finer scales still for smaller organisms. Much of our understanding of the mechanisms regulating ecosystem processes has been gained at the organism scale in small systems. But in order to develop models to predict the behaviour of large, coarse-scale systems, specific detail may be aggregated to the point where the community characteristics of biological controls are generally lost. We know that the biological processes are proximate determinants of decomposition rates but community variables, other than microbial biomass, are not required to simulate observed trends in carbon pools in grassland, forest and agricultural systems at field scales. For example, there is extensive information on the roles of invertebrates in organic matter decomposition and soil organic matter turnover, but there are no biologically explicit variables in the CENTURY model (Parton *et al.*, 1987) of soil organic matter dynamics which has been validated at field scale. The problems of retaining detail and resolution of plant processes in system models while avoiding the errors of aggregating variation in a large number of components are considered in detail by Rastetter *et al.* (1992).

The questions addressed here are, therefore, what determines the scales at which species and food web effects are apparent as process controls, and how can we link the often disparate scales at which organisms operate and system properties are measured?

Ecosystem Functioning: Seeing the Wood for the Trees

Ecosystems and processes have no inherent dimensions. The units we arbitrarily impose in space and time can determine the perceived linkages between species and function. Increasing the dimensions of process measurements from mg m^{-2} to kg ha^{-1} does not convert a micro-plot into an ecosystem.

Ecosystems have been defined as the total community of organisms in a given space interacting with their physical environment (Tansley, 1935; Odum, 1968) and can therefore be considered as a functional unit at any scale from a soil crumb to the biosphere (Coleman *et al.*, 1998). Catchments (watersheds) are one of the most useful operational ecosystem units (Bormann and Likens, 1979; Swank and Crossley, 1988) and have provided much insight into the biogeochemical controls over system functioning and responses to management and climatic perturbations. This system approach led to the definition of an ecosystem during the International Biological Programme (IBP) as 'a unit containing plant, herbivore and decomposer sub-systems integrated by nutrient cycles which have larger internal fluxes than across the boundaries' (Swift *et al.*, 1979). In fact, these community and system concepts of ecosystem functioning can involve fundamentally different perspectives of process controls by community and system ecologists.

Population and community ecologists consider organisms to be proximate factors regulating processes with rates determined by population sizes, resource availability and physical environmental constraints. Understanding population dynamics and species interactions requires sampling at spatial and temporal scales appropriate to the ecology of the species. Process measurements (e.g. CO_2 production, ammonification, decomposition) made at the same scale as these activities predictably reflect the particular ecological characteristics of the species and their interactions. However, the functioning of ecosystems as landscape units also involves abiotic buffering of energy, mineral, sediment and water fluxes by biophysical components of the system such as micro-relief, woody tissues, soil organic matter, ion exchange capacity and weathering minerals. Scaling up community processes from moss tussock and microcosms to plot and landscape scales requires the integration of these biophysical parameters which may not increase along the same dimensions as community structure. For example, increasing horizontal measurements by a metre might not alter the species composition of plant and animal communities and the functioning of food webs. But increasing measurements to 1 m depth involves large organic matter and mineral pools with different dynamics which largely determine how the soil subsystem functions. Hence, it is generally easier to identify the importance of community interactions as process controls, relative to biophysical parameters, when scaling down than when scaling up.

Legacies of the Past and Patterns of the Present

Current distributions and activities of organisms, and their effects on processes, are often determined by antecedent events. The bulk of soil organic matter (SOM), and the associated biota, changes at slower rates than vegetation cover. Land use and management often involves stepwise changes in vegetation cover which can, therefore, result in anomalous associations of plants, soil organisms and soil properties. Reforestation of old grassland with deciduous or coniferous trees results in different patterns of soil organic matter distribution and dynamics. The use of lignin derivatives as biochemical tracers enabled Sanger *et al.* (1997) to show that over a period of 45 years, Norway spruce (*Picea abies*) accumulated SOM only in the top 15 cm of the soil compared with material derived from ash (*Fraxinus excelsior*), which was distributed throughout the profile. Hence soil organism communities under spruce were functioning in two components of the system with different dynamics: the surface horizons accumulating nutrients and organic matter, while organic matter pools in the underlying soil, derived from grass, were probably degrading. Interpretation of the effects of invertebrates on C or N mineralization in different soil horizons (e.g. Anderson *et al.*, 1985) could be confounded by these SOM pools having biochemical properties derived from different parent plant materials. Dendooven and Anderson (1995) showed shorter-term antecedent effects of weather events on denitrification in pasture soils. *De novo* induction of microbial reductase enzymes took 18–24 h following the onset of anaerobic conditions in water-logged soils. It was found, however, that nitrate reductase persisted in aerobic soils as a legacy of previous anaerobic events and accounted for the almost instantaneous flushes of N_2O in response to summer rainfall events, causing transitory saturation of soil aggregates. The magnitude and N_2O/N_2 gas ratio of these fluxes could not be explained without understanding what had pre-determined the microbial responses to rainfall events.

Higher-order events also obscure the effects of food web interactions on system processes. Beare *et al.* (1992) showed that soil and residue management practices of conventional and no-till systems resulted in food webs with very different structural characteristics. Under conventional till there were greater flows of C and N via bacteria, protozoa and bacterivorous nematodes, and five times fewer earthworms, than under no-till. There were, however, no significant differences in crop yields between tillage treatments (House *et al.*, 1984) because cultivation, mulching by crop residues and fertilizer applications usually over-ride the effects of the soil biota on soil structure and nutrient turnover. For example, the elimination of earthworms by fungicides from fertilized pasture over 20 years resulted in changes in surface accumulations of grass litter, lower SOM content, increased bulk density and reduced hydraulic conductivity (Clements *et al.*, 1991). Sward production increased, however, because fungal pathogens rather than soil structure and nutrient availability were constraints to plant growth. Conversely, the introduction of earthworms to pastures with low fertilizer inputs, slow organic matter turnover and poor soil structure, can significantly enhance production. While plant production is potentially a good measure of soil fertility, it may be difficult to relate to soil community processes at plot scale because other factors affect nutrient availability; initial effects of soil fauna on seedling populations may be obscured by compensatory plant population processes of tillering or self-thinning; while growth and yield are also affected by herbivory.

It is evident that the interactions of organisms regulating carbon and nutrient fluxes through food webs are difficult to relate to system-level fluxes because biophysical pools and processes often buffer these effects. If, however, the activities of the organisms are synchronized, or affect biophysical parameters, they can reinforce pathways for effects that influence ecosystem processes on larger scales and over longer periods.

Metabolic and Modulating Effects of Organisms

Organisms contribute to ecosystem processes through direct and indirect mechanisms. The direct mechanisms are mainly the consequences of metabolic processes dissipating complex molecules to release energy, carbon, nitrogen and minerals through respiration, excretion and chemotrophy. Indirect mechanisms include many 'top down' feedback effects which have been described as 'ecosystem engineering' (Jones and Lawton, 1995), whereby organisms influence ecosystem functioning as a consequence of their physical presence (trees or coral reefs structuring the system) or catalytic activities (dam building by beavers). In order to avoid confusion over the use of this term it is simpler to define these indirect effects as 'modulating' functions (Anderson, 1995), reflecting the nature of their activities in mediating much larger energy or mass fluxes than their metabolic requirements. An important difference between metabolic and modulating functions is that the former are directly proportional to the density and physiological state of the organisms, while the latter create durable artefacts, such as soil structures or nutrient-rich patches, which have cumulative effects on the activity states of other organisms. Modulating effects of vegetation include habitat structure for other organisms, microclimate, entrainment of sediment, soil stabilization by roots, soil organic matter accumulation and partitioning of incident precipitation and nutrient capture. When herbivore and decomposer communities modify these functions (Table 1.1) their effects may be amplified up the hierarchy of process controls to effect changes at a system level.

Decomposer communities

Soil organisms can be classified into microflora, and micro-, meso- and macrofauna according to thallus or body width (Swift *et al.*, 1979). These sizes broadly reflect the scales, or domains (see below), at which the organisms function and

Table 1.1. Metabolic and modulating processes of herbivores and soil organisms.

Metabolic processes	Modulating processes		
	Herbivores	Soil microorganisms	Soil fauna
Respiration Excretion Chemotrophy	Canopy structure (light, heat, water, nutrients) Leaf/litter quality Plant species balances Soil compaction Trace gases (methane)	Aggregate stability Soil crusting, runoff Symbionts Trace gas production	Aggregate formation Macropores, infiltration Transport processes SOM stabilization Microbial community structure Microsites for trace gases (methane and nitrous oxide)

the extent to which their activities are constrained by, or are able to modify, the structure of soil and litter. Hence protozoa and nematodes are able to feed fairly specifically on bacteria and fungi within the litter and soil matrix but these interactions may be constrained by the pore size of soils (Elliott *et al.*, 1980). Similarly, the extent to which mycophagous mesofauna are able to access active fungal tissues changes as hyphae ramify in plant tissues as decomposition progresses, and the physical structure of the matrix can modify the dynamics of these interactions (Leonard and Anderson, 1991). Large organisms, such as earthworms and millipedes, are rarely able to feed selectively on microbial tissues but conditioning by fungal species can influence food selection and hence feedback to microbial community structure, activities and dispersal (Clegg *et al.*, 1994; Dighton *et al.*, 1997; Maraun *et al.*, 1998).

Small organisms (e.g. bacteria and protozoa) have higher mass-specific metabolic rates than large organisms (earthworms and millipedes). Models of carbon, energy or nutrient flows through food webs therefore tend to be dominated by trophic links between bacteria, fungi, protozoa and nematodes and are relatively insensitive to the composition and activities of meso- and macrofauna populations (de Ruiter *et al.*, 1997). However, as Macfadyen (1963) first pointed out, the saprotrophic fauna may respire only 10–15% of the carbon flux through soils but have much larger indirect (modulating) effects on microbial processes because high consumption rates and low assimilation efficiencies result in changes in the physical and chemical properties of organic matter. In general terms, increasing body size is associated with a shift in the relative importance of metabolic to modulating functions (Table 1.2), which reflects decreasing direct interactions between animals and microorganisms (grazing, predation), a shift to indirect effects on microbial populations (comminution and geophagy), and increasing importance of fauna effects on soil physical processes (infiltration, redox, physical stabilization of soil organic matter). Hence, while bacteria–protozoa links in food chains may constitute most of the metabolic N flux mediated by invertebrates, earthworms can modulate comparable N fluxes by optimizing conditions for denitrification (Fig. 1.1) or enhancing microbial decomposition rates as a consequence of bioturbation.

Meso- and macrofauna are involved in the active and passive transport of organic matter into locations where decomposition is determined by different combinations of biotic and physical variables in the soil surface including physical stabilization in mineral soils (Swift *et al.*, 1979). This 'cascade process' of litter decomposition, with different organisms acting as 'gates' for physical transport processes in soils, is closely analogous to the sequential roles of invertebrate guilds processing organic matter in stream systems (Anderson, 1987; Wagener *et al.*, 1998). But, because the processes operate over small distances in soils, the passive roles of comminuters in structuring soils are less recognized than active transfers by anecic lumbricid earthworms and mound-building termites (Anderson, 1988). Unlike stream systems, soil animal and microbial activities also have reciprocal effects on hydrologic fluxes through the formation of surface crusts and macropores which affect infiltration rates and

Table 1.2. Functional groups of soil fauna (see text for details).

Group:	Microflora	Microfauna	Mesofauna	Macrofauna
Size (ø):		<100–200 µm	<200 mm	2–20 mm
Functions:	Catabolism →	Predation →	Grazing → Comminution →	Bioturbation

Metabolic: ←——

Increasing contribution to community metabolism
Increasing indirect effects on microbial and physical processes

Modulating: ——→

hydraulic conductivity. Because the organisms are creating relatively persistent structures which can determine the direction of water fluxes and modulate the high kinetic energy of precipitation and runoff, their effects can influence vegetation dynamics at landscape scales (Elkins *et al.*, 1986; Tongway *et al.*, 1989).

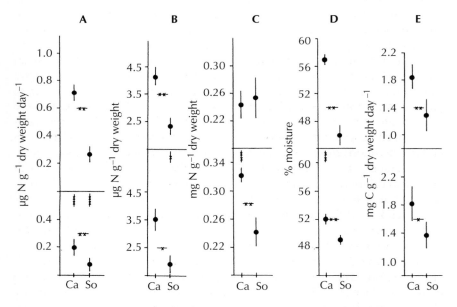

Fig. 1.1. Modulating effects of earthworms on denitrification in pasture soils. A: Mean rates of N_2O production during an autumn (fall) period are shown for casts (Ca) and adjacent mini-cores of soil (So) with concentrations of nitrate (B), ammonium (C), moisture (D) and available carbon measured as cumulative CO_2 production (E). The upper figures are for a pasture receiving 200 kg N ha^{-1} as inorganic fertilizer and the lower figures for unfertilized pastures. The enhanced rates of denitrification in the fertilized plot are a consequence of higher stocking levels increasing inputs of high quality dung into the soil and not direct losses of fertilizer-N. The vertical or horizontal bars marked with crosses indicate significant differences within and between treatments. After Elliott *et al.* (1991).

Herbivores

The magnitude of the contributions by grazing animals to community metabolism are comparable to those of fauna in decomposer communities (Phillipson, 1973). Similarly, their major effects on system functioning are manifested through the modulation of much larger fluxes as a consequence of the effects of structural effects on the system such as beaver dams, grazing, trampling, mass transfers of green plant material to soils, changing plant community structure, decomposition rates and nutrient cycling. The subject area has been comprehensively reviewed elsewhere (Milchunas and Lauenroth, 1993; Jeffries *et al.*, 1994; Dyer *et al.*, 1998) and only two aspects of grazing will be considered here. Firstly, the effects of frequency and intensity of feeding events on plant communities (Table 1.3) and, secondly, the impact of insect outbreaks on the resistance and resilience of ecosystem processes.

The physiological responses of plants to grazing, as with invertebrate–microbial interactions, depend on the magnitude of tissue damage in relation to their regenerative capacity. In grasslands, plant communities often show non-linear responses to increasing grazing intensity, with production (and biodiversity) enhanced up to a threshold, 'break even' point beyond which plants are unable to compensate for tissue removal (Hilbert *et al.*, 1981). A function describing the optimized level of herbivory has been derived using four variables: mean shoot relative growth rate, change in relative growth rate after grazing, grazing intensity and recovery time after grazing (Dyer *et al.*, 1998). Woody plants tend to show different patterns of responses to vertebrate browsing and insect herbivores which range from tolerance, through the induction of chemical defences and regeneration responses following major defoliation events (Haukioja *et al.*, 1983; Lambers, 1993; Jeffries *et al.*, 1994). The frequency and intensity of these events, in relation to damage to meristems and apical dominance of shoots (Danell *et al.*, 1997), act cumulatively to exceed the compensation point of plant responses and result in changes in plant or community function. At the finest scale, intensive feeding by fungivorous nematodes on the hyphae of vesicular arbuscular mycorrhizae impaired phosphorus uptake, N fixation and growth of soybeans (Table 1.2). Recent reports that the coexistence of plant species in complex assemblages can be determined by the diversity of arbuscular mycorrhizae (van der Heijden *et al.*, 1998), suggests a potential mechanism for community changes similar to the effects of root-feeding invertebrates, showing that aboveground and belowground herbivory could respectively accelerate or retard plant succession. In moorlands the balance between grass and heather is maintained when heather maintains a dominant canopy cover and the vigour of grass growth is constrained by low N availability. Severe defoliation by insects (or overgrazing by domestic stock) can open up the heather canopy, reduce competitiveness with grass and increase rates of organic matter turnover (Table 1.3). Similarly, occasional but intensive browsing by moose can result in the replacement of aspen by less palatable spruce trees producing low quality litter, increasing SOM accumulation and decreasing rates of nutrient cycling.

Table 1.3. Responses of vegetation to herbivory. The compensation capacity of plants to tissue removal, and the frequency and intensity of feeding events, determines whether the plant community accommodates or changes in response to herbivory (the stress and scale parameters were not specified in the studies cited).

System	Dominant species/type	Plant responses	Stressor	Stress intensity	Scale (nominal)	Change in system	Reference
Boreal	Spruce Aspen	Slow Slow	Moose on aspen	Infrequent severe	$10 \text{ m}^2 \text{ year}^{-1}$	Increased spruce Increased SOM	Pastor et al. (1993)
Moors	Heather Grass	Slow Fast	Heather beetle	Infrequent severe	$1 \text{ m}^2 \text{ year}^{-1}$	More grass More SOM	Scandrett and Gimmingham (1991)
Meadow	Grass Forbs	Medium Medium	Invertebrates Above ground Below ground	Frequent moderate Frequent moderate	$10 \text{ cm}^2 \text{ day}^{-1}$ $10 \text{ cm}^2 \text{ day}^{-1}$	Accelerates succession Retards succession	Brown and Gange (1992)
Pot experiment	Soybean Mycorrhiza	Medium Fast	Nematodes	Frequent Severe	mm h^{-1}	Reduced N fixation and soybean growth	Salawu and Estey (1979)

These examples suggest that the stresses from herbivory can build up to threshold values through different patterns of events varying in frequency and intensity. The level of this threshold, and the magnitude of change, depends on different properties of resistance (inertia to change) and resilience (rate of recovery) which are characteristic of all ecosystems. When the stress threshold exceeds these homeostatic mechanisms the system may degrade or shift into new equilibrium states. For example, consumption of 5–10% of leaf production in forests by herbivorous insects constitutes a minor nitrogen flux that has an insignificant effect on system functioning. In contrast, Zlotin and Khodashova (1980) have detailed the impact of severe, periodic outbreaks of *Tortrix viridiana* which completely defoliate oak (*Quercus robur*) forests in spring over extensive areas in the Urals of Eastern Europe. During a period of 6–8 weeks it was shown that the reduction in the canopy cover allowed greater light and heat transfer to the forest floor than under the intact tree canopies and increased surface temperatures by up to 3.5°C. Grass production was stimulated by the higher light intensity and temperature, and increased nitrogen supply from the rapid decomposition of green leaf fragments, frass and insect remains. In July grass biomass was reduced by canopy closure and by grazing animals moving into the defoliated areas to feed on the high quality sward. Leaf area index increased as a consequence of adventitious growth in the outbreak areas, and higher N concentrations in leaves resulted in later leaf fall than in non-defoliated areas. As a consequence of these compensatory processes there were no differences in tree-ring increments between defoliated and non-defoliated forests. Given the high exchange capacity and organic matter content of chernozem soils, there were also likely to have been insignificant nutrient losses from the system.

In contrast, localized outbreaks of defoliating insects in mixed deciduous forest in the Coweeta catchment (North Carolina) resulted in increased N leaching from the system (Swank, 1986). Apparently, the ground flora, soil organic matter and mineral exchange pools in this system did not have the capacity to retain the mobilized N which was rapidly leached below tree rooting depth as a consequence of the high hydraulic conductivity of these coarse-textured soils. In this case it is likely that the N concentrations in stream water caused by herbivory would increase as a function of the area extent and intensity of canopy defoliation. Hence, the heterogeneity of the tree species mosaic influences the extent of such outbreaks and the localization of effects on the associated forest floor and soil communities.

Patches in Nests and Nested Patches

Plants (Campbell *et al.*, 1991) and animals (Kotilar and Wiens, 1990) have strategies for exploiting resources which are heterogeneously distributed. These activities create patches of different dynamic states that dissipate or aggregate energy and mass. The net balance of these sources and sinks determines whether fluxes emerge at higher levels of organization (Rastetter *et al.*, 1992) and determine the overall dynamics of the system (Addiscott, 1995). Figure 1.2

illustrates interactions between organisms of different sizes through the creation of biological and biophysical sources and sinks at different scales.

Dominance and domains

The habitat space influenced by an organism can be referred to as its domain (Anderson, 1995). For static organisms the domain is the immediate zone

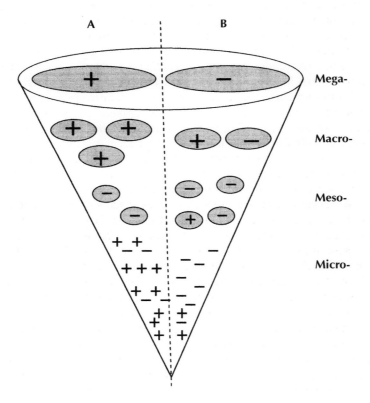

Fig. 1.2. Interaction of functional domains at different scales acting as sources (+) or sinks (−) for nitrogen. Scenario A illustrates the cascade effects of an event such as litterfall (mega-scale) activating soil fauna, which could inhibit mesofauna activities as a consequence of their feeding activities, but may activate microfauna which mobilize nutrients from the microflora. The net effect of the cascade is N mobilization at an earlier phase in decomposition processes than with microbes alone. Scenario B illustrates the situation where there is extensive microbial immobilization of N in fresh plant litter that is not overcome by mesofauna activities which, at low densities, can stimulate microbial growth. Here macrofauna have no net effect on N mineralization and the plant (mega-scale) is temporarily N-limited.

around the organism ranging from a few microns for bacteria to many square metres under the canopy of a tree. The species domain is determined by the density and distribution pattern of individuals. For mobile organisms the domain constitutes the home range and may be contiguous or overlapping for individuals. These patterns may be highly dynamic in the case of animals in comparison with the dispersal mechanisms of plants but both create patches that create a cascade of interactive effects on smaller organisms. Conversely, the activities of small organisms can feed back to influence the domain of larger organisms. A few examples will illustrate these points.

We have already considered how the distribution of tree species influences the type and distribution of soil organic matter, and the localization of insect outbreaks. Mast fruiting by trees also focuses the foraging activity of pigs and other vertebrates under the canopy umbrella. Disturbance to soils, fertilization with dung, and seed and seedling predation may affect tree succession through Markovian processes, resulting in changes in soil organic matter pools and associated communities. Similar patch enrichment by animal activities has been shown where nutrient accessions to rook colonies in woodlands during the breeding season far exceed atmospheric inputs (Wier, 1969), and where islands of acacias in areas of broad-leaved savannas in South Africa mark ancient (Iron Age) settlements where the soil is enriched with P (Woomer and Swift, 1994, p. 236). The same phenomenon operates in the formation of nutrient-rich patches by soil fauna and their exploitation by plant roots (Hodge *et al.*, 1998). Tree canopy architecture also determines spatial patterns of throughfall (Whelan and Anderson, 1996); N and S deposition (Whelan *et al.*,1998); underlying patterns of microbial community structure (Wilkinson and Anderson, unpublished observations); decomposition (Stöckli, 1991); the activity of nitrifying bacteria, ionic balances and aluminium leaching (Berg *et al.*, 1997); and denitrification (Willison and Anderson, 1991). A similar cascade can be constructed for localized areas of earthworm activity produced by inputs of high or low quality litter under clover or grass, trees of different species (Van Hoof, 1983) or crop residues (Tian *et al.*, 1993). As a consequence of different litter types and earthworm activities (and other macrofauna) there are differences in levels of bioturbation (Van Hoof, 1983), and reciprocal effects on microarthropods, nematodes and microbial communities analogous to the tillage effects on food web structure described by Beare *et al.* (1992).

Sinks and sources

The dynamics of aggregative and dissipative processes are a characteristic of all ecosystems involving redistribution of materials within the system and exchanges between systems (Addiscott, 1995; Coleman *et al.*, 1998). The net balance between sources and sinks operating at all scales determines whether the dynamics of the system are aggrading, degrading or steady state (Fig. 1.2).

Undisturbed forests have approximately steady-state carbon balances (production/respiration $\cong$ 1) but contain a mosaic of degrading and aggrading patches of about 0.1 ha where biomass fluctuates between 0 and 400 t ha^{-1} several times within a few hundred years (Shugart and West, 1981). Following commercial logging, massive sediment export from catchments reflects extensive areas of soil erosion but not significant redistribution of nutrient-rich fines within the system which can influence vegetation recovery (Anderson and Spencer, 1991). Within small areas of mixed woodland, Van Hoof (1983) showed that the removal of ash litter by earthworms under ash trees increased splash erosion and surface wash from exposed soil but patches of less palatable beech leaves intercepted water and sediment resulting in no net sediment losses from the system. Similarly, at a finer scale, earthworm casts and burrows form sources and sinks for suspended sediment in close proximity (Syers and Springett, 1983). Because these structures are durable, events such as litterfall, dung deposition or rain after drought can synchronize burrowing, casting and litter removal by individual earthworms so that the sum of these sink/source processes in the species domain results in net surface water infiltration or runoff over larger areas.

Nitrogen dynamics illustrate how the balance of microsite processes are linked to plot-scale responses (Fig. 1.2). Net mineralization occurs when gross N mobilization exceeds immobilization. The shift in the balance from microbial N immobilization to mineralization occurs over time in decomposing litter, or between microsites in soils, and can be changed by soil fauna feeding activities. These continue when there are fresh inputs of plant residues with a high C/N ratio, such as straw, but the N immobilization capacity by microorganisms decomposing the resource may exceed exogenous sources of mineral N and result in N immobilization at plot scale. Conversely, when the litter cohort moves into the mineralization phase the same level of faunal activity may significantly enhance N mobilization from microbial pools. Plant roots ramify the soil matrix with a rooting density which is a function of the species and rates of N mineralization (Bowen, 1983). In organic soils with slow rates of ammonium-N mineralization at the microsite scale, the domains of microbial and invertebrate activities are innervated by the extramatrical hyphae of mycorrhizae which are more persistent than root hairs. In mineral soils, nitrate-N is not only associated with higher flux rates but its mobility requires a relatively low density of fine roots which occupy relatively protected sites in the soil matrix. The architecture of roots, both at microsites and at the level of plant domains, is indicative of the scales at which plants are integrating animal, microbial and biophysical processes regulating nutrient and water availability.

Patterns of Species and Processes in Space and Time

A major problem of linking community structure and soil processes is that the two components have different relationships to increasing dimensions in space or time (Fig. 1.3). Species number increases as a log-linear function of sample number with increments where communities of small organisms are determined by larger vegetation patterns (or animal activities). Flux measurements, however, are the integrated product of sinks and sources produced by both biological and biophysical processes. As the scale of process measurements increases, the effects of smaller organisms are subsumed under patterns determined by the distribution and activities of larger organisms, soil variables and climate controls. Two approaches to analysing these relationships are to quantify the covariance between species assemblages and process rates within these spatial patterns and, secondly, to use measures of the cumulative, modulating effects of communities on the biophysical complexity of soils which are determinants of system functioning.

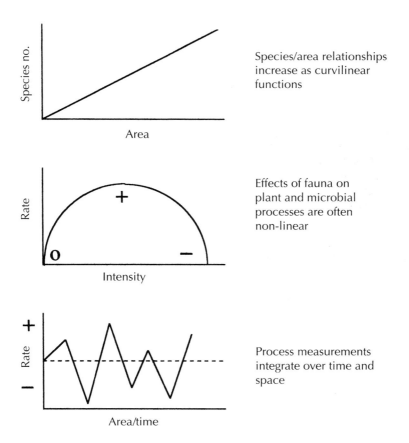

Fig. 1.3. Issues involved in linking community structure and ecosystem processes over increasing spatial scales.

Spatial heterogeneity

The effects of food web interactions within species domains are optimized in patches of different sizes which are heterogeneously distributed, but these aggregations and activities are often related to higher-order patterns and events. Geostatistics provides a means for both defining spatial autocorrelation of variables and for using the degree and scale of autocorrelation to interpolate distributions of species, or species assemblages, across an area (Robertson, 1994). Jackson and Caldwell (1993) carried out one of the first ecological studies using this approach to map the heterogeneous patterns of nutrient distribution associated with desert plants. Concentrations of ammonium, nitrate, phosphorus and potassium showed that the patch size established by the plant domain was generally less than a metre. Hence analysis of plant–soil interactions needs to be investigated below this scale, and scaled up by integrating these patches into larger spatial (and temporal) scales rather uniformly across plots (as implicitly occurs when measurements made at fine scales are converted into larger units; e.g. from $g\ m^{-2}$ to $kg\ ha^{-1}$). While this may appear intuitively obvious in the case of extremely patchy vegetation patterns, the principles apply at all scales where there are species assemblages affecting plant–soil interactions.

Robertson and Freckman (1995) used a similar approach to investigate the distribution of nematodes in a maize field. Sampling was carried out before planting and showed that bacteriophagous nematodes had a very patchy, large-scale distribution with very high concentrations in localized areas of the plot (possibly related to soil moisture) while plant-feeding nematodes showed no spatial distribution patterns. Although the system was not re-sampled after planting it is likely that the distribution of the plant-feeders, and associated rhizosphere communities, would be imposed by the spacing interval of the maize plants. The centres of activity for bacterial feeders could also be related to variation in plant growth (e.g. by precision harvesting) and soil conditions producing these responses. By using spatial heterogeneity as a source of information on species assemblages and their effects on plant/soil relationships we may gain better understanding of the scales at which species interactions are driving system processes.

Biophysical diversity

The consequence of severe stress and disturbance to biotic communities is generally a reduction in species richness and an increase in the dominance (lower equitability) of physiologically robust species. These relationships are widely used in ecology as diagnostics of system responses and recovery from natural and anthropogenic perturbations. Analogous components and responses can be identified in the components of biophysical diversity (Table 1.4) and can similarly be used to relate 'community structure' to system functioning.

Table 1.4. Attributes of biodiversity and biophysical diversity of soil organic matter (SOM) and soil physical structure.

Parameter	Biodiversity	Biophysical diversity
Component	Functional group, species, genotype	SOM, aggregates, porosity
Richness	Number of species, etc.	Number of fractions or void classes
Evenness	Equitability of species	Equitability of mass or voids
Resilience	Size and physiological tolerance	Chemical/physical stability
Disturbance	Dominance by a few species Loss of species	Dominance of resistant fractions Loss of structure
Recovery	Increased diversity Restoration of functional complexity	Increased diversity Restoration of functional complexity

The resistance and resilience of soils to anthropogenic (pollutants, cultivation) and natural perturbations is mainly determined by the mass of SOM, buffering by exchange pools and soil structure. Operationally, these constituents are often considered as single pools, or large aggregated groups (fast and slow SOM pools; sand, silt and clay fractions; micropores and macropores) but actually contain a very large diversity of components with different functional properties and turnover times. Soil structure, for example, may exhibit more or less steady-state characteristics in undisturbed systems for pore and aggregate size distributions (e.g. Azooz *et al.*, 1996) but these are highly dynamic entities with life spans determined by soil animal activities, plant root dynamics and the slaking of soil pores. When soils are disturbed or degraded the diversity of physical or chemical components is reduced and process rates are determined by the characteristics of the dominant components. For example, when compaction eliminates macropores, rates of infiltration and hydraulic conductivity are determined by the properties of meso- and micropores until biotic activity (or tillage) restores structure. Similarly, the loss of SOM components under continuous cultivation affects different attributes of soil structure, water holding capacity, ion exchange and nutrient cycling. The degradation of soil properties resulting from intensive cultivation, the recovery of soil properties under conservation management practices, and the role of the biota in soil restitution, is generally recognized (Logan *et al.*, 1991; Juma, 1994). The quantification of changes in soil biophysical diversity from the condition of natural systems, or undegraded reference plots such as fallows is, therefore, an attribute of the whole soil community function, including plant roots, and potentially a useful measure of soil health.

Conclusions

The dynamics of carbon or nutrient fluxes through food webs reveals the complexities of the mechanisms controlling ecosystem processes. As we scale up to process measurements over larger areas and longer time periods the characteristics of species are generally subsumed into progressively larger functional groups. Consequently their roles as rate determinants are rarely apparent more than one or two orders of magnitude above the scale of their functional domains. The difficulty of linking species and processes appears to be a consequence of three attributes of these systems.

The first is that it is difficult to measure, model or understand the contributions of large numbers of species in a functional group, particularly for modulating processes which have residual effects. As the number of species increases, their contribution to processes becomes progressively harder to detect for more than about five species as a consequence of niche differentiation. Defining natural assemblages comprising higher numbers of species either requires that the operational definition of a functional group is broadened, or that the area of process measurements is increased. Secondly, measurements of decomposition, nitrogen mineralization or gas fluxes are integrating measures of community activity. Mass losses from litter bags, for example, are the cumulative effects of a succession of antecedent processes affecting decomposition, just as the composition and production of a grass sward is affected by the legacy of herbivory. Thirdly, as the dimensions of a system are expanded, processes are increasingly dominated by the biophysical properties of SOM and mineral pools which buffer the effects of organism activities in soil microsites.

Two approaches considered for linking microsite and macroscale processes were to analyse the activities of organisms within functional repetitive units, which may be associated with the spatial patterns of plants, litter or animal activities, or to use changes in soil biophysical diversity as a link between soil functioning and community activity. Until we have resolved these disparities between scaling of species effects and process measurement there will continue to be a disparity between the perceptions of community and system ecologists on the functional importance of food web interactions in ecosystem functioning.

References

Addiscott, T.M. (1995) Entropy and sustainability. *European Journal of Soil Science* 46, 161–168.

Anderson, J.M. (1987) Forest soils as short dry rivers. Effects of invertebrates on transport processes. *Verhandlung Gesellschaft für Ökologie* 17, 33–46.

Anderson, J.M. (1988) Invertebrate-mediated transport processes in soils. *Agriculture, Ecosystems and Environment* 24, 5–19.

Anderson, J.M. (1993) Functional attributes of biodiversity in land use systems. In: Greenland, D.J. and Szabolcs, I. (eds) *Soil Resilience and Sustainable Land Use*. CAB International, Wallingford, pp. 267–290.

Anderson, J.M. (1995) Soil organisms as engineers: microsite modulation of macroscale processes. In: Jones, C.J. and Lawton, J.H. (eds) *Linking Species and Ecosystems.* Chapman and Hall, New York, pp. 94–106.

Anderson, J.M. and Spencer, T. (1991) *Carbon, Nutrient and Water Balances of Tropical Rain Forest Ecosystems Subject to Disturbance.* MAB Digest No. 7, UNESCO, Paris.

Anderson, J.M., Leonard, M.A., Ineson, P. and Huish, S. (1985) Faunal biomass: a key component of a general model of nitrogen mineralization. *Soil Biology and Biochemistry* 17, 735–737.

Azooz, R.H., Arshad, M.A. and Franzluebbers, A.J. (1996) Pore size distribution and hydraulic conductivity affected by tillage in northwestern Canada. *Soil Science Society of America Journal* 60, 1197–1201.

Beare, M.H., Parmelee, R.W., Hendrix, P.F., Cheng, W., Coleman, D.C. and Crossley, D.A. (1992) Microbial and faunal interactions and the effects on litter nitrogen and decomposition in agroecosystems. *Ecological Monographs* 62, 569–591.

Berg, M.P., Verhoef, H.A., Bolger, T., Anderson, J.M., Beese, F., Couteaux, M.M., Ineson, P., McCarthy, F., Palka, L., Raubuch, M., Splatt, P. and Willison, T. (1997) Effects of air pollutant–temperature interactions on mineral-N dynamics and cation leaching in replicate forest soil transplantation experiments. *Biogeochemistry* 39, 295–326.

Bormann, F.H. and Likens, G.E. (1979) *Pattern and Process in a Forested Ecosystem.* Springer-Verlag, New York.

Bowen, G.D. (1983) Tree roots and soil nutrients. In: Bowen, G.D. and Nambiar, E.K.S. (eds) *Nutrition of Plantation Forests.* Academic Press, London, pp. 147–179.

Brown, V.K. and Gange, A.C. (1992) Secondary plant succession: how is it modified by insect herbivory? *Vegetatio* 101, 3–13.

Campbell, B.D., Grime, J.P. and Mackey, J.M.L. (1991) A trade-off between scale and precision in resource foraging. *Oecologia* 87, 532–538.

Clegg, C.D., van Elsas, J.D., Anderson, J.M. and Lappin-Scott, H.M. (1994) Assessment of the role of a terrestrial isopod in the survival of a genetically modified pseudomonad and its detection using polymerase chain reaction. *FEMS Microbiology Ecology* 15, 161–168.

Clements, R.O., Murray, P.J. and Sturdy, R.G. (1991) The impact of 20 years' absence of earthworms and three levels of N fertilizer on a grassland soil environment. *Agriculture, Ecosystems and Environment* 36, 75–85.

Coleman, D.C., Hendrix, P.F., Beare, M.H., Crossley, D.A., Hu, S. and van Vliet, P.C.J. (1994) The impacts of management and biota on nutrient dynamics and soil structure in sub-tropical agroecosystems: impacts on detritus food webs. In: Pankhurst, C.E., Doube, B.E., Gupta, V.V.S.R. and Grace, P.R. (eds) *Soil Biota: Management in Sustainable Farming Systems.* CSIRO, Melbourne, pp. 133–143.

Coleman, D.C., Hendrix, P.F. and Odum, E.P. (1998) Ecosystem health: an overview. In: *Soil Chemistry and Ecosystem Health.* Soil Science Society of America, Madison, Wisconsin, pp. 1–20.

Danell, K., Haukioja, E. and HussDanell, K. (1997) Morphological and chemical responses of mountain birch leaves and shoots to winter browsing along a gradient of plant productivity. *Ecoscience* 4, 296–303.

Dendooven, L. and Anderson, J.M. (1995) Maintenance of denitrification potential in pasture soil following anaerobic events. *Soil Biology and Biochemistry* 27, 1251–1260.

Dighton, J., Jones, H.E., Robinson, C.H. and Beckett, J. (1997) The role of abiotic factors, cultivation practices and soil fauna in the dispersal of genetically modified micro-organisms in soils. *Applied Soil Ecology* 5, 109–131.

Dodd, M.E., Silvertown, J., McConway, K., Potts, J. and Crawley, M. (1994) Stability in the plant communities of the Park Grass Experiment: the relationship between species richness, soil pH and biomass variability. *Philosophical Transactions of the Royal Society London* 346, 185–193.

Dyer, M.I., Turner, C.L. and Seastedt, T.R. (1998) Biotic interactivity between grazers and plants: relationships contributing to atmospheric boundary layer dynamics. *Journal of Atmospheric Sciences* 55, 1247–1259.

Ehrlich, P.R. (1993) Biodiversity and ecosystem function: need we know more? In: Greenland, D.J. and Szabolcs, I. (eds) *Soil Resilience and Sustainable Land Use.* CAB International, Wallingford, pp. vii–xi.

Elkins, N.Z., Sabol, G.V., Ward, T.J. and Whitford, W.G. (1986) The influence of subterranean termites on the hydrological characteristics of a Chiuhuahuan desert ecosystem. *Oecologia* 68, 521–528.

Elliott, E.T., Anderson, R.V. and Coleman, D.C. (1980) Habitable pore space and microbial trophic interactions. *Oikos* 35, 327–335.

Elliott, P.W., Knight, D. and Anderson, J.M. (1991) Variables controlling denitrification from earthworm casts and soil in permanent pastures. *Biology and Fertility of Soils* 11, 24–29.

Haukioja, E., Kapiainen, K., Niemelä, P. and Tuomi, J. (1983) Plant availability hypothesis and other explanations of herbivore cycles. *Oikos* 40, 419–432.

Hilbert, D.W., Swift, D.M., Detling, J.K. and Dyer, M.I. (1981) Relative growth rates and the grazing optimization hypothesis. *Oecologia* 51, 14–18.

Hodge, A., Stewart, J., Robinson, D., Griffiths, B.S. and Fitter, A.H. (1998) Root proliferation, soil fauna and plant nitrogen capture from nutrient-rich patches in soil. *New Phytologist* 139, 479–494.

House, G.J., Stinner, B.R., Crossley, D.A. and Odum, E.P. (1984) Nitrogen cycling in conventional and no-tillage agro-ecosystems. *Journal of Applied Ecology* 21, 991–1012.

Jackson, R.B. and Caldwell, M.M. (1993) The scale of nutrient heterogeneity around individual plants and its quantification with geostatistics. *Ecology* 74, 612–614.

Jeffries, R.L., Klein, D.R. and Shaver, G.R. (1994) Vertebrate herbivores and northern plant communities: reciprocal influences and processes. *Oikos* 71, 193–206.

Johnson, K.H., Vogt, K.A., Clark, H.J., Schmitz, O.J. and Vogt, D.J. (1996) Biodiversity and the productivity of ecosystems. *Trends in Ecology and Evolution* 11, 372–377.

Jones, C.G. and Lawton, J.H. (1995) *Linking Species and Ecosystems.* Chapman and Hall, New York.

Juma, N.G. (1994) A conceptual framework to link carbon and nitrogen cycling to soil structure formation. *Agriculture, Ecosystems and Environment* 51, 257–267.

Kareiva, P.M. and Anderson, M. (1988) Spatial aspects of species interactions: the wedding of models and experiments. In: Hastings, A. (ed.) *Community Ecology.* Springer-Verlag, Berlin, pp. 35–50.

Kotilar, N.B. and Wiens, J.A. (1990) Multiple scales of patchiness and patch structure: a hierarchic framework for the study of heterogeneity. *Oikos* 59, 253–260.

Lambers, H. (1993) Rising CO_2, secondary plant metabolism, plant–herbivore interactions and litter decomposition. *Vegetatio* 104/105, 263–271.

Lawton, J.H. and Brown, V.K. (1993) Redundancy in Ecosystems. In: Greenland, D.J. and Szabolcz, I. (eds) *Soil Resilience and Sustainable Land Use.* CAB International, Wallingford, pp. 255–270.

Leonard, M.A. and Anderson, J.M. (1991) Growth dynamics of Collembola (*Folsomia candida*) and a fungus (*Mucor plumbeus*) in relation to nitrogen availability in spatially simple and complex systems. *Pedobiologia* 35, 163–173.

Logan, T.J., Lal, R. and Dick, W.A. (1991) Tillage systems and soil properties in North America. *Soil and Tillage Research* 20, 241–270.

Macfadyen, A. (1963) The contribution of soil fauna to total soil metabolism. In: Doeksen, J. and van der Drift, J. (eds) *Soil Organisms*. North-Holland, Amsterdam, pp. 3–17.

Maraun, M., Visser, S. and Scheu, S. (1998) Oribatid mites enhance the recovery of the microbial community after a strong disturbance. *Applied Soil Ecology* 9, 175–181.

McGrady-Steed, J., Harris, J. and Morin, P.J. (1997) Biodiversity regulates ecosystem predictability. *Nature* 390, 162–165.

Milchunas, D.G. and Lauenroth, W.K. (1993) Quantitative effects of grazing on vegetation and soils over a global range of environments. *Ecological Monographs* 63, 327–366.

Odum, E.P. (1968) Energy flow in ecosystems: a historical review. *American Zoologist* 8, 11–18.

Parton, W.J., Schimel, D.S., Cole, C.V. and Ojima, D.S. (1987) Analysis of factors controlling soil organic matter levels in Great Plain grasslands. *Soil Science Society of America Journal* 51, 1173–1179.

Pastor, J., Dewey, B., Naiman, R.J., McInnes, P.F. and Cohen, Y. (1993) Moose browsing and soil fertility in the boreal forests of Isle Royale National Park. *Ecology* 74, 467–480.

Phillipson, J. (1973) The biological efficiency of protein production by grazing and other land-based systems. In: Jones, J.G.W. (ed.) *The Biological Efficiency of Protein Production*. Cambridge University Press, Cambridge, pp. 217–235.

Rastetter, E.B., King, A.W., Cosby, B.J., Hornberger, G.M., O'Neill, R.V. and Hobbie, J.E. (1992) Aggregating fine-scale ecological knowledge to model coarser-scale attributes of ecosystems. *Ecological Applications* 2, 55–70.

Robertson, G.P. (1994) The impact of soil and crop management practices on soil spatial heterogeneity. In: Pankhurst, C.E., Doube, B.M., Gupta, V.V.S.R. and Grace, P.R. (eds) *Soil Biota*. CSIRO, Melbourne, pp. 156–161.

Robertson, G.P. and Freckman, D.W. (1995) The spatial distribution of nematode trophic groups across a cultivated ecosystem. *Ecology* 76, 1425–1432.

de Ruiter, P.C., Neutel, A.-M. and Moore, J.C. (1997) Soil food web interactions and modelling. In: Benckiser, G. (ed.) *Fauna in Soil Ecosystems*. Dekker, New York, pp. 363–386.`

Salawu, E.O. and Estey, R.H. (1979) Observations on the relationships between a vesicular-arbuscular fungus, a fungivorous nematode, and the growth of soybeans. *Phytoprotection* 60, 99–102.

Sanger, L.J., Anderson, J.M., Little, D. and Bolger, T. (1997) Phenolic and carbohydrate signatures of organic matter in soils developed under grass and forest plantations following changes in land use. *European Journal of Soil Science* 48, 311–317.

Scandrett, E. and Gimmingham, C.H. (1991) The effect of heather beetle (*Lochmaea suturalis*) on vegetation in a wet heath in NE Scotland. *Holarctic Ecology* 14, 24–30.

Shugart, H.H. and West, D.C. (1981) Long term dynamics of forest ecosystems. *American Science* 66, 647–652.

Stöckli, H. (1991) Influence of stemflow upon the decomposer system in two beech stands. *Revue d'Écologie et de Biologie du Sol* 28, 265–268.

Swank, W.T. (1986) Biological control of solute losses from ecosystems. In: Trudgil, S.T. (ed.) *Solute Processes*. John Wiley & Sons, Chichester, pp. 85–139.

Swank, W.T. and Crossley, D.A. (1988) *Forest Hydrology and Ecology at Coweeta*. Springer-Verlag, New York.

Swift, M.J., Heal, O.W. and Anderson, J.M. (1979) *Decomposition in Terrestrial Ecosystems*. Blackwell Scientific, Oxford.

Syers, J.K. and Springett, J.A. (1983) Earthworm ecology in grassland soils. In: Satchell, J. (ed.) *Earthworm Ecology*. Chapman and Hall, London, pp. 67–83.

Tansley, A.G. (1935) The use and abuse of vegetational concepts and terms. *Ecology* 16, 284–307.

Tian, G., Brussaard, L. and Kang, B.T. (1993) Biological effects of plant residues with contrasting chemical compositions under humid tropical conditions – effect on soil fauna. *Soil Biology and Biochemistry* 25, 731–737.

Tilman, D. and Downing, J.A. (1994) Biodiversity and stability in grasslands. *Nature* 367, 363–365.

Tongway, D.J., Ludwig, J.A. and Whitford, W.G. (1989) Mulga log mounds: fertile patches in the semi-arid woodlands of eastern Australia. *Australian Journal of Ecology* 14, 263–268.

Van der Heijden, M.G., Klironomos, J.N., Ursic, M., Moutoglis, P., Streitwolf-Engel, R., Boller, T., Wiemken, A. and Sanders, I.R. (1998) Mycorrhizal fungal diversity determines plant biodiversity, ecosystem variability and productivity. *Nature* 396, 69–71.

Van Hoof, P. (1983) Earthworm activity as a cause of splash erosion in a Luxembourg forest. *Geoderma* 31, 195–204.

Verhoef, H.A. and Brussaard, L. (1990) Decomposition and nitrogen mineralization in natural and agroecosystems: the contribution of soil animals. *Biogeochemistry* 11, 175–211.

Wagener, S.M., Oswood, M.W. and Schimel, J.P. (1998) Rivers and soils: parallels in carbon and nutrient processing. *BioScience* 48, 104–108.

Weir, J.S. (1969) Importation of nutrients into woodlands by rooks. *Nature* 221, 487–488.

Whelan, M.J. and Anderson, J.M. (1996) Modelling spatial patterns of throughfall and interception losses in a Norway spruce (*Picea abies*) plantation at plot scale. *Journal of Hydrology* 186, 335–354.

Whelan, M.J., Sanger, L.J., Baker, M. and Anderson, J.M. (1998) Spatial patterns of throughfall and mineral ion deposition in a lowland Norway spruce (*Picea abies*) plantation at plot scale. *Atmospheric Environment* 32, 3493–3501.

Willison, T.W. and Anderson, J.M. (1991) Denitrification potentials, controls and spatial patterns in a Norway spruce plantation. *Forest Ecology and Management* 44, 69–76.

Woomer, P.L. and Swift, M.J. (eds) (1994) *The Biological Management of Tropical Soil Fertility*. John Wiley & Sons, Chichester.

Zlotin, R.I. and Khodashova, K.S. (1980) *The Role of Animals in Biological Cycling of Forest-Steppe Ecosystems*. Dowden, Hutchinson and Ross, Stroudsburg, Pennsylvania.

2

Keystone Arthropods as Webmasters in Desert Ecosystems

W.G. Whitford

*US Environmental Protection Agency, Environmental Sciences Division, Las Vegas, Nevada, USA**

There are several arthropod taxa that meet the criteria for keystone taxa in desert ecosystems. Although the keystone concept was based on the effect of predators on the structure of herbivore–detritivore communities (Paine, 1966, 1969), the concept has been broadened to include species or groups of animals that structure animal and/or plant communities, regulate rates of ecosystem processes, structure landscapes or serve as essential links in food webs (Brown and Heske, 1990). In deserts, different groups of arthropods fit into all of these keystone roles. Indeed, it can be argued that most of the structural and functional properties of desert ecosystems are dependent upon a few taxa of arthropods.

Hurlbert (1997) argues that the keystone metaphor should be dropped in favour of functional importance. He provides a mathematical definition of functional importance as the change in production which would occur on the removal of a particular species from the biocoenosis. Keystone taxa affect the productivity of other taxa in numerous direct and indirect ways which are not described by the simple mathematical relationship of functional importance presented by Hurlbert (1997). Here the evidence for keystone roles of two arthropod taxa in desert ecosystems and landscapes is reviewed. The arthropod taxa, for which there is sufficient evidence in the literature to ascribe keystone status, are social insects (termites and ants).

In deserts, some arthropods fill functional roles that are the domain of annelids and other invertebrates in more mesic environments. Arthropods fill many of the important functional roles in deserts because they are less constrained by low water availability and extreme thermal environments than are most other kinds of animals (Whitford, 1991; Heatwole, 1996). Arthropods

* See p. viii for mailing address.

© CAB *International* 2000. *Invertebrates as Webmasters in Ecosystems* (eds D.C. Coleman and P.F. Hendrix) play keystone roles as regulators of ecosystem processes and as engineers creating landscape patterns. Despite the paucity of literature on patch generation in desert landscapes, there are several studies that provide evidence for the keystone role of ants and termites as agents which effect landscape heterogeneity at the different spatial and temporal scales reviewed here. Here I also describe how termites and ants control the spatial and temporal distribution of critical resources such as water and nutrients, thereby affecting essential ecosystem processes such as productivity and nutrient cycling. Because keystone arthropods control the flow of critical resources and modify ecosystem structure, they are webmasters in desert ecosystems.

Pedogenesis

An important component of the development of a soil profile is soil turnover. Ants and termites are important contributors to soil turnover in arid and semi-arid environments. In addition to physical disturbance, ants and termites frequently change the chemical composition of the soil. Rates of soil turnover resulting from the construction of feeding gallery sheeting by subterranean termites are variable because this behaviour varies with recent climatic conditions. The published estimates of soil turnover by subterranean termites are impressive (Table 2.1). Soil turnover by ants can also be very high. Most of the estimates of soil turnover by ants are low because most of the data are for species that build large permanent nests and/or are for single species and not the entire ant community. In northern Chihuahuan Desert grasslands, the quantity of soil moved by ants in the cleaning and construction of ephemeral nests ranged from 21.3 to 85.8 kg ha^{-1} year^{-1} (Whitford *et al.*, 1995). These are conservative estimates based on samples collected in early summer. The soil materials of the nest discs are friable and easily eroded by wind and water. In the Chihuahuan Desert, nests of most ant species are surrounded by fresh spoil material following intense summer rainstorms. A conservative estimate of the annual soil turnover by ants in these ecosystems would be two to three times that of the single early summer estimate. Extrapolating from the conservative averages for annual soil turnover by ants and termites on a Chihuahuan Desert watershed, these insects are transporting the equivalent of 2 cm of soil ha^{-1} in a decade.

These estimates of soil movement are similar to those reported for the ant community in a semiarid shrubland in Australia. Briese (1982) estimated that the ant community moved soil at a rate of 0.03 mm year^{-1} on a clay soil where densities of soil nesting ants are relatively low. It was estimated that funnel ants (*Aphaenogaster barbigula*) brought an average of 336 g m^{-2} year^{-1} of soil profile material to the surface (Table 2.1). This is equivalent to the annual deposition of a layer of soil 0.28 mm thick (Eldridge and Pickard, 1994). Rates of soil transport by ant communities in Western Australia were of the same order of magnitude as those reported by Briese (1982), i.e. 10–37 g m^{-2} year^{-1} (Lobry de Bruyn and Conacher, 1994). Lobry de Bruyn and Conacher found several instances where

Table 1.1. Estimated quantities of soil (kg ha^{-1} year^{-1}) brought to the surface by desert ants and termites during the construction of feeding galleries and nest chambers, and in nest repair.

Ant species	Location	Turnover rate
Ant community	*Atriplex vesicaria* shrub steppe	350–420 (1)
Aphaenogaster barbigula (funnel ants)	*Calitris–Eucalyptus* open woodland	3360 (2)
Ant community	Heath, Western Australia	310 (3)
Ant community	Wandoo woodland, Western Australia	200 (3)
Ant communities	Variety of Chihuahuan Desert shrublands and grasslands	21.3–85.8 (4)
Termite species		
Heterotermes aureus	Sonoran Desert, Arizona,	
Gnathamitermes perplexus	USA	750 (5)
Macrotermes subhyalinus	Senegal	675–950 (6)
Gnathamitermes tubiformans	Chihuahuan Desert, USA	
	Mixed grassland–shrubland	4095 (7)
	Creosotebush shrubland	801
	Black-grama grassland	981
	Watershed	2600

References: (1) Briese (1982), (2) Eldridge and Pickard (1994), (3) Lobry de Bruyn and Conacher (1994), (4) Whitford *et al.* (1995), (5 and 6) Lobry de Bruyn and Conacher (1990), (7) MacKay and Whitford (1988).

the ant spoil had a higher proportion of clay-sized particles in comparison with soil not influenced by ant nest excavation. This was attributed to ants excavating into a clay subsoil or being size selective in the material they excavated.

Macropores and Water

By definition, deserts are water-limited systems. It is axiomatic that organisms that enhance water availability play a keystone role in those systems. Several arthropod taxa affect water infiltration and water storage at spatial scales ranging from the small patch to whole landscape units. The effect of arthropods on spatial and temporal patterns of soil-water availability structures the species composition and productivity patterns of the vegetation.

It is only in the past 20–25 years that soil scientists have rediscovered the importance of soil macropores in the spatial distribution of water in soil and the effects of these pores on water infiltration (Phillips *et al.*, 1989). Macropores are continuous tubes or spaces (voids) in the soil. Macropores transport flowing water to the deeper parts of the soil profile faster than predicted by infiltration models of soils of various particle-size distributions. Since the number of macropores need not be large to affect water infiltration, burrow and pore constructions by invertebrates are important contributors to the availability of water in arid ecosystems.

Most macropores are biological in origin. Several groups of arthropods construct burrows in the soils of arid ecosystems. These include ants, termites, burrowing spiders, cicadas and beetles. Of these, only ants and termites are sufficiently numerous and widespread in deserts to qualify as keystone taxa with respect to water infiltration. In deserts, social insects (termites and ants) dominate the semiarid and arid regions where earthworms are rare or absent.

In Australian arid rangelands, subterranean termites produce numerous small galleries and voids (small hollow chambers) in the surface soil. In a mallee area there was an average of 137.5 voids m^{-2} to a depth of 20 cm (Whitford *et al.*, 1992). Soil voids produced by subterranean termites contribute directly to the high infiltration rates that characterize these soils. Macropores created by termites enhance infiltration in the red earths of the mulga region of eastern Australia, where Greene *et al.* (1990) found that relatively high infiltration rates on unburned plots coincided with the presence of preferential flow paths produced by termites. Greene *et al.* (1990) reported that annual burning resulted in a marked reduction in biological activity and the filling of termite-produced channels by sand and silt.

In central Australia, banded or striped vegetation patterns are characteristic of some landscapes (Tongway and Ludwig, 1990). In these landscapes vegetation is concentrated in stripes which are perpendicular to the slope of the landscape. The large patches of virtually barren ground between stripes are erosion slopes that shed water during intense rains. Runoff water from the erosion slopes is slowed by vegetation and infiltrates within the vegetation bands. Colonies of ants and termites are concentrated within the vegetation bands providing macropores that enhance infiltration and storage of water. There are significantly more gallery tunnels produced by foraging subterranean termites in mulga (*Acacia aneura*) groves than on the erosion slopes (Whitford *et al.*, 1992). Termites and ants also produce macropores in the soils of log mounds formed around dead trees on the erosion slopes. These macropores are responsible for the higher infiltration rates recorded on log mounds in comparison with rates on adjacent unmodified soils (Tongway *et al.*, 1989). The activities of the arthropods also contribute to the higher nutrient levels in the soils of the log mounds. The log mounds produce scattered productive patches on the barren erosion slopes. The productivity of the log mound patches and of the banded mulga groves results from the high rates of water infiltration and storage via macropores produced by termites and ants. This is an example of how termites and ants establish positive feedback loops where increased soil moisture produces more suitable conditions for the insects thereby reinforcing the water–nutrient enrichment of the patch.

Soil-nesting ants produce concentrations of macropores in the form of tunnels near the surface entrance(s) of the nests. Studies of infiltration rates near the centre and at the margins of harvester ant (*Pogonomyrmex rugosus*) nests revealed that both saturated hydraulic conductivity and the differences between saturated and unsaturated conductivity were highest near the nest centres (Herrick and Whitford, 1995). This study illustrates that ants create macropores capable of conducting water during overland flow. Overland flow occurs frequently as a result of intense, short duration rains in deserts. Lobry de Bruyn and Conacher (1994) report a similar finding for a semiarid environment in Western Australia. They found that ant biopores only transmit water down the soil profile when the surface layer of soil is saturated and water is ponding on the soil surface.

In the Chihuahuan Desert, the abundance of soil-nesting ants and subterranean termites decreases dramatically on fine-textured soils (Whitford *et al.*, 1995). The paucity of these insects in fine-textured soils contributes to the virtual absence of infiltration into these soils. In the Chihuahuan Desert banded vegetation develops on fine-textured soils on low-gradient slopes (Montana, 1990). Ant colonies and termite foraging galleries are virtually absent on the erosion slopes of these landscapes but are numerous within the vegetation bands (W.G. Whitford, unpublished data). This pattern of nest distribution of ants and foraging activity of subterranean termites results in concentrations of macropores within the vegetation bands. The macropores enhance infiltration and water storage within the bands thereby producing a positive feedback mechanism that contributes to the dynamics of this landscape.

In studies in the Sahel comparing plots from which termites were excluded with plots with termites present, Mando and Miedema (1997) reported an average of 88 ± 25 large voids m^{-2} (0.8–1.2 cm diameter), but no voids in the termite-excluded plots. They reported that termites accounted for more than 60% of the macropores in this arid region. Mando (1998), in a study of water infiltration on plots with termites excluded, found significant increases ($P < 0.01$) in infiltration on plots with termites (Fig. 2.1). In Mando's study an average of 51.3% of the precipitation infiltrated plots with termites but only 36.3% of the annual precipitation infiltrated plots from which termites were excluded. Field experiments in the northern Chihuahuan Desert, USA, in which subterranean termites were chemically excluded from plots, provided evidence that macropores created by foraging termites were important for water infiltration. Water infiltration was significantly higher on the plots with termites than on plots with termites excluded (Elkins *et al.*, 1986) (Fig. 2.1). This difference in infiltration was directly attributable to the absence of termite-produced macropores in the soils of the termite-excluded plots.

Bulk density (mass of soil per unit volume) is a soil property which is directly related to the total porosity of the soil and, hence, to infiltration and water storage. The effects of ants and termites on soil bulk density were reviewed in detail by Lobry de Bruyn and Conacher (1990). Most of the data on the effects of termites on bulk density were from studies of mound-building termites. In most instances the bulk density of mounds was greater than that of the surrounding soil. There are no quantitative data on the bulk density effects

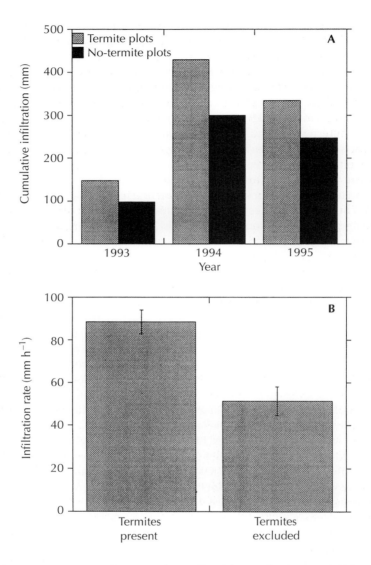

Fig. 2.1. (A) Cumulative infiltration of rainfall on plots with termites and plots with termites excluded in the African Sahel of Burkina Faso. The cumulative infiltration differences among years is the result of variation in rainfall (data from Mando, 1998). (B) Final infiltration rates resulting from rainfall simulation on a northern Chihuahuan Desert watershed on plots with termites and plots with termites excluded. The error bars are ± one standard deviation (data from Elkins *et al.*, 1986).

of subterranean termites. However, the large quantity of foraging gallery sheeting produced by subterranean termites may serve to decrease the bulk density of soil in areas with high concentrations of foraging galleries (MacKay and Whitford, 1988). Bulk densities of ant-nest soils are generally lower than those of surrounding soils (Lobry de Bruyn and Conacher, 1990).

Most of the studies on the effects of ants on water infiltration have been conducted in the Australian arid zone. Eldridge (1993) reported that funnel ants (*Aphaenogaster barbigula*) produced nest entrances at densities of up to 37 m^{-2}. Steady-state water infiltration on soils with entrances averaged 23.3 mm min^{-1} which was approximately four times higher than that on nest-free soils. Infiltration rates on ant nests were much higher ($120.5 \text{ mm min}^{-1}$) than in soil away from ant nests (36.8 mm min^{-1}) (Majer *et al.*, 1987).

Other species of arthropods may also indirectly contribute to water infiltration and storage as a result of their position within food webs. For example, burrowing spiders indirectly contribute to increased water storage in the mulga groves of the banded vegetation landscapes in Australia. Monitor lizards (*Varanus gouldii*) dig the spiders from their burrows leaving excavated pits that average 93 cm in depth. The highest densities of excavated pits were in the interception zone of the mulga groves (77 ha^{-1}). These pits become organic matter- and water-rich germination sites (Whitford, 1999). However, burrowing spiders do not fulfil the criteria for keystone taxa.

Soil Nutrients

In deserts, as in most other kinds of ecosystems, the spatial and temporal patterns of nutrient availability are nearly as important as water as a factor governing vegetation structure and productivity. Nitrogen availability is generally the second most important factor limiting productivity in arid ecosystems (Whitford *et al.*, 1987). Keystone arthropods affect both the temporal and spatial patterns of nitrogen availability in desert soils. The same arthropod taxa that enhance water availability are the most important taxa affecting the spatial and temporal availability of nitrogen.

Termites are abundant in most arid and semiarid regions of the world (Lee and Wood, 1971) and where they are numerous they consume a large fraction of the dead plant material. Plant material that is decomposed by the gut symbionts of termites has a very different fate to that of plant material decomposed by free-living microflora and microfauna. The release of mineral nitrogen from the organic nitrogen of dead plant material and dung is the most important source of soil nitrogen that supports plant growth. In the Chihuahuan Desert, two groups of arthropods, termites and omnivorous prostigmatid mites, are responsible for mineralization of most of the organic nitrogen. Since the organic carbon in desert soils limits the growth of the soil microflora, the way in which these two groups of arthropods affect soil organic carbon has important ramifications for nitrogen availability.

Studies in the Chihuahuan Desert on plots from which termites were excluded but in which the remainder of the soil microflora and microfauna were intact showed that termites consumed a large proportion of the annual standing crop of dead plant parts and animal dung (Table 2.2) (Whitford, 1991). Observations of termite harvesting activity in arid regions of Australia and South Africa suggest that termites consume a large proportion of the dead plant material. Data from Tanzania reported by Jones (1989) provide some additional quantitative evidence for this conclusion.

Because the hindgut microflora of many species of termites are capable of decomposing complex molecules such as hemicellulose and lignin, their faecal material contains almost no recalcitrant carbon compounds. Thus, their excretory products do not contribute to the slow breakdown pool of organic carbon and nitrogen. The breakdown of complex organic molecules in the hindgut of termites is the mechanism responsible for the negative relationship between termite abundance and soil organic matter (Nash and Whitford, 1995). While some of the mineralized nitrogen is returned to the soil as faecal material in gallery cement, most is probably lost as volatiles. The other minerals in plant materials processed by termites are returned to the soil in faecal matter which is either deposited in the soil or used as cement in foraging gallery construction. In the Chihuahuan Desert there is a strong correlation between available nitrogen and soil organic matter (Whitford *et al.*, 1987). In this desert soil, organic matter varies inversely with the abundance and activity of subterranean termites ($r = -0.97$) (Nash and Whitford, 1995). Nash and Whitford (1995) concluded that in the northern Chihuahuan Desert, subterranean termites are responsible for most of the variation in soil organic matter. In a detailed study of soil properties in semi-arid Tanzania, Jones (1989) reported that density of termites and vegetation type accounted for most of the variation in soils. Termites contributed to the depletion of organic C, total N and associated

Table 2.2. Range of consumption of dead plant parts and dung by subterranean termites in the northern Chihuahuan Desert. Consumption is expressed as a percentage of the annual standing crop (from Whitford, 1991).

Material	Percentage consumed
Annual plants, above-ground parts	40–90
Annual plants, dead roots	50–70
Perennial grass, above-ground parts	60–90
Perennial grass, dead roots	50–70
Shrub leaves	0–90[a]
Dead wood	<1–5[b]
Cattle dung	60–100
Rabbit dung	15–50

[a] Variation dependent upon shrub species, landscape position and availability of other forage for termites.
[b] In the Chihuahuan Desert, subterranean termites only scrape the surface material of dead wood that has been softened by fungi.

nutrients throughout the soils of the region. Nash and Whitford (1995) argue that organic carbon is more rapidly and completely metabolized by the hindgut microflora and microfauna of termites than is organic matter processed by free-living microflora and microfauna. The environment of the termite gut is stable and suitable for a mix of microbes that are capable of breaking down most natural organic molecules to carbon dioxide and water. Thus most of the ingested organic matter is released in gaseous or liquid form and little is eliminated as faeces. Nash and Whitford (1995) conclude that the consumption of organic input (detritus) by subterranean termites is the primary determinant of soil organic matter (soil organic carbon) in arid and semiarid environments where termites are dominant soil animals.

The large fraction of net plant production consumed by termites (Table 2.2) suggests that these arthropods are responsible for much of the nutrient cycling in deserts (Schaefer and Whitford, 1981) (Fig. 2.2). In a detailed field study of how subterranean termites (*Gnathamitermes tubiformans*) affected the pools of nitrogen, phosphorus and sulphur in a shrub-dominated community, they found that the total estimated termite population turned over in approximately one-third of a year. Most of the minerals in the considerable fractions of dead plant materials consumed by termites were cycled back into the ecosystem by predators that fed on termites (Fig. 2.2). Other avenues by which nitrogen returned to the system included the gallery sheeting constructed around food items. Schaefer and Whitford concluded that Chihuahuan Desert termites regulate the rates of recycling of nutrients, especially nitrogen and phosphorus (Fig. 2.2). They argue that alate flights, foraging activity and the high mortality rate (close to 100%) of the large number of alates, contribute to the critical timing of release of nutrients since they are coincident with large rainfalls. In deserts, timing of release of nutrients may be as critical as the forms in which the nutrients become available to plants. Nutrients must be available immediately following large rain events in order for rapid plant growth to occur. The emergence and subsequent loss to predation of termite alates coincident with the first intense rains is therefore essential for the continued functioning of these ecosystems by providing a pulse of nutrients at a critical time.

Most studies of the effects of ants on soil nutrients have focused on single locations (Lobry de Bruyn and Conacher, 1990). The published data indicate that some species of ants affect soil chemistry, while other species do not. Ants that transport fruits and seeds, plant parts and dead insects to the nest, concentrate organic matter in the vicinity of that nest. Ant species that have an effect on soil chemistry are characterized by long-lived colonies (> 10 years) and milling seeds or processing plant parts at the nest. Long-lived colonies of *Pogonomyrmex barbatus* and *Pogonomyrmex rugosus* mill seeds within the nest and deposit chaff in midden piles which are usually located within a metre of the nest entrance (Wagner *et al.*, 1997). *P. rugosus* was reported to discard 60% of the total energy content of materials collected by foragers and returned to the nests. Soil nutrients that are concentrated in the soils of long-lived *Pogonomyrmex* spp. include N, P and K (Whitford, 1988; Whitford and DiMarco,

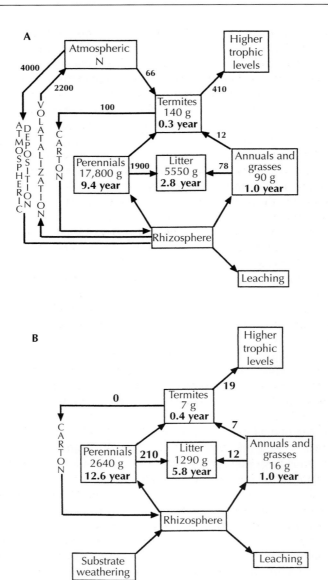

Fig. 2.2. (A) The role of subterranean termites, *Gnathamitermes tubiformans,* in nitrogen cycling in a northern Chihuahuan Desert shrubland ecosystem. Numbers adjacent to arrows are mass flux values in g ha⁻¹. The bold face numbers in the boxes are the estimated turnover time in years for the estimated standing stock of nitrogen in the pool presented as mass in grams. The flux from atmospheric N to termites was estimated from rates of N fixation by termite gut symbionts. (B) The role of subterranean termites, *Gnathamitermes tubiformans,* in phosphorus cycling in a northern Chihuahuan Desert shrubland ecosystem. The fluxes, standing crops and turnover times are presented as in (A) (from Schaefer and Whitford, 1981).

1995; Wagner *et al.*, 1997). In a study of soils and vegetation associated with harvester ant (*Pogonomyrmex rugosus*) nests on a Chihuahuan Desert watershed, Whitford and DiMarco (1995) reported that there were few differences in soil and vegetation associated with the nests in shrub–grass mosaic sites in the basin, but large differences between soil nutrients and vegetation associated with nests and soils at a distance from nests at the midslope locations. It is likely that the relative effects of ants on soils vary with landscape position and the nutrient status of the soils on a particular landscape unit.

Long-lived nests of termites are reported to have effects similar to those of ant nests on soil nutrients. For example, the mounds of mound-nesting termites frequently have nutrients concentrated in the soils adjacent to the base of the mounds similar to the pattern reported for harvester ant nests. The soils of such mounds are modified by the addition of excreta and/or saliva, so that the soils contain higher concentrations of plant nutrients relative to the nutrient content of undisturbed soil. These mound soils are enriched in organic matter, N, P, Ca and Mg (Malaka, 1977; Arshad, 1982; Pomeroy, 1983; Wood *et al.*, 1983). Hence, soil eroded from nest mounds produces a nutrient-rich area around the base of the mound. The relative importance of mound-building termites on the patch distribution of soil nutrients is probably as dependent upon landscape position and nutrient status of the matrix soil as is the case for ant nests.

Although not keystone taxa, omnivorous microarthropods play a critical role in decomposition and nitrogen mineralization of materials not consumed by termites. The primary source of soil nitrogen and soil organic carbon in desert soils is from the decomposition of that fraction of dead roots and litter not eaten by termites. The rates of decomposition and mineralization of this material are regulated by soil mites (Santos and Whitford, 1981; Santos *et al.*, 1981; Parker *et al.*, 1984). Soil mites of the family *Tydeidae* feed on nematodes and fungi. In the early stages of decomposition, tydeid mites regulate the numbers of microbivorous nematodes, which keeps the nematodes from overexploiting the microbial decomposers (Santos *et al.*, 1981). In later stages of decomposition, fungi are the dominant decomposers. Small fungivorous mites, such as tydeids and tarsonemids grazing on the fungi are responsible for most of the nitrogen mineralization and for maintaining the balance between nutrient immobilization by fungal biomass and release of mineral to the soil via their excretory products (Parker *et al.*, 1984).

In deserts where soils dry rapidly, bacteria, protozoans and nematodes that are active only in water films on soil particles are in a cryptobiotic or anhydrobiotic state most of the time (Whitford, 1989) (Fig. 2.3). Whitford (1989) calculated that 50% of the soil protozoan population was encysted at a soil water potential of -0.1 MPa, and that virtually the entire protozoan population was inactive at -0.4 MPa. Freckman *et al.* (1987) found that the soil water potential at which 50% of the population was anhydrobiotic was -0.4 MPa and that 99% of the soil nematodes were anhydrobiotic at water potentials between -3.0 and -5.0 MPa. Desert soils are drier than -6.0 MPa much of the year

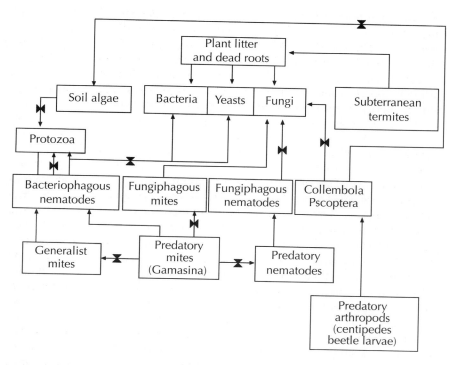

Fig. 2.3. Belowground food web based on data from the Chihuahuan Desert in North America. Web linkages that cease to function at soil water potentials drier than −3.0 MPa are shown as feeding arrows with a black valve. Arrows without a black valve show feeding relationships that persist in soils drier than −3.0 MPa (figure modified from Whitford, 1989).

and under these soil moisture conditions the active soil microflora is reduced to species of fungi that are capable of growth and reproduction in dry soils. In dry soils, species of soil acari are the only active members of the soil microfauna (MacKay *et al.*, 1986, 1987). Therefore, in arid and semiarid environments, a few taxa of soil mites regulate rates of decomposition and nitrogen mineralization. Since nitrogen availability is second only to water as the factor limiting productivity in deserts (Whitford *et al.*, 1987), these taxa of soil mites are regulators of nitrogen availability in North American hot deserts.

There is surprising convergence in the soil microarthropod faunas of arid regions. The soil microarthropod fauna of the Australian arid lands is dominated by many of the same genera of prostigmatid mites that dominate the soils of North American deserts (Noble *et al.*, 1996). Noble *et al.* (1996) reported densities of soil microarthropods ranging from 2600 m^{-2} to 14,000 m^{-2}. The high abundance and taxonomic similarity suggests that small prostigmatid mites have similar roles in decomposition and nitrogen mineralization in the

Australian arid region as in North American deserts. There are very limited data on the soil fauna of other deserts. However, studies in Chile (Cepeda-Pizarro, personal communication) and in southern Africa (Coineau and Massoud, 1977) show that in these deserts the soil acarines are predominantly small prostigmatids. These limited data suggest that the small omnivorous prostigmatid mites may have a role in decomposition and nitrogen mineralization similar to that reported in North American deserts.

Plant Production

The importance of the effects of subterranean termites on soil properties and ecosystem processes has been documented in studies comparing plant production on plots with termites and plots without termites. While growth of the dominant shrub, *Larrea tridentata*, was not significantly affected by the elimination of termites (Gutierrez and Whitford, 1989), biomass production of the annual plants on termite-excluded plots (11.2 g m^{-2}) was less than half that recorded on plots with termites present (24.6 g m^{-2}) (Gutierrez and Whitford, 1987). Not only was overall productivity affected by the presence of termites, there were significant differences in species composition of the annual plant community. The differences in annual plant productivity were directly related to the differences in water infiltration and soil nutrient concentrations on the termite-excluded and termites-present plots (Gutierrez and Whitford, 1989). These keystone attributes of subterranean termites translate directly into productivity differences that Hurlbert (1997) defines as functional importance.

The nests of seed-harvesting ants may affect water and nitrogen availability in soils adjacent to the nest discs. The effects of large harvester ant nests on plant production varies both temporally and spatially (Whitford, 1988; Whitford and DiMarco, 1995). The large plants and high biomass per unit area of *Erodium texanum* reported at the edges of harvester ant nests by Whitford (1988) were not present in the 3 years subsequent to the year of the study. The large plants and high biomass of Indian rice grass, *Orysopsis hymenoides*, associated with *Pogonomyrmex owyheei* nest mounds in unburned Great Basin desert did not occur in an area that had been burned 3 years prior to the study (Nowak *et al.*, 1990). Nowak *et al.* (1990) explained these differences as indirect allelopathy by ants defoliating shrub seedlings that were potential competitors with the grass. However, they did not sample soils and could not address the question of spatial or temporal variability in soil nutrients potentially resulting from the activity of the harvester ants. These studies demonstrate that some species of ants enhance plant production in a temporally and spatially variable manner.

The examples of the effects of termites and ants on the spatial and temporal heterogeneity of soil resources serves to support my contention that these are keystone taxa. Heterogeneity of soil resources directly and indirectly affects the productivity, abundance and species composition of plants. This supports the argument that these taxa are keystones in desert ecosystems. In addition, the experiments in which termites were removed showed reduced plant production as a result of that removal. This fits Hurlbert's (1997) definition of keystone. These examples show that the simple mathematical relationship of functional importance proposed by Hurlbert (1997) is inadequate to describe the many direct and indirect effects of ants and termites on numerous other species in the community and on ecosystem properties and processes. This review shows that Hurlbert's functional importance is not equivalent to the concept of keystone taxa.

Conclusion

Although the data summarized in this review are predominantly for the Chihuahuan Desert in North America and a few desert areas in Australia, it is logical that ants and termites fit the webmaster concept in other deserts. It is certainly clear that these arthropods directly affect the heterogeneity of desert ecosystems by their effects on soil properties. Spatial and temporal heterogeneity are important features affecting the biological diversity and sustainability of ecosystems, especially those ecosystems of extreme desert environments. These keystone arthropods are adapted to life in dry soil, and fluctuations in that environment which exceed the tolerance limits of many other animals. As has been discussed in this review, termites and ants affect virtually all ecosystem processes, i.e. infiltration, runoff, sediment yield, nutrient cycling and plant production.

In some of these processes, termites and ants are the major players and in others they share a major role with another group of arthropods such as omnivorous prostigmatid mites. Not only do these arthropods affect processes on the ecological time-scale, they also affect processes on the geological time-scale through the movement of soil materials to the surface where they are subjected to physical and biological weathering. These arthropods are truly webmasters in desert ecosystems.

Acknowledgement

The US Environmental Protection Agency (EPA) through its Office of Research and Development funded the preparation of this contribution. It has been subjected to the Agency's peer and administrative review and has been approved as an EPA publication. The US Government has a non-exclusive, royalty-free licence in and to any copyright covering this article.

References

Arshad, M.A. (1982) Influence of the termite *Macrotermes michaelseni* (Sjost) on soil fertility and vegetation in a semi-arid savannah ecosystem. *Agro-Ecosystems* 8, 47–58.

Briese, D.T. (1982) The effects of ants on the soil of a semi-arid salt bushland habitat. *Insectes Sociaux* 29, 375–386.

Brown, J.H. and Heske, E.J. (1990) Control of a desert–grassland transition by a keystone rodent guild. *Science* 225, 1705–1707.

Coineau, Y. and Massoud, Z. (1977) Decouverte d'un noveau peuplement psammique de microarthropodes du milieu interstitiel aluen des sables fin. *Comptes Rendus des Seances l'Academie des Sciences (Paris), Series D* 285, 1073–1074.

Eldridge, D.J. (1993) Effect of ants on sandy soils in semi-arid eastern Australia: local distribution of nest entrances and their effect on infiltration of water. *Australian Journal of Soil Research* 31, 509–518.

Eldridge, D.J. and Pickard, J. (1994) Effects of ants on sandy soils in semi-arid eastern Australia: II. Relocation of nest entrances and consequences for bioturbation. *Australian Journal of Soil Research* 32, 323–333.

Elkins, N.Z., Sabol, G.V., Ward, J.J. and Whitford, W.G. (1986) The influence of subterranean termites on the hydrological characteristics of a Chihuahuan Desert ecosystem. *Oecologia* 68, 521–528.

Freckman, D.W., Whitford, W.G. and Steinberger, Y. (1987) Effect of irrigation on nematode population dynamics and activity in desert soils. *Biology and Fertility of Soils* 3, 3–10.

Greene, R.S.B., Chartres, C.J. and Hodgkinson, K.C. (1990) The effects of fire on the soil in degraded semi-arid woodland. I. Cryptogram cover and physical and micro-morphological properties. *Australian Journal of Soil Research* 28, 755–777.

Gutierrez, J.R. and Whitford, W.G. (1987) Chihuahuan Desert annuals: importance of water and nitrogen. *Ecology* 68, 2032–2045.

Gutierrez, J.R. and Whitford, W.G. (1989) Effect of eliminating subterranean termites on the growth of creosotebush, *Larrea tridentata. Southwestern Naturalist* 34, 549–551.

Heatwole, H. (1996) *Adaptations of Desert Organisms.* Springer-Verlag, Berlin.

Herrick, J.E. and Whitford, W.G. (1995) Contributions of ant activity to spatial variability in hydrologic properties in the Chihuahuan Desert. *Agronomy Abstracts* 309.

Hurlbert, S.H. (1997) Functional importance *vs* keystoneness: Reformulating some questions in theoretical biocenology. *Australian Journal of Ecology* 22, 369–382.

Jones, J.A. (1989) Environmental influences on soil chemistry in central semiarid Tanzania. *Soil Science Society of America Journal* 53, 1748–1758.

Lee, K.E. and Wood, T.G. (1971) *Termites and Soils.* Academic Press, London.

Lobry deBruyn, L.A. and Conacher, A.J. (1990) The role of termites and ants in soil modification. *Australian Journal of Soil Research* 28, 55–93.

Lobry deBruyn, L.A. and Conacher, A.J. (1994) The bioturbation activity of ants in agricultural and naturally vegetated habitats in semi-arid environments. *Australian Journal of Soil Research* 32, 555–570.

MacKay, W.P. and Whitford, W.G. (1988) Spatial variability of termite gallery production in Chihuahuan Desert plant communities. *Sociobiology* 14, 281–289.

MacKay, W.P., Silva, S., Lightfoot, D.C., Pagani, M.I. and Whitford, W.G. (1986) Effect of increased soil moisture and reduced soil temperature on a desert soil arthropod community. *The American Midland Naturalist* 116, 45–56.

MacKay, W.P., Silva, S. and Whitford, W.G. (1987) Diurnal activity patterns and vertical migration in desert soil microarthropods. *Pedobiologia* 30, 65–71.

Majer, J.D., Walker, T.C. and Berlandier, F. (1987) The role of ants in degraded soils within Dyandra state forest. *Mulga Research Centre Journal* 9, 15–16.

Malaka, S.L. (1977) A study of the chemistry and hydraulic conductivity of mound materials and soils from different habitats of some Nigerian termites. *Australian Journal of Soil Research* 15, 878–891.

Mando, A. (1998) Soil-dwelling termites and mulches improve nutrient release and crop performance on Sahelian crusted soil. *Arid Soil Research and Rehabilitation* 12, 153–164.

Mando, A. and Miedema, R. (1997) Termite-induced change in soil structure after mulching degraded (crusted) soil in the Sahel. *Applied Soil Ecology* 6, 261–263.

Montana, C. (1990) A floristic-structural gradient related to land forms in the southern Chihuahuan Desert. *Journal of Vegetation Science* 1, 669–674.

Nash, M.H. and Whitford, W.G. (1995) Subterranean termites: regulators of soil organic matter in the Chihuahuan Desert. *Biology and Fertility of Soils* 19, 15–18.

Noble, J.C., Whitford, W.G. and Kaliszweski, M. (1996) Soil and litter microarthropod populations from two contrasting ecosystems in semi-arid eastern Australia. *Journal of Arid Environments* 32, 329–346.

Nowak, R.S., Nowak, C.L., DeRocher, T., Cole, N. and Jones, M.A. (1990) Prevalence of *Oryzopsis hymenoides* near harvester ant mounds: indirect facilitation by ants. *Oikos* 58, 190–198.

Paine, R.T. (1966) Food web complexity and species diversity. *The American Naturalist* 100, 65–75.

Paine, R.T. (1969) A note on trophic complexity and community stability. *The American Naturalist* 103, 91–93.

Parker, L.W., Santos, P.F., Phillips, J. and Whitford, W.G. (1984) Carbon and nitrogen dynamics during the decomposition of litter and roots of a Chihuahuan Desert annual, *Lepidium lasiocarpum*. *Ecological Monographs* 54, 339–360.

Phillips, R.E., Quisenberry, V.L., Zeleznik, J.M. and Dunn, G.H. (1989) Mechanism of water entry into simulated macropores. *Soil Science Society of America Journal* 53, 1629–1635.

Pomeroy, D.E. (1983) Some effects of mound-building termites on the soils of a semi-arid area of Kenya. *Journal of Soil Science* 34, 555–570.

Santos, P.F. and Whitford, W.G. (1981) The effects of microarthropods on litter decomposition in a Chihuahuan Desert ecosystem. *Ecology* 62, 654–663.

Santos, P.F., Phillips, J. and Whitford, W.G. (1981) The role of mites and nematodes in early stages of buried litter decomposition in a desert. *Ecology* 62, 664–669.

Schaefer, D.A. and Whitford, W.G. (1981) Nutrient cycling by the subterranean termite *Gnathamitermes tubiformans* in a Chihuahuan Desert ecosystem. *Oecologia* 48, 277–283.

Tongway, D.J. and Ludwig, J.A. (1990) Vegetation and soil patterning in semi-arid mulga lands of eastern Australia. *Australian Rangeland Journal* 11, 5–20.

Tongway, D.J., Ludwig, J.A. and Whitford, W.G. (1989) Mulga log mounds: fertile patches in the semi-arid woodlands of eastern Australia. *Australian Journal of Ecology* 14, 263–268.

Wagner, D., Brown, M.J.F. and Gordon, D.M. (1997) Harvester ant nests, soil biota, and soil chemistry. *Oecologia* 112, 232–236.

Whitford, W.G. (1988) Effects of harvester ant, *Pogonomyrmex rugosus*, nests on soils and a spring annual, *Erodium texanum*. *Southwestern Naturalist* 33, 482–485.

Whitford, W.G. (1989) Abiotic controls on the functional structure of soil food webs. *Biology and Fertility of Soils* 8, 1–6.

Whitford, W.G. (1991) Subterranean termites and long-term productivity of desert rangelands. *Sociobiology* 19, 235–243.

Whitford, W.G. (1998) Contributions of pits dug by Goannas (*Varanus gouldii*) to the dynamics of banded mulga landscapes in eastern Australia. *Journal of Arid Environments* 40, 453–457.

Whitford, W.G. and DiMarco, R. (1995) Variability in soils and vegetation associated with harvester ant (*Pogonomyrmex rugosus*) nests on a Chihuahuan Desert watershed. *Biology and Fertility of Soils* 20, 169–173.

Whitford, W.G., Reynolds, J.F. and Cunningham, G.L. (1987) How desertification affects nitrogen limitation of primary productivity on Chihuahuan Desert watersheds. In: Aldon, E.F., Gonzalez, V., Carlos, E. and Moir, W.H. (eds) *Strategies for Classification and Management of Native Vegetation for Food Production in Arid Zones.* USDA Forest Service, Rocky Mountain Forest and Range Experiment Station Report. Fort Collins, Colorado, pp. 143–153.

Whitford, W.G., Ludwig, J.A. and Noble, J.C. (1992) The importance of subterranean termites in semi-arid ecosystems in south-eastern Australia. *Journal of Arid Environments* 22, 87–91

Whitford, W.G., Forbes, G.S. and Kerley, G.I. (1995) Diversity, spatial variability, and functional roles of invertebrates in desert grassland ecosystems. In: McClaran, M.P. and VanDevender, T.R. (eds) *The Desert Grassland.* University of Arizona Press, Tucson, pp. 152–195.

Wood, T.G., Johnson, R.A. and Anderson, J.M. (1983) Modification of soils in Nigerian savanna by soil-feeding *Cubitermes* (Isoptera, Termitidae). *Soil Biology and Biochemistry* 15, 575–579.

3

Responses of Grassland Soil Invertebrates to Natural and Anthropogenic Disturbances

J.M. Blair[1], T.C. Todd[2] and M.A. Callaham, Jr[1]

[1]Division of Biology and [2]Department of Plant Pathology, Kansas State University, Manhattan, KS 66506, USA

Introduction

Grassland ecosystems possess many fundamental characteristics which make them well suited for studies of belowground communities and processes. Although net primary productivity of grasslands varies across geographic localities, largely as a function of climate (Sala *et al.*, 1988), virtually all grasslands allocate a large proportion of primary production belowground in the form of rhizomes, roots and root exudates. Root-to-shoot biomass ratios in temperate grasslands worldwide average 3.7 and are typically an order of magnitude greater than in forest ecosystems, and exceeded only by cold deserts and tundra (Jackson *et al.*, 1996). Further, temperate grasslands exceed all other biomes in biomass of both total and living fine roots (≤ 2 mm in diameter) (Jackson *et al.*, 1997). Living roots are a direct resource for belowground herbivores, while exudates, sloughed cells and dead roots are important substrates for the soil microflora, microbivores and detritivores. In addition, the roots of most grassland plants include arbuscular mycorrhizal fungal symbionts (Rice *et al.*, 1998), which may support additional trophic pathways belowground (Lussenhop, 1992). In many ways, the large contribution of roots to belowground food webs, relative to aboveground detritus in grasslands, probably increases their importance in affecting soil invertebrate community structure and function in these ecosystems.

While herbivory is obviously important in grasslands, much of the carbon fixed by grassland producers enters the detrital food web, either directly (litterfall, root senescence, sloughing of root cells, exudation) or indirectly after being processed by herbivores (Elliott *et al.*, 1988). However, the climatic conditions associated with grasslands (high climatic variability, seasonal and longer-term periods of drought stress) and low tissue quality of plant detritus result in relatively slow decomposition rates. This pattern of high C allocation belowground and slow decomposition has led to the large stores of soil organic matter and nutrients characteristic of most grasslands (Seastedt, 1995). In the central USA, soils of tallgrass prairie are particularly noted for their high belowground productivity and biomass, large accumulations of organic matter and nutrients, and an abundant, diverse assemblage of soil biota (Rice *et al.*, 1998).

A second general characteristic of many grasslands is the occurrence of frequent disturbances that affect plant community composition, primary productivity and nutrient cycling (Blair, 1997; Collins and Steinauer, 1998). Mesic grasslands, in particular, occur under climatic conditions suitable for the development of alternative vegetative states, including shrublands or woodlands. In many instances, the persistence and dominance of grasses in these ecosystems is determined by various disturbance regimes which favour graminoid plant species. This is evident in the semihumid tallgrass prairies of the Great Plains where woody vegetation can replace grasslands in the absence of periodic 'disturbances' such as fire and grazing by large ungulates. Although a substantial body of literature is available on the effects of fire and grazing on plant and soil responses in these grasslands (Risser *et al.*, 1981; Knapp *et al.*, 1998a), less is known about responses of soil invertebrates to these disturbances, or the potential consequences of changes in belowground communities for ecosystem processes. This is unfortunate, since the combination of a diverse assemblage of soil invertebrates and natural disturbance regimes that alter plant and soil community composition, soil microclimate, and key ecosystem processes makes grasslands ideal for testing hypotheses about controls on soil invertebrate abundances and composition (microclimate, resource quantity and quality, plant productivity and diversity, etc.), as well as the relationships between soil community structure and ecosystem functioning. For example, what are the relationships between the effects of various disturbances on resource quality and quantity and changes in the soil biota? How are these linked to process-level responses? Further, a better understanding of how soil communities and processes respond to natural disturbances may aid in predicting and understanding the potential impacts of future environmental changes, such as altered climatic conditions, elevated atmospheric CO_2 concentrations or increased N deposition.

Our goals in this paper are: (i) to summarize the major factors influencing invertebrate abundance and distribution in tallgrass prairie soils, focusing on the responses of selected soil invertebrate groups to natural disturbances (fire, grazing, drought); (ii) to identify potential linkages between changes in soil communities and the effects of disturbances on key plant and soil characteristics or processes; and (iii) to discuss potential effects of novel anthropogenic perturbations on soil communities and processes. We focus on North American tallgrass prairies and draw upon studies undertaken at the Konza Prairie Long-Term Ecological Research (LTER) site (Knapp *et al.*, 1998a), with results from other grasslands where appropriate.

Characteristics of Grassland Soil Invertebrate Communities

In spite of the superficial simplicity and uniformity of most grassland ecosystems, grassland soil invertebrate communities are surprisingly diverse and
complex (Elliott *et al.*, 1988; Stanton, 1988; Curry, 1994). Most major soil
invertebrate groups are present in grasslands, although total belowground
invertebrate biomass and the relative abundance of some meso- (e.g. acari) and
macrofaunal (e.g. earthworms) groups generally decline with decreasing
average annual precipitation (Stanton, 1988). Studies of soil food web structure
and function in an arid shortgrass steppe ecosystem (Hunt *et al.*, 1987; Elliott
et al., 1988) illustrate the organization of grassland detrital food webs into
bacterial- and fungal-based compartments, as well as the importance of roots
for both herbivore- and decomposer-based pathways. The structure of belowground food webs is somewhat different in more mesic grasslands, where
certain macrofauna (e.g. earthworms) are more abundant, microarthropod
densities and composition are different, and the ratios of herbivorous and
fungal-feeding nematodes to bacterivores are higher (Rice *et al.*, 1998).
Estimates of the average density of major invertebrate groups at Konza Prairie
are presented in Table 3.1, along with similar estimates for a more arid mixed-
grass prairie site. Nematodes dominate the soil invertebrates of tallgrass prairie
in terms of abundance, but earthworms and macroarthropods compose most
of the invertebrate biomass.

Much of the research on soil invertebrates to date at Konza Prairie has
focused on the factors influencing the abundance and composition of the soil
invertebrate community, especially the effects of fire, grazing (or mowing) and
climatic variability. These factors, and their interactions with one another, alter
the relative availability of several key resources (water, light and nutrients) in
ways that affect the structure and functioning of these grasslands (Knapp *et al.*,
1998a). The same three factors also affect the soil physical environment
(temperature, water content, compaction), resource supply rates (plant
productivity, detrital inputs, availability of surface litter) and quality of available
resources (plant species composition, tissue N content), all of which may
influence soil communities and food webs. Therefore, we begin by discussing the
effects of these natural disturbances on tallgrass prairie structure and function
and responses of the soil invertebrate community, focusing primarily on soil
arthropods and nematodes.

Effects of fire

Fire is a common, large-scale disturbance in mesic and sub-humid grasslands
throughout the world. In North America, periodic fires were integral to the
development and maintenance of tallgrass prairie ecosystems (Anderson,
1990), and today fire is widely used as a management tool to limit invasion by
woody plants and to maintain the high productivity of the dominant warm-

Table 3.1. Estimates of mean abundance of major groups of soil invertebrates in two Kansas grasslands. Konza Prairie is a semi-humid tallgrass prairie (mean annual precipitation = 835 mm) and Hays is a more arid mixed-grass prairie (mean annual precipitation = 580 mm). Where possible, data from multiple years, sites and fire treatments were combined.

		Density (no m^{-2})	
Invertebrate group	Sample depth (cm)	Konza Prairie	Hays
Earthworms[a,b,c]	30	150	ND
Macroarthropods (total)[b,c,d,e]	30	84	ND
Herbivores	30	46	ND
Detritivores	30	22	ND
Predators	30	16	ND
Microarthropods (total)[f,g]	5	54,100	32,800
Prostigmatid mites	5	30,200	24,600
Mesostigmatid mites	5	2,000	900
Oribatid mites	5	16,300	5,300
Collembola	5	4,100	1,600
Others	5	1,500	400
Nematodes (total)[h]	40	6,414,000	5,908,000
Herbivores	40	2,372,000	2,186,000
Fungivores	40	2,598,000	2,130,000
Microbivores	40	1,117,000	1,339,000
Omnivore/predators	40	327,000	253,000

Data summarized from: [a]James, 1982; [b]Seastedt *et al.*, 1987; [c]Todd *et al.*, 1992; [d]Seasteadt, 1984a; [e]Seasteadt *et al.*, 1986; [f]Seastedt, 1984b; [g]O'Lear and Blair, 1999; [h]Todd *et al.*, 1999. ND, not determined.

season (C$_4$) grasses (Bragg and Hulbert, 1976). The effects of fire on ecosystem characteristics and processes in tallgrass prairie have been fairly well documented and are manifold (Collins and Wallace, 1990). They include changes in the soil microclimate, patterns of productivity, plant tissue chemistry, plant species composition, microbial activity and soil nutrient availability, all of which have the potential to impact on soil food webs.

Fire results in the removal of aboveground detritus which, in tallgrass prairie, can accumulate up to 1000 g m^{-2} and depths of 30 cm in the absence of fire (Knapp and Seastedt, 1986). This accumulation of surface litter and standing dead plant material limits light penetration, insulates the soil and minimizes evaporation. Fire results in an essentially complete removal of this detrital layer, altering the energy environment and microclimate of the soil, and resulting in substantially warmer and often drier, soil conditions. Soil temperatures in burned prairie can average up to 10°C warmer than on comparable unburned prairie (Blair, 1997), with potential consequences for soil microbial and faunal activity and rates of seasonal plant growth. One important effect of these changes in soil microclimate is to promote the rapid phenological development and growth of the C_4 grasses (Old, 1969; Knapp, 1985), which increase in abundance and biomass, while forbs and cool-season (C_3) grasses decline with repeated fires (Collins, 1987). Thus, frequent burning tends to increase total aboveground net primary productivity (Briggs and Knapp, 1995) while simultaneously increasing the dominance of a few C_4 grasses and reducing plant species diversity (Collins, 1992). Similar changes occur belowground, as greater root production by the C_4 grasses leads to increased belowground plant biomass (Table 3.2). Studies of root dynamics using a root window approach indicate greater root lengths in burned, compared with unburned, prairie (Fig. 3.1). Increased root lengths and biomass appear to result more from increased productivity than from changes in rates of root disappearance (Seastedt and Ramundo, 1990). Higher root productivity on burned prairie has also been documented using root in-growth bags (Johnson, unpublished observations). Concomitant with increased belowground productivity is a decrease in the N content of roots and rhizomes (Fig. 3.2). Wider C:N ratios of roots and rhizomes increase N immobilization in soils of frequently burned prairie (Ojima *et al.*, 1994; Blair *et al.*, 1998).

Results regarding the effects of fire on soil microbial biomass in tallgrass prairie are equivocal, with potential differences between the short-term responses to fire and the long-term effects of repeated, frequent burning. Ojima *et al.* (1994) found no effects of fire on microbial biomass within the first 1–2 years of burning plots on Konza prairie, although longer-term annual burning in a nearby site reduced microbial biomass C and N, and resulted in a small (2.5%) but significant increase in microbial biomass C:N ratio. In contrast, Garcia and Rice (1994) did not detect a significant consistent response in microbial biomass C and N after 5

Table 3.2. Comparisons of root and rhizome biomass of burned and unburned tallgrass prairie (Rice *et al.*, 1998, modified from Seastedt and Ramundo, 1990).

Depth of sample (cm)	Study site	Burned prairie (g m^{-2})	Unburned prairie (g m^{-2})	Reference
35 cm	Illinois	1064	839	Hadley and Kieckhefer (1963)
5 cm	Missouri	956	669	Kucera and Dahlman (1968)
100 cm	Illinois	2107	1908	Old (1969)
30 cm	Kansas	1002	790	Ojima *et al.* (1994)[a,b]
30 cm	Kansas	1086	859	Seastedt and Ramundo (1990)[a,b]
20 cm	Kansas	1618	1362	Garcia (1992)
20 cm	Kansas	960	838	Benning (1993)[b]

[a] Calculated from C data, assuming mass = 2.5 times the amount of C.
[b] Rhizomes not included.

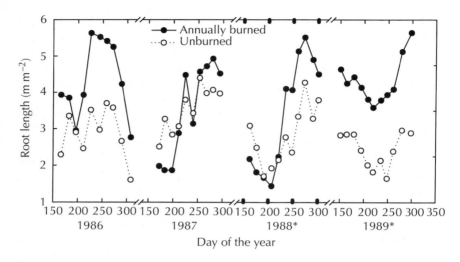

Fig. 3.1. Seasonal changes in root length in burned and unburned tallgrass prairie (from Rice *et al.*, 1998). Asterisks represent years with below average precipitation.

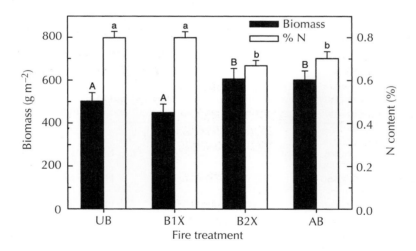

Fig. 3.2. Effects of fire frequency on live root biomass (solid bars) and N concentration in live roots (open bars). Small bars are 1 SE. Fire treatments were UB, unburned; B1X, burned once in the year of measurement; B2X, burned twice before measurement; AB, annually burned. Different letters represent significant differences among fire treatments ($P < 0.05$) in root biomass (upper case) or root N content (lower case). Data from Blair (1997).

years of annual burning. In summary, the effect of repeated frequent fire in tall-grass prairie is to alter the soil environment, increase root inputs but lower the quality and heterogeneity of these inputs, and possibly lower microbial biomass. How do these changes affect the soil invertebrate community?

Soil invertebrate groups differ in their responses to fire in tallgrass prairie (Rice *et al.*, 1998). In general, herbivorous taxa in both micro- and macrofaunal groups appear to be most consistent in their response to fire, and tend to increase in abundance and/or biomass in spite of greater fluctuations in soil moisture and temperature, and generally lower N content of roots. For example, densities of some herbivorous macroarthropods including root-grazing scarab beetle (*Phyllophaga* spp.) larvae (Seastedt, 1984a) and xylem-feeding cicada nymphs (Seastedt *et al.*, 1986) have increased following fire (Fig. 3.3). Populations of some herbivorous nematodes, such as graminoid-feeding *Helicotylenchus* spp., also are typically greater in burned, compared with unburned, prairie (Todd, 1996). Patterns of herbivorous nematode abundance can often be correlated with changes in live-root mass over time in burned prairie (Fig. 3.4). The generally positive response of belowground herbivores to fire in tallgrass prairie appears to be a consequence of enhanced root produc-tivity, and occurs in spite of reduced root-N content. This suggests that resource availability (i.e. root productivity) is a key factor for belowground herbivores as a group in these grasslands, although other factors such as a reduction in predation or changes in microclimate cannot be ruled out.

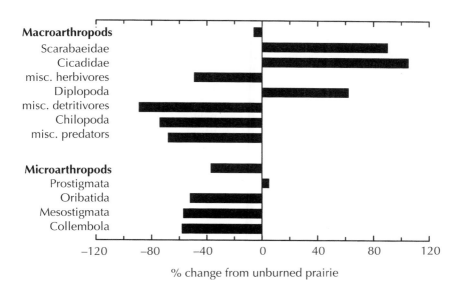

Fig. 3.3. Percentage change in mean abundance of selected arthropod groups on burned, relative to unburned prairie. Data from Seastedt *et al.* (1986) and Seastedt (1984b).

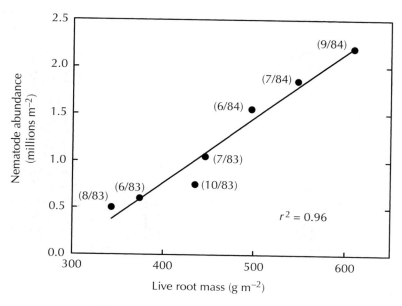

Fig. 3.4. Correlation between the abundance of herbivorous nematodes and biomass of live roots in annually burned tallgrass prairie over a 2-year period. Sample dates are indicated in parentheses (from Rice *et al.,* 1998).

Changes in plant species composition associated with different burning regimes are also important, and can be manifest in the composition and relative abundance of belowground herbivores. Nine years of annual burning or fire exclusion in a replicated field plot experiment resulted in substantial changes in the relative biomass of grasses and forbs (Fig. 3.5A), with C_4 grasses composing the vast majority of total aboveground plant biomass (98%) on burned plots, and forbs composing a much greater proportion of total plant biomass (44%) on unburned plots (Collins *et al.,* 1998). These changes in the relative abundance of grasses and forbs were correlated with a large shift in the composition of the herbivorous nematode community (Fig. 3.5B). After 9 years of treatment, the abundance of the dominant graminoid-feeding *Helicotylenchus* spp. was 137% greater in annually burned plots, while nematodes in the subfamily Paratylenchinae became the dominant herbivores in the unburned plots (Todd, 1996), possibly reflecting the non-graminoid feeding preferences of some taxa in the latter group. Responses of non-herbivorous nematodes to fire have been variable, with microbivores (primarily bacterial-feeding species in the Cephalobidae) often more abundant in annually burned compared with unburned prairie, and predator/omnivores generally more abundant in unburned prairie (Todd, 1996).

Measurement of soil nematode populations in a series of watersheds with different fire histories, but all burned in the year of measurement, provided some

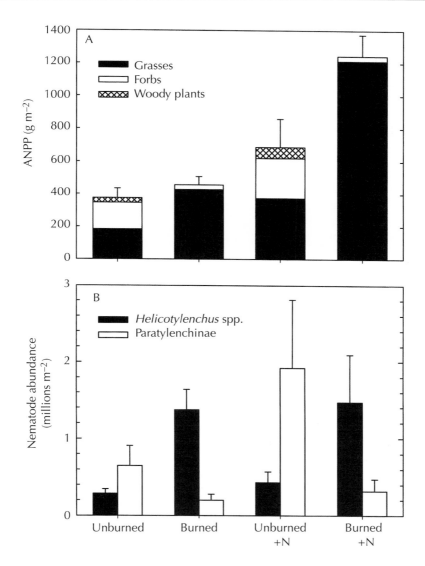

Fig. 3.5. Change in the annual aboveground net primary productivity (ANPP) of plants by major life form (A) and the composition of herbivorous nematodes (B) in response to 9 years of fire treatments (unburned or annually burned) and N addition (control or +10 g N m^{-2}). Small bars are 1 SE, and refer to total ANPP in (A).

intriguing data on long-term responses to different fire frequencies (Fig. 3.6). Abundances of both fungivores and microbivores were greatest in watersheds burned recently (mostly annually burned watersheds), and both groups declined in abundance as time since burning increased. The highest densities

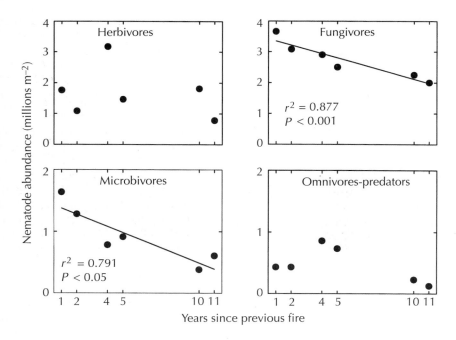

Fig. 3.6. Mean abundance of nematode trophic groups in burned watersheds as a function of time since previous fire. All watersheds were burned in the spring of 1991, and samples were collected in the autumn of 1991.

of the omnivore-predator group occurred at intermediate fire frequencies. There were no significant differences in abundances of total herbivores with fire frequency; this is not surprising since these watersheds included plant communities with varied species composition. The responses of fungivores and microbivores were not likely to be a result of short-term microclimatic changes induced by fire, since all watersheds were burned in the year of measurement. Instead, these data suggest that long-term responses by microbial-feeding nematodes to burning are due to biotic interactions, such as changes in root productivity or changes in microbial populations which accrue over time.

Experimental additions of C can increase populations of microbivore and omnivore nematodes in tallgrass prairie (Seastedt *et al.*, 1988), and it seems likely that greater root-derived C inputs in burned prairie may be responsible for increased populations of microbial-feeding nematodes. Interestingly, microbial biomass has been reported to decrease under long-term annual burning (Ojima *et al.*, 1994), although frequent burning appears to enhance numbers of microbial-feeding nematodes. One possible explanation is that increased C inputs from higher root productivity in burned prairie stimulates activities of soil microbes which, in turn, support greater populations of microbial-feeding nematodes. In this case, the apparent lack of relationship between microbial biomass and numbers of microbivore and fungivore nematodes may simply result from the feeding activities of the nematodes which can reduce standing stocks of microbial biomass (e.g. Wardle and Yeates, 1993). This hypothesis warrants further investigation.

Soil microarthropods, which include a variety of functional groups, exhibit variable responses to fire in tallgrass prairie. Lussenhop (1976, 1981) reported greater densities of soil microarthropods in a burned Illinois prairie, compared with unburned prairie, which he attributed to increased root productivity and belowground inputs. However, research at Konza Prairie has indicated lower abundances of total microarthropods in burned prairie, relative to unburned sites, at least in the surface 5 cm of soil (Fig. 3.3; Seastedt, 1984b). These differences were primarily due to lower numbers of oribatid mites and Collembola, groups which are largely detritivores/fungivores and which were, apparently, adversely affected by reduced amounts of surface litter and/or changes in soil microclimate. Numbers of prostigmatid mites, which included predacious and many small euedaphic, fungivorous taxa, were unaffected by burning. Burning also had no effect on microarthropod densities in the 5–10 cm deep soil layer.

These results suggest that soil microarthropod abundances in tallgrass prairie soils are not correlated with increased root C inputs. It may be that microarthropods as a group are affected by changes in habitat associated with loss of surface litter, or that microarthropods in these carbon-rich soils are affected more by resource quality (e.g. N content, C:N ratio) than resource quantity (e.g. inputs of new C). The latter hypothesis is supported by experiments in which adding inorganic N increased both belowground tissue quality and micro-arthropod abundances (Fig. 3.7). Surprisingly, additions of C as sugar in the same experiment also increased the abundance of microarthropods and bacterial-feeding nematodes, and resulted in an unexpected increase in plant production and belowground plant tissue N content (Seastedt *et al.*, 1988). Presumably, a pulsed addition of labile C may have stimulated soil N mineralization (a priming effect) resulting in higher N availability later in the growing season. In any case, it appears that manipulations which increase resource quality (plant N content) have a positive impact on soil microarthropods in these grasslands.

Reduced abundances of microarthropods in burned prairie may also be related to reduced resource heterogeneity (lower plant species diversity). Increased heterogeneity of leaf litter due to inputs from multiple species can alter microarthropod abundances and composition in forests (Blair *et al.*, 1990; Chapter 10, this volume), but the effect of plant species diversity on soil microarthropods in grasslands remains to be investigated. An alternative hypothesis is that lower microarthropod abundances in burned prairie are due to a less favourable soil microclimate (drier and warmer, more variable), but this is not supported by other studies in which soil moisture has been manipulated experimentally in burned prairie (O'Lear and Blair, 1999).

There has been substantial research on the effects of fire on soil processes in these grasslands, and we can ask how these changes in soil processes relate

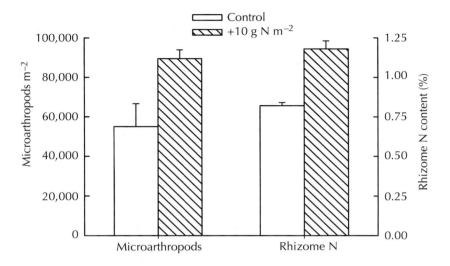

Fig. 3.7. Changes in the abundance of total soil microarthropods and rhizome N concentrations in response to added N (from Seastedt *et al.*, 1988). Small bars are 1 SE.

to changes in the soil invertebrate community. Several lines of evidence suggest increased rates of C flow belowground and decreased N availability in burned tallgrass prairie. Root biomass and productivity are generally enhanced in burned tallgrass prairie (Seastedt and Ramundo, 1990), indicating greater C inputs. O'Lear *et al.* (1996) also found that a uniform substrate (wooden dowels) decayed more rapidly in soils of burned, compared with unburned, prairie (Fig. 3.8). In that study, a wildfire on one of the unburned watersheds did not affect decay rates, suggesting that short-term microclimatic changes were not responsible for differences in decay rates under different fire treatments. Instead, they suggested that longer-term changes in the soil biota may underlie the faster decay rates in frequently burned prairie.

Knapp *et al.* (1998b) estimated that total annual soil CO_2 flux (root + heterotrophic respiration) was 33% greater in burned, relative to unburned, prairie. An interesting result of that study was that CO_2 fluxes from previously unburned plots that were burned for the first time were intermediate compared with annually burned and unburned plots (Fig. 3.9). The fact that soil microclimates were similar in both the newly burned and annually burned plots suggests that enhanced soil CO_2 flux from annually burned prairie is not simply a result of microclimate (i.e. warmer soils), and that changes in biotic factors (root productivity, soil microbial and faunal activity) are key to explaining increased soil respiration in frequently burned prairie. Another important effect of frequent burning in these grasslands is reduced net N mineralization rates (Fig. 3.9) and the potential for enhanced N immobilization (Ojima *et al.*, 1994; Blair, 1997;

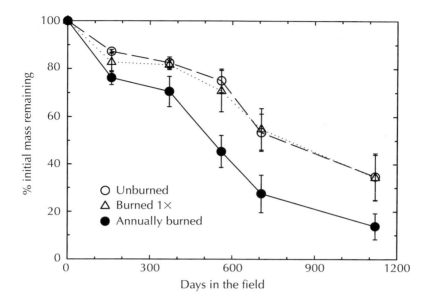

Fig. 3.8. Percentage mass remaining (mean ± 1 SE) over time of buried wood dowels on burned (closed circles) and unburned (open circle) watersheds. Open triangles are an unburned watershed that was burned once in a wildfire during the course of the study (from O'Lear *et al.*, 1996).

Turner *et al.*, 1997). Greater plant and microbial demand for N in burned prairie seems likely to increase the importance of invertebrate–microbe–plant interactions. However, the significance of changes in soil food web structure and function on C and N flux in these grasslands remains to be determined.

Effects of grazing

Grazing by large herbivores is another common natural disturbance in grassland ecosystems, often with substantial effects on ecosystem properties and processes (McNaughton, 1985; Collins *et al.*, 1998; Frank *et al.*, 1998; Knapp *et al.*, 1999). These effects include removal of aboveground biomass, reductions in root productivity and/or biomass, return of nutrients to the soil in more labile form (i.e. dung and urine), physical alteration of the soil (e.g. compaction) and modulation of plant species composition, all of which can affect soil invertebrate communities. However, few studies have examined the effects of grazing on soil invertebrate communities in tallgrass prairie, and many of these have focused on one particular component of grazing: aboveground biomass removal. We include results from both grazing and mowing (an experimental surrogate) studies here.

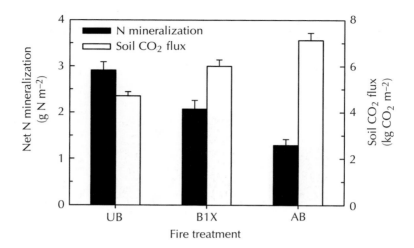

Fig. 3.9. Effects of fire on *in situ* measurements of growing season (May–October) net N mineralization (Blair, 1997) and annual soil CO_2 flux (Knapp *et al.*, 1998b). Fire treatments are unburned (UB), burned once in the year of measurement (B1X), or burned annually (AB). Small bars are 1 SE.

Aboveground biomass removal by either mowing or grazing tends to reduce root growth and biomass in tallgrass prairie (Turner *et al.*, 1993; Knapp *et al.*, 1999; Johnson, unpublished observations), and alter the N content of below-ground plant tissue (Fig. 3.10). Several studies on Konza Prairie indicate an increase in root or rhizome N concentrations (% N) in response to moderate levels of mowing or grazing, although total amounts of N (g N m^{-2}) in roots and rhizomes may be reduced (Seastedt *et al.*, 1988; Todd *et al.*, 1992; Turner *et al.*, 1993). Changes in either productivity or resource quality of roots may affect populations of belowground invertebrates, with potentially conflicting results. Both mowing and grazing have produced negative responses of herbivorous nematodes in tallgrass prairie. Todd *et al.* (1992) found that 2 years of mowing reduced herbivore densities by 54%, while 9 years of annual mowing and raking to remove aboveground biomass resulted in a 29% reduction in populations of herbivorous nematodes in tallgrass prairie (Todd, 1996). Similarly, comparisons of nematode populations inside and outside bison grazing exclosures indicated 45% reduction in total nematode abundance ($P = 0.01$) and a 50% reduction in herbivorous nematode abundance ($P = 0.05$) in grazed areas (Fig. 3.11). This is consistent with the hypothesis that root productivity is a primary determinant of herbivorous nematode abundances (Yeates and Coleman, 1982; Todd *et al.*, 1992).

In contrast, clipping at intermediate levels has been reported to increase populations of belowground herbivores in shortgrass steppe (Smolik and Dodd, 1983; Stanton, 1983). Increases in the abundance of belowground arthropod herbivores, including scarab beetle larvae (Seastedt *et al.*, 1986) and homopterans

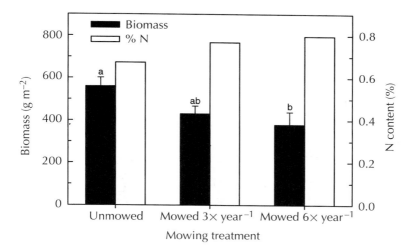

Fig. 3.10. Effects of aboveground biomass removal on live root biomass and root-N concentration (calculated from Turner *et al.*, 1993). Mowing treatments are: unmowed; mowed three times per year; mowed six times per year. Different letters represent significant differences among mowing treatments.

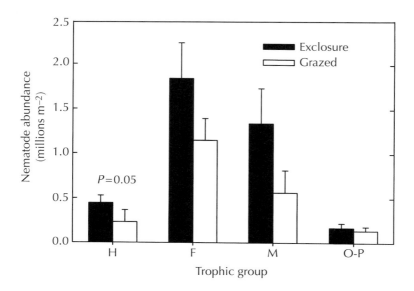

Fig. 3.11. Effects of bison grazing on the abundance of nematodes by trophic group. Trophic groups are: H, herbivores; F, fungivores; M, microbivores; O-P, omnivore-predators. Grazing significantly reduced total nematode densities ($P = 0.001$). Small bars are 1 SE.

(Seastedt and Reddy, 1991) have also been reported in response to aboveground biomass removal (Fig. 3.12). For these herbivores, any increases in resource quality (e.g. N content) in response to aboveground herbivory (Seastedt *et al.*, 1988) may compensate for decreases in root productivity. This would explain the apparently conflicting results of different studies, or responses of different invertebrate groups. In addition, as with fire (Blair, 1997), there may be differences between the short- and long-term effects of aboveground biomass removal, with responses of belowground invertebrates varying accordingly. For example, microbivore nematode numbers increased as a short-term response to mowing (Todd *et al.*, 1992), but were unaffected by 9 years of aboveground biomass removal (Todd, 1996). The short-term response was attributed to a transient increase in detrital inputs due to higher root mortality in the mowed treatments. Other studies have reported reduced microbivore densities in the presence of clipping (Stanton, 1983) and grazing (Fig. 3.11). The responses of microarthropods to mowing or grazing have not been addressed in tallgrass prairie. However, in a comprehensive study of the composition and distribution of soil microarthropods in shortgrass steppe, Leetham and Milchunas (1985) found very little change in soil microarthropod communities in areas subjected to light versus heavy grazing.

Effects of mowing or grazing on ecosystem processes in tallgrass prairie are substantial (Turner *et al.*, 1993; Collins *et al.*, 1998; Knapp *et al.*, 1999). Unlike fire, grazing often reduces C inputs belowground and lowers rates of soil CO_2 flux (Knapp *et al.*, 1998b), but enhances rates of N transformation including net mineralization and nitrification (Knapp *et al.*, 1999). Changes in the soil invertebrate community have not been definitively linked to these process-level changes, although the reductions in belowground herbivores, particularly nematodes, may play a role in reduced C flux belowground in grazed prairie.

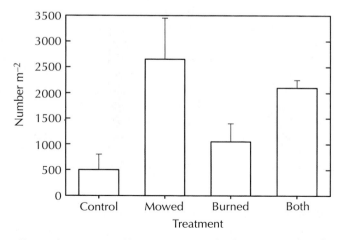

Fig. 3.12. Effects of mowing and burning on root-feeding *Sternorrhyncha* (Homoptera) (from Seastedt and Ready, 1991). Small bars are 1 SE.

Effects of drought

Grassland ecosystems are subject to frequent dry intervals during the growing season, and to longer-term periodic droughts. These dry periods can affect soil invertebrate communities directly, through changes in the soil environment, and indirectly through effects on primary producers and microbial decomposers. Effects of seasonal or longer drought periods include both reduced root productivity (Hayes and Seastedt, 1987) and changes in tissue N content. Hayes (1985) documented decreased root and rhizome N concentrations under drought, although the proportion of shoot N translocated to roots was higher during drought. In contrast, Schimel *et al.* (1991) found higher root N concentrations at more xeric upland prairie sites, relative to mesic lowland prairie, which they attributed to a higher plant nitrogen use efficiency as water limitation increased.

Nematodes might be expected to be more responsive than microarthropods to periods of water stress, since their activities are largely restricted to water films and water-filled pore spaces. In fact, nematode densities in tallgrass prairie do exhibit large seasonal fluctuations in abundance, with a general trend for increasing numbers over the growing season punctuated by population depressions during dry summer months. Similarly, nematode abundances in drought years can be as much as 60% lower than in years with near average precipitation (Todd, 1996). However, these responses to seasonal and annual droughts are not uniform across functional groups, and herbivorous taxa usually display the largest reductions (55–67%) in response to drought periods, while microbivorous taxa (primarily bacterivores) display the smallest (8–38%) responses. We interpret the relatively large response of herbivorous nematodes as a consequence of reduced root productivity during drought periods (Hayes and Seastedt, 1987). The reduced response of other functional groups, which should have similar moisture requirements, suggests that the direct effects of drought on the soil environment (i.e. reduced soil water content) are less important than production responses in these grasslands. In contrast to nematodes, many groups of microarthropods (e.g. many acari) are tolerant of dry soil conditions (Coleman and Crossley, 1996). The responses of microarthropods to drought have not been explicitly addressed at Konza, but can be inferred from comparisons of microarthropod densities at upland and lowland sites on Konza Prairie, and from the studies presented in the following section on altered precipitation regimes.

Responses of Grassland Invertebrates to Anthropogenic Disturbances

Research summarized in the previous sections indicates the importance of natural disturbances in these grassland ecosystems, and the many varied responses of different components of the soil-food web. In the face of increasing anthropogenic impacts on terrestrial ecosystems, it seems useful to consider whether we can use information garnered from studies of 'natural' disturbances to predict, or at least explain, the consequences of novel anthropogenic changes on grassland ecosystems and their soil-food webs. In this section we consider three potentially important anthropogenic disturbances – altered amounts of precipitation, elevated CO_2 concentrations and increased N inputs – and summarize results available to date regarding the effects of these factors on soil invertebrates in tallgrass prairie.

Altered precipitation amounts

Ecosystem processes (e.g. production, decomposition and nutrient cycling) in grasslands are strongly influenced by water availability, making them especially vulnerable to the altered patterns of rainfall predicted by global climate change models. For the Central Plains these predictions generally include decreased summer precipitation (i.e. Karl *et al.*, 1991; Gregory *et al.*, 1997) and increased variability in the amounts and timing of rainfall events (Houghton *et al.*, 1996). The potential impacts of changes in precipitation patterns and amounts on the soil biota of grasslands are largely unknown, although some inferences can be made based on studies across naturally occurring soil moisture gradients. We recently used two experimental approaches to address the potential effects of altered precipitation patterns and amounts on grassland soil invertebrates. The first was an irrigation experiment utilizing control and irrigated plots along a topographic gradient at Konza Prairie to examine the effects of eliminating seasonal soil-water deficits in tallgrass prairie (Knapp *et al.*,1994). The second was a reciprocal core transplant across a regional precipitation gradient, in which large, intact soil–plant cores were transplanted between a more mesic tallgrass site (Konza Prairie) and a more arid mixed-grass site (Hays, Kansas) to determine the responses of grassland ecosystems to different soil-water regimes (O'Lear and Blair, 1999). Here we summarize our major conclusions regarding the effects of altered precipitation amounts and soil-water availability on soil microarthropods and nematodes. Details regarding experimental design, sampling procedures and analyses for microarthropods and nematodes are presented in O'Lear and Blair (1999) and Todd *et al.* (1999).

Responses of both nematodes and microarthropods were more complex than expected. Nematodes and microarthropods differed in their responses to changes in water availability, although results from both the irrigation and reciprocal core transplant experiment were fundamentally similar for each group. Total nematode numbers increased with increasing water availability. As expected, based on responses to drought, herbivorous nematodes were most responsive to changes in water availability, and averaged 71% greater abundance under wetter conditions (irrigated plots, more mesic sites) across both experiments. For example, in the reciprocal transplant experiment, herbivorous nematodes were most abundant at the mesic site, regardless of core origin (Fig. 3.13). Responses of other nematode trophic groups were variable, with microbivores, in some

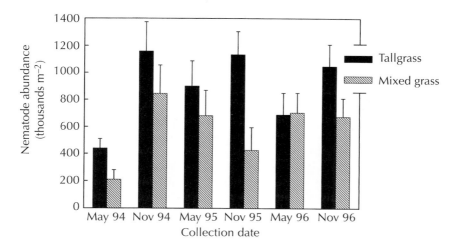

Fig. 3.13. Abundances of herbivorous nematodes in soil cores incubated at an arid mixed-grass site (hatched bars) or a more mesic tallgrass prairie site (solid bars). Values are averaged across cores originating from both sites, and small bars are 1 SE.

cases, inversely related to soil-water content. Microarthropod responses to soil-water availability were also complex. Depth distributions of microarthropods at a given location suggested positive responses to water availability in the soil profile (i.e. highest abundances at lower soil depths). However, experimentally increasing water availability, by irrigation or by transplanting cores to a wetter environment, generally decreased total microarthropod numbers (Fig. 3.14). In addition, the responses of specific groups of microarthropods varied, resulting in shifts in composition of microarthropod assemblages. Interestingly, for both nematodes and microarthropods, the short-term changes measured over the course of these manipulative experiments (2–4 years) were not consistent with differences in community structure and abundance developed at sites subject to long-term climatic differences. This suggests the need for caution in extrapolating results from short-term manipulative studies to predict the consequences of long-term climate changes.

Elevated atmospheric CO_2

Much effort has been directed towards assessing the impacts of rising levels of atmospheric CO_2 on the structure and functioning of terrestrial ecosystems. To date, most elevated CO_2 studies have focused on plant responses, or ecosystem processes, with few specifically examining responses of the soil invertebrate community. Those that have done so have produced mixed results (Yeates *et al.*,

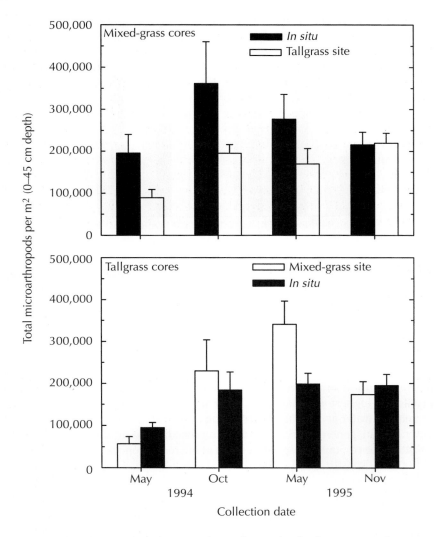

Fig. 3.14. Abundances of total microarthropods to a depth of 45 cm in soil cores originating from the more arid mixed-grass site or the more mesic tallgrass prairie site. Soil cores either remained at their site of origin until sampling (solid bars) or were transplanted to the alternative site in the autumn of 1993 (open bars). Small bars are 1 SE. (From O'Lear and Blair, 1999.)

1997; Jones *et al.*, 1998; Lussenhop *et al.*, 1998). The recent completion of a long-term field experiment utilizing open-top chambers to study the effects of elevated CO_2 in tallgrass prairie (Owensby *et al.*, 1993, 1997) provided us with the opportunity to look for longer-term changes in the soil invertebrate community. A one-time sampling of soil invertebrates in the autumn of 1996, following 8 years of exposure to twice ambient CO_2, indicated no significant changes in soil arthropod or earthworm abundances or gross faunal composition (Table 3.3). Abundances of total nematodes tended to be greater under elevated CO_2 (Fig. 3.15), although the effect was not statistically significant ($P = 0.14$), possibly due to the high spatial variability of nematodes and the low number of replicate chambers per treatment ($n = 3$).

Herbivore and omnivore-predator nematodes tended to exhibit the greatest increases. Responses of the herbivores are consistent with measured increases in root productivity in tallgrass prairie under elevated CO_2 (Owensby *et al.*, 1993) and the known responses of herbivorous nematodes to root productivity in these ecosystems. The lack of response by other soil invertebrates suggests that they may be relatively insensitive to the effects of elevated CO_2 in tallgrass prairie. These results should be viewed as preliminary, however, since they represent only one sample date and analysis at only a coarse taxonomic

Table 3.3. Soil arthropods and earthworms sampled from ambient and elevated CO_2 (2× ambient) chambers on 11 November 1996, after 8 years of CO_2 treatment. Two samples were taken from each of three replicate chambers per treatment. Sampling depths were 30 cm for macroarthropods and earthworms and 5 cm for microarthropods. Results are expressed as number per m², with standard error in parentheses.

Taxon	Control (no chamber)	Ambient CO_2	Elevated CO_2
Macroarthropods			
Araneae	35 (15)	35 (12)	37 (18)
Cicadidae	37 (13)	96 (51)	32 (20)
Chilopoda	51 (17)	72 (22)	59 (21)
Coleoptera	352 (79)	408 (119)	301 (109)
Diplopoda	74 (75)	13 (15)	13 (7)
Diplura	114 (58)	157 (48)	171 (55)
Diptera larvae	40 (15)	5 (6)	5 (4)
Isopoda	21 (14)	184 (99)	168 (88)
Lepidoptera larvae	66 (49)	56 (33)	37 (34)
Microarthropods			
Prostigmata	55,905 (6,866)	36,987 (6,405)	37,412 (10,432)
Mesostigmata	3,309 (1,113)	3,733 (1,426)	4,666 (1,103)
Oribatida	37,666 (9,209)	30,455 (6,763)	43,604 (15,046)
Collembola	5,260 (1,565)	7,041 (2,878)	8,229 (3,111)
Others	3,648 (1,103)	2,630 (1,243)	1,951 (608)
Earthworms			
Lumbricidae	128 (25)	176 (76)	67 (26)
Megascolecidae	80 (44)	171 (72)	208 (97)
Cocoons	171 (61)	192 (80)	59 (35)

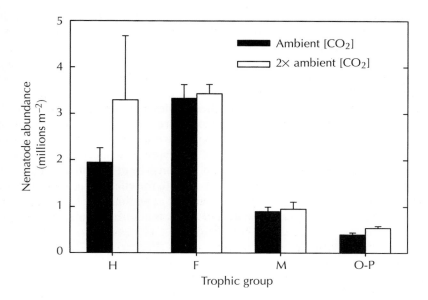

Fig. 3.15. Nematode responses, by trophic group, to 8 years of elevated CO_2 treatment. Trophic groups are: H, herbivores; F, fungivores; M, microbivores; O-P, omnivore-predators. Small bars are 1 SE.

scale. Some recent studies have suggested changes in the distribution of some soil invertebrates in response to elevated CO_2 (Lussenhop *et al.*, 1998), while others have reported individual species responses even where total numbers of major invertebrate groups, such as Collembola, did not change (Jones *et al.*, 1998). Clearly, more studies are needed before generalizations can be made regarding soil invertebrate responses to elevated CO_2.

Nitrogen fertilization

Human activities have more than doubled the amount of available nitrogen in the biosphere (Vitousek *et al.*, 1997), and elevated N deposition has been identified as a serious anthropogenic threat to many terrestrial ecosystems (Fenn *et al.*, 1998). Given the potential for N limitation in tallgrass prairie (Seastedt *et al.*, 1991; Blair *et al.*, 1998), it is not surprising that N loading in long-term fertilization experiments has produced shifts in plant species composition (Collins *et al.*, 1998), with potential consequences for ecosystem functioning (Wedin and Tilman, 1996). It seems likely that N additions should alter belowground food webs as well, but this has not been well studied in these grasslands. We can, however, use data from fertilization experiments to explore the potential consequences of N enrichment in tallgrass prairie.

Fertilization with N has produced some of the most dramatic and consistent responses by soil biota in field experiments at Konza Prairie. As previously noted, microarthropod numbers increased substantially in response to a single fertilizer application (Seastedt *et al.*, 1988). Chronic N additions over a 9-year period also resulted in an average increase of 51% for herbivorous nematodes (Fig. 3.16). This is somewhat surprising, since most studies of nematode responses to fire, grazing and drought indicate stronger correlations with root productivity or biomass, than with root N content. However, in all of those instances, there was a negative correlation between the effects of the disturbance on root biomass and root N content. In contrast, 4 years of fertilizer N addition resulted in both modest increases in live root biomass (15% greater than non-fertilized control plots), and substantial increases in live-root N content (77% greater than non-fertilized control plots, Benning and Seastedt, 1997). Apparently, resource quality does become an important factor with chronic N enrichment (Todd, 1996), and these findings have been incorporated into a response surface model of the interactive effects of resource quantity and quality for belowground herbivores (Rice *et al.*, 1998). Based on this model, factors that increase either root productivity or root quality independently may increase populations of belowground herbivores, but factors that simultaneously do both should have the greatest effects.

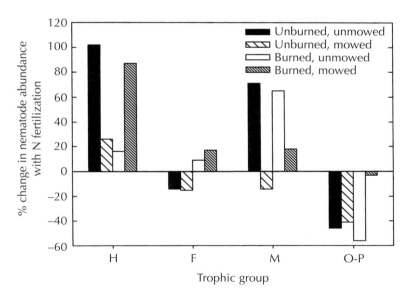

Fig. 3.16. Nematode responses, by trophic group, to 9 years of N addition (+10 g N m^{-2} year^{-1}) relative to unfertilized plots of the same fire/mowing treatment. Trophic groups are: H, herbivores; F, fungivores; M, microbivores; O-P, omnivore-predators. Treatments were either annually burned in the spring or not burned, and either annually mowed and raked to remove aboveground biomass or not mowed.

Fertilization also greatly increased the abundance of microbivorous nematodes in unmowed prairie (Fig. 3.16), with little change in the abundance of fungivores. The net effect was a decreased ratio of fungivore : bacterivore nematodes, suggesting fundamental changes in the detrital food web of N-enriched prairie. Microbial activity in unfertilized tallgrass prairie is fungal-dominated (Rice *et al.*, 1998), and this is reflected by typically high ratios of fungivores to microbivores in the nematode community (Table 3.1). Nitrogen fertilization in tallgrass prairie has been shown to increase microbial biomass N, while simultaneously decreasing microbial biomass C (Garcia and Rice, 1994). These changes coincided with an increase in the proportion of microbial respiration attributable to bacterial versus fungal respiration (Garcia, 1992), suggesting a shift in microbial community composition and activity from one dominated by fungal activity to one with a greater bacterial component. Together, these results suggest that an important impact of increased N deposition may be a shift to a more bacterially based food web. The long-term consequences of this change for soil N dynamics and organic matter storage have not yet been determined, but they may be substantial.

Conclusions

Disturbances are common in tallgrass prairie, and their role in driving community structure and aboveground ecosystem processes has received considerable recognition (Collins and Steinauer, 1998). The studies summarized here indicate that disturbance also plays a significant role in affecting belowground processes and structuring soil invertebrate communities. Natural disturbances, such as fire and grazing, alter specific components of the soil community in different ways. Many of these responses to disturbance appear to be indirect, and are mediated by changes in the quantity or quality of belowground inputs. Root productivity is a key variable affecting populations of herbivorous nematodes and some herbivorous arthropods, and enhanced C inputs associated with frequent fires support large numbers of microbivores and fungivores. In other cases, the quality of belowground resources (C/N ratios) appears to exert a greater influence than total C inputs. Interestingly, the novel anthropogenic disturbances discussed here (climate change, elevated CO_2, increased N deposition) produced some unexpected responses by the soil invertebrate community, suggesting the complexities of belowground interactions and the need for longer-term studies to evaluate the impacts of anthropogenic change. The responses of soil invertebrate communities to specific disturbances were often correlated with changes in ecosystem processes, such as C processing and N cycling, but determining to what extent these changes are related remains a challenge and an area for further research.

References

Anderson, R.C. (1990) The historic role of fire in the North American grassland. In: Collins, S.L. and Wallace, L.L. (eds) *Fire in North American Tallgrass Prairies*. University of Oklahoma Press, Norman, Oklahoma, pp. 8–18.

Benning, T.L. (1993) Fire frequency and topoedaphic controls of net primary productivity in the tallgrass prairie. Dissertation, University of Colorado, Boulder, Colorado.

Benning, T.L. and Seastedt, T.R. (1997) Effects of fire, mowing and nitrogen addition on root characteristics in tall-grass prairie. *Journal of Vegetation Science* 8, 541–546.

Blair, J.M. (1997) Fire, N availability, and plant response in grasslands: a test of the transient maxima hypothesis. *Ecology* 78, 2359–2368.

Blair, J.M., Parmelee, R.W. and Beare, M.H. (1990) Decay rates, nitrogen fluxes and decomposer communities of single- and mixed-species foliar litter. *Ecology* 71, 1976–1985.

Blair, J.M., Seastedt, T.R., Rice, C.W. and Ramundo, R.A. (1998) Terrestrial nutrient cycling in tallgrass prairie. In: Knapp, A.K., Briggs, J.M., Hartnett, D.C. and Collins, S.C. (eds) *Grassland Dynamics: Long-Term Ecological Research in Tallgrass Prairie*. Oxford University Press, New York, pp. 222–243.

Bragg, T.B. and Hulbert, L.C. (1976) Woody plant invasion of unburned Kansas bluestem prairie. *Journal of Range Management* 29, 19–23.

Briggs, J.M. and Knapp, A.K. (1995) Interannual variability in primary production in tallgrass prairie: climate, soil moisture, topographic position and fire as determinants of aboveground biomass. *American Journal of Botany* 82, 1024–1030.

Coleman, D.C. and Crossley, D.A., Jr (1996) *Fundamentals of Soil Ecology*. Academic Press, San Diego.

Collins, S.L. (1987) Interaction of disturbances in tallgrass prairie: a field experiment. *Ecology* 68, 1243–1250.

Collins, S.L. (1992) Fire frequency and community heterogeneity in tallgrass prairie vegetation. *Ecology* 73, 2001–2006.

Collins, S.L. and Steinauer, E.M. (1998) Disturbance, diversity, and species interactions in tallgrass prairie. In: Knapp, A.K., Briggs, J.M., Hartnett, D.C. and Collins, S.C. (eds) *Grassland Dynamics: Long-Term Ecological Research in Tallgrass*. Oxford University Press, New York, pp. 140–156.

Collins, S.L. and Wallace, L.L. (eds) (1990) *Fire in North American Tallgrass Prairies*. University of Oklahoma Press, Norman, Oklahoma.

Collins, S.L., Knapp, A.K., Briggs, J.M., Blair, J.M. and Steinauer, E. (1998) Modulation of diversity by grazing and mowing in native tallgrass prairie. *Science* 280, 745–747.

Curry, J.P. (1994) *Grassland Invertebrates. Ecology, Influences on Soil Fertility and Effects on Plant Growth*. Chapman and Hall, New York.

Elliott, E.T., Hunt, H.W. and Walter, D.E. (1988) Detrital foodweb interactions in North American grassland ecosystems. *Agriculture, Ecosystems and Environment* 24, 41–56.

Fenn, M.E., Poth, M.A., Aber, J.D., Baron, J.S., Bormann, B.T., Johnson, D.W., Lemly, A.D., McNulty, S.G., Ryan, D.F. and Stottlemyer, R. (1998) Nitrogen excess in North American Ecosystems: predisposing factors, ecosystem responses, and management strategies. *Ecological Applications* 8, 706–733.

Frank, D.A., McNaughton, S.J. and Tracy, B.F. (1998) The ecology of the earth's grazing ecosystems. *BioScience* 48, 513–521.

Garcia, F.O. (1992) Carbon and nitrogen dynamics and microbial ecology in tallgrass prairie. Dissertation, Kansas State University, Manhattan, Kansas.

Garcia, F.O. and Rice, C.W. (1994) Microbial biomass dynamics in tallgrass prairie. *Soil Science Society of America Journal* 58, 816–823.

Gregory, J.M., Mitchell, J.F.B. and Brady, A.J. (1997) Summer drought in northern mid-latitudes in a time-dependent CO_2 climate experiment. *Journal of Climate* 10, 662–686.

Hadley, E.B. and Kieckhefer, B.J. (1963) Productivity of two prairie grasses in relation to fire frequency. *Ecology* 44, 389–395.

Hayes, D.C. (1985) Seasonal nitrogen translocation in big bluestem during drought conditions. *Journal of Range Management* 38, 406–410.

Hayes, D.C. and Seastedt, T.R. (1987) Root dynamics of tallgrass prairie in wet and dry years. *Canadian Journal of Botany* 65, 787–791.

Houghton, J.T., Meiro Filho, L.G., Callander, B.A., Harris, N., Kattenberg, A. and Maskell, K. (eds) (1996) *Climate Change 1995: the Science of Climate Change*. Contribution of Working Group I to the Second Assessment Report of the Intergovernmental Panel on Climate Change. Cambridge University Press, Cambridge, Massachusetts.

Hunt, H.W., Coleman, D.C., Ingham, E.R., Elliott, E.T., Moore, J.C., Rose, S.L., Reid, C.P.P. and Morley, C.R. (1987) The detrital food web in a shortgrass prairie. *Biology and Fertility of Soils* 3, 57–68.

Jackson, R.B., Canadell, J., Ehleringer, J.R., Mooney, H.A., Sala, O.E. and Schulze, E.-D. (1996) A global analysis of root distributions for terrestrial biomes. *Oecologia* 108, 389–411.

Jackson, R.B., Mooney, H.A. and Schulze, E.-D. (1997) A global budget for fine root biomass, surface area, and nutrient contents. *Proceedings of the National Academy of Sciences USA* 94, 7362–7366.

James, S.W. (1982) Effects of fire and soil type on earthworm populations in a tallgrass prairie. *Pedobiologia* 24, 37–40.

Jones, T.H., Thompson, L.J., Lawton, J.H., Bezemer, T.M., Bardgett, R.D., Blackburn, T.M., Bruce, K.D., Cannon, P.F., Hall, G.S., Hartley, S.E., Howson, G., Jones, C.G., Kampichler, C., Kandeler, E. and Ritchie, D.A. (1998) Impacts of rising atmospheric carbon dioxide on model terrestrial ecosystems. *Science* 280, 441–443.

Karl, T.R., Heim, R.R. and Quayle, R.G. (1991) The greenhouse effect in central North America: If not now, when? *Science* 251, 1058–1061.

Knapp, A.K. (1985) Effect of fire and drought on the ecophysiology of *Andropogon gerardii* and *Panicum virgatum* in a tallgrass prairie. *Ecology* 66, 1309–1320.

Knapp, A.K. and Seastedt, T.R. (1986) Detritus accumulation limits productivity in tallgrass prairie. *BioScience* 36, 662–668.

Knapp, A.K., Koelliker, J.K., Fahnestock, J.T. and Briggs, J.M. (1994) Water relations and biomass responses to irrigation across a topographic gradient in tallgrass prairie. In: Wickett, R.G., Lewis, P.D., Woodcliffe, A. and Pratt, P. (eds) *Proceedings of the Thirteenth North America Prairie Conference*. Preney Print and Litho, Windsor, Ontario, pp. 215–220.

Knapp, A.K., Briggs, J.M., Hartnett, D.C. and Collins, S.C. (eds) (1998a) *Grassland Dynamics: Long-Term Ecological Research in Tallgrass Prairie*. Oxford University Press, New York.

Knapp, A.K., Conard, S.L. and Blair, J.M. (1998b) Determinants of soil CO_2 flux from a sub-humid grassland: effect of fire and fire history. *Ecological Applications* 8, 760–770.

Knapp, A.K., Blair, J.M., Briggs, J.M., Collins, S.L., Hartnett, D.C., Johnson, L.C. and Towne, E.G. (1999) The keystone role of bison in North American tallgrass prairie. *BioScience* 49, 39–50.

Kucera, C.L. and Dahlman, R.C. (1968) Root-rhizome relationships in fire-treated stand of big bluestem, *Andropogon gerardii* Vitman. *American Midland Naturalist* 80, 268–271.

Leetham, J.W. and Milchunas, D.G. (1985) The composition and distribution of soil microarthropods in the shortgrass steppe in relation to soil water, root biomass, and grazing by cattle. *Pedobiologia* 28, 311–325.

Lussenhop, J. (1976) Soil arthropod response to prairie burning. *Ecology* 57, 88–98.

Lussenhop, J. (1981) Microbial and microarthropod detrital processing in a prairie soil. *Ecology* 62, 964–972.

Lussenhop, J. (1992) Mechanisms of microarthropod–microbial interactions in soil. *Advances in Ecological Research* 23, 1–33.

Lussenhop, J., Trionis, A., Curtis, P.S., Teeri, J.A. and Vogel, C.S. (1998) Responses of soil biota to elevated atmospheric CO_2 in poplar model systems. *Oecologia* 113, 247–251.

McNaughton, S.J. (1985) Ecology of a grazing ecosystem: the Serengeti. *Ecological Monographs* 55, 259–294.

Ojima, D.S., Schimel, D.S., Parton, W.J. and Owensby, C.E. (1994) Long- and short-term effects of fire on nitrogen cycling in tallgrass prairie. *Biogeochemistry* 24, 67–84.

Old, S.M. (1969) Microclimate, fire and plant production in an Illinois prairie. *Ecological Monographs* 39, 355–384.

O'Lear, H.A. and Blair, J.M. (1999) Responses of soil microarthropods to changes in soil water availability in tallgrass prairie. *Biology and Fertility of Soils* 29, 207–217.

O'Lear, H.A., Seastedt, T.R., Briggs, J.M., Blair, J.M. and Ramundo, R.A. (1996) Fire and topographic effects on decomposition rates and nitrogen dynamics of buried wood in tallgrass prairie. *Soil Biology and Biochemistry* 28, 323–329.

Owensby, C.E., Coyne, P.I., Ham, J.M., Auen, L.M. and Knapp, A.K. (1993) Biomass production in a tallgrass prairie ecosystem exposed to ambient and elevated CO_2. *Ecological Applications* 3, 644–653.

Owensby, C.E., Ham, J.M., Knapp, A.K., Rice, C.W., Coyne, P.I. and Auen, L.M. (1997) Ecosystem-level responses of tallgrass prairie to elevated CO_2. In: Koch, G. and Mooney, H. (eds) *Carbon Dioxide and Terrestrial Ecosystems*. Physiological Ecology Series, Academic Press, New York, pp. 175–193.

Rice, C.W., Todd, T.C., Blair, J.M., Seastedt, T.R., Ramundo, R.A. and Wilson, G.W.T. (1998) Belowground biology and processes. In: Knapp, A.K., Briggs, J.M., Hartnett, D.C. and Collins, S.C. (eds) *Grassland Dynamics: Long-Term Ecological Research in Tallgrass Prairie*. Oxford University Press, New York, pp. 244–264.

Risser, P.G., Birney, C.E., Blocker, H.D., May, S.W., Parton, W.J. and Wiens, J.A. (1981) *The True Prairie Ecosystem*, US/IBP Synthesis Series 16. Hutchinson Ross, Stroudsburg, Pennsylvania.

Sala, O.E., Parton, W.J., Joyce, L.A. and Lauenroth, W.K. (1988) Primary production of the central grassland region of the United States. *Ecology* 69, 40–45.

Schimel, D.S., Kittel, T.G.F., Knapp, A.K., Seastedt, T.R., Parton, W.J. and Brown, V.B. (1991) Physiological interactions along resource gradients in a tallgrass prairie. *Ecology* 72, 672–684.

Seastedt, T.R. (1984a) Belowground macroarthropods of annually burned and unburned tallgrass prairie. *American Midland Naturalist* 111, 405–408.

Seastedt, T.R. (1984b) Microarthropods of burned and unburned tallgrass prairie. *Journal of the Kansas Entomological Society* 57, 468–476.

Seastedt, T.R. (1995) Soil systems and nutrient cycles on the North American prairie. In: Joern, A. and Keeler, K.H. (eds) *The Changing Prairie*. Oxford University Press, Oxford, pp. 157–174.

Seastedt, T.R. and Ramundo, R.A. (1990) The influence of fire on belowground processes of tallgrass prairies. In: Collins, S.L. and Wallace, L.L. (eds) *Fire in North American Tallgrass Prairies*. University of Oklahoma Press, Norman, Oklahoma.

Seastedt, T.R. and Reddy, M.V. (1991) Fire, mowing and insecticide effects on soil Stenorrhyncha (Homoptera) densities in tallgrass prairie. *Journal of the Kansas Entomological Society* 62, 238–242.

Seastedt, T.R., Hayes, D.C. and Petersen, N.J. (1986) Effects of vegetation, burning and mowing on soil macroarthropods of tallgrass prairie. In: Clambey, G.K. and Pemble, R.H. (eds) *Proceedings of the Ninth North American Prairie Conference*. Tri-College Press, Fargo, North Dakota, pp. 99–102.

Seastedt, T.R., Todd, T.C. and James, S.W. (1987) Experimental manipulations of arthropod, nematode and earthworm communities in a North American tallgrass prairie. *Pedobiologia* 30, 9–17.

Seastedt, T.R., James, S.W. and Todd, T.C. (1988) Interactions among soil invertebrates, microbes and plant growth in the tallgrass prairie. *Agriculture, Ecosystems and Environment* 24, 219–228.

Seastedt, T.R., Briggs, J.M. and Gibson, D.J. (1991) Controls of nitrogen limitation in tallgrass prairie. *Oecologia* 87, 72–79.

Smolik, J.D. and Dodd, J.L. (1983) Effects of water and nitrogen, and grazing on nematodes in a shortgrass prairie. *Journal of Range Management* 36, 744–748.

Stanton, N.L. (1983) The effect of clipping and phytophagous nematodes on net primary production of blue grama, *Bouteloua gracilis*. *Oikos* 40, 249–257.

Stanton, N.L. (1988) The underground in grasslands. *Annual Review of Ecology and Systematics* 19, 573–589.

Todd, T.C. (1996) Effects of management practices on nematode community structure in tallgrass prairie. *Applied Soil Ecology* 3, 235–246.

Todd, T.C., James, S.W. and Seastedt, T.R. (1992) Soil invertebrate and plant responses to mowing and carbofuran application in a North American tallgrass prairie. *Plant and Soil* 144, 117–124.

Todd, T.C., Blair, J.M. and Milliken, G.A. (1999) Effects of altered soil water availability on a tallgrass prairie nematode community. *Applied Soil Ecology* 13, 45–55.

Turner, C.L., Seastedt, T.R. and Dyer, M.I. (1993) Maximization of aboveground grassland production: the role of defoliation frequency, intensity, and history. *Ecological Applications* 3, 175–186.

Turner, C.L., Blair, J.M., Schartz, R.J. and Neel, J.C. (1997) Soil N availability and plant response in tallgrass prairie: effects of fire, topography and supplemental N. *Ecology* 78, 1832–1843.

Vitousek, P.M., Aber, J.D., Howarth, R.W., Likens, G.E., Matson, P.M., Schindler, D.W., Schlesinger, W.H. and Tilman, D.G. (1997) Human alteration of the global nitrogen cycle: sources and consequences. *Ecological Applications* 7, 737–750.

Wardle, D.A. and Yeates, G.W. (1993) The dual importance of competition and predation as regulatory forces in terrestrial ecosystems: evidence from decomposer food webs. *Oecologia* 93, 303–306.

Wedin, D.A. and Tilman, D. (1996) Influence of nitrogen loading and species composition on the carbon balance of grasslands. *Science* 274, 1720–1723.

Yeates, G.W. and Coleman, D.C. (1982) Role of nematodes in decomposition. In: Freckman, D.W. (ed.) *Nematodes in Soil Ecosystems.* University of Texas Press, Austin, Texas, pp. 55–80.

Yeates, G.W., Tate, K.R. and Newton, P.C.D. (1997) Response of the fauna of a grassland soil to doubling of atmospheric carbon dioxide concentration. *Biology and Fertility of Soils* 25, 307–315.

Effects of Invertebrates in Lotic Ecosystem Processes

4

J.B. Wallace[1,2] and J.J. Hutchens, Jr[1]

[1]*Institute of Ecology and* [2]*Department of Entomology, University of Georgia, Athens, GA 30602, USA*

Introduction

Freshwater invertebrates perform many roles in ecosystem processes (Palmer *et al.*, 1997) and these roles are frequently associated with a diverse array of feeding habits which have been organized into functional feeding groups (FFGs). Wallace and Webster (1996) reviewed many roles of FFGs in stream ecosystems. Streams differ markedly from most ecosystems in that the unidirectional flow of water through areas of different relief, lithology, runoff and large woody debris generates an array of channel forms (Brussock *et al.*, 1985). These various channel forms result in many diverse habitats (Frissell *et al.*, 1986), which place many constraints on organisms and the type and form of their food resources. This physical heterogeneity, including substrate and current velocity, is important in that it influences many aspects of stream ecology including nutrient dynamics, biotic diversity of animals, functional feeding groups, predator–prey interactions, refugia from disturbances, micro- and macroflora, organic matter retention and transport, as well as local secondary production (see Wallace and Webster, 1996).

Most streams have a highly diverse macroinvertebrate assemblage which is represented by several FFGs that have evolved a diverse array of morpho-behavioural mechanisms for exploiting foods. Throughout this chapter we will follow the functional classification of Merritt and Cummins (1996), which is based on mechanisms used by invertebrates to acquire foods. These functional groups are as follows: *scrapers*, animals adapted to graze or scrape materials from mineral and organic substrates; *shredders*, organisms that comminute primarily coarse particulate organic matter (CPOM ≥ 1 mm diameter); *gatherers*, animals that feed primarily on fine particulate organic matter (FPOM ≤ 1 mm diameter) deposited within streams; *filterers*, animals that have specialized anatomical structures (e.g. setae, mouthbrushes or fans, etc.) or silk and silk-like secretions that act as sieves to remove particulate matter from suspension (Wallace and Merritt, 1980); and *predators*, those organisms that feed primarily on animal tissue.

These FFGs refer primarily to *modes* of feeding or the food acquisition system (*sensu* Cummins, 1986) and not type of food *per se*. For example, many filter-feeding insects in streams are primarily carnivores (e.g. Benke and Wallace, 1980, 1997). Scrapers also consume quantities of what must be characterized as epilithon (FPOM, accompanying microbiota, algae) and not solely attached algae. Likewise, although shredders may select materials that have been 'microbially conditioned' by colonizing fungi and bacteria (e.g. Cummins and Klug, 1979), these shredders also ingest attached algal cells, protozoans and various other components of the fauna during feeding (Merritt and Cummins, 1996). While it appears valid to separate taxa according to the mechanisms used to obtain foods, many questions remain concerning the ultimate sources of assimilated energy for each of these functional groups (e.g. Mihuc, 1997).

In most streams draining forested regions in eastern North America, allochthonous inputs from the surrounding forest far exceed inputs from within-stream primary production (Webster *et al.*, 1995). Only tundra and arid land streams appear to have levels of autochthonous production exceeding those of allochthonous inputs (Webster and Meyer, 1997). Thus, it is not surprising that animal assemblages of streams draining forested regions have long been viewed as relying primarily on allochthonous inputs from the surrounding forest (Cummins, 1973). Most of the particulate organic matter entering streams draining forested regions is primarily in the form of CPOM, leaves and woody debris (Fig. 4.1), which usually compose a large portion of the total particulate organic matter (POM) standing crop (Jones, 1997). In the absence of storms, very little of the organic matter stored in headwater streams is exported as CPOM. Most carbon loss at baseflow is in the form of FPOM and dissolved organic carbon (DOC) (Naiman and Sedell, 1979; Webster and Patten, 1979; Wallace *et al.*, 1982). Thus, shallow headwater streams can be thought of as sites of input, storage, transformation and export of organic matter, since DOC and FPOM are much more amenable to entrainment and downstream transport compared with CPOM. Although DOC does enter the stream through groundwater, some of this DOC is derived from leaf litter stored within the wetted channel (Meyer *et al.*, 1998). Invertebrates promote downstream transport by bioturbation and conversion of CPOM to more easily transported FPOM (Fig. 4.1). Downstream organisms, especially filterers, promote retention of organic matter, whereas other functional groups such as grazers, gatherers and predators often enhance transport through their feeding activities. In larger downstream areas, most organic matter is available as either suspended or deposited FPOM, and these regions are dominated by filterers, gatherers, predators and, to a lesser extent, shredder-herbivores when macrophytes are available (Fig. 4.1).

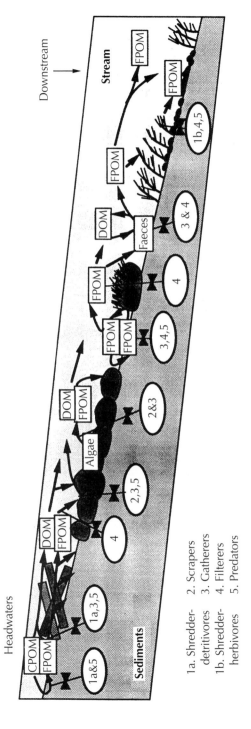

Fig. 4.1. A conceptual diagram showing how invertebrate functional feeding groups affect different ecosystem processes along a hypothetical stream gradient. Note how different reaches along the gradient are dominated by different resources and abiotic conditions. Upstream reaches (left) often have much stored CPOM behind woody debris dams. Middle reaches can be active sites of algal production on cobble and boulders, while downstream reaches (right) can have abundant aquatic macrophytes and large inputs of FPOM from upstream. Also, notice how FPOM and DOM are important resources along the entire stream length. Processes affected by functional feeding groups (1–5) are indicated by lines with hourglass symbols and are described in the text.

1a. Shredder-
    detritivores
1b. Shredder-
    herbivores

2. Scrapers
3. Gatherers
4. Filterers
5. Predators

This chapter focuses on various processes that are influenced by invertebrate activities rather than on individual functional groups themselves. Invertebrates influence lotic ecosystem processes by several mechanisms, which we will discuss in the following sections. First, invertebrates alter resource quantity, size and shape. Second, invertebrates enhance downstream movement of carbon and associated nutrients. Third, their activity can also enhance retention of carbon and nutrients. Fourth, predators, through their influence on standing crop of prey, may affect all processes. Finally, we address how large-scale physical changes along stream gradients affect invertebrate assemblages and their primary roles in stream ecosystems.

Invertebrate Activities that Alter Resource Quantity, Size and Shape

Invertebrates have been shown to alter organic resources in streams by various mechanisms, including grazing of periphyton assemblages by scrapers (=grazers), feeding on macrophytes, shredding of leaf detritus and processing of woody debris. These activities can affect ecosystem structure by reducing standing crops and modifying assemblage structure of both primary producers and heterotrophs. Ecosystem processes can also be changed, for example primary production.

Since early reviews by Gregory (1983) and Lamberti and Moore (1984) there have been numerous studies of periphyton–scraper interactions. A classic example dramatically demonstrating grazing effects on algae was conducted in a California stream by Lamberti and Resh (1983), where grazing caddisflies greatly reduced the standing crop of algae and increased algal turnover rate. The important role of grazers in many streams was highlighted by a meta-analysis of 89 stream experiments, which indicated that grazers maintained at ambient densities reduced periphyton biomass 70% of the time (Feminella and Hawkins, 1995). Furthermore, comparison of periphyton community structure with and without grazers indicated that >80% of the experiments showed grazers altered algal taxonomic and physiognomic structure (Feminella and Hawkins, 1995).

Although macrophytes in streams are generally assumed to be rarely fed upon, invertebrate consumption of living macrophytes may be important (see review by Newman, 1991). For example, floating-leaf macrophytes, such as *Nuphar*, can be heavily grazed and contribute high quality FPOM to the detrital food web throughout the growing season (Wallace and O'Hop, 1985). Some macroinvertebrates have been used successfully as biological control agents of introduced aquatic macrophytes such as alligatorweed and *Hydrilla* in Coastal Plain streams (Buckingham, 1994). However, feeding on submerged macrophytes in streams is poorly studied in North America, although some European studies do indicate that grazing can be significant at certain times during the growing season (Jacobsen, 1993; Jacobsen and Sand-Jensen, 1994a, b).

Overall, the impact of aquatic invertebrates on submerged macrophytes and the ecosystem consequences of shredder-herbivores are not well known (Newman, 1991).

Invertebrate shredders in laboratory streams increase the rate of conversion of CPOM to FPOM (Petersen and Cummins, 1974). Shredders generally have low assimilation efficiencies and high ingestion rates (McDiffett, 1970; Golladay *et al.*, 1983), resulting in much FPOM generation (McDiffett, 1970). A direct relationship was also shown between shredder activity, leaf litter breakdown and subsequent ingestion of fine particles by filter-feeding insects (Short and Maslin, 1977). Furthermore, laboratory feeding studies indicate that shredders facilitate leaching of DOC from CPOM (Meyer and O'Hop, 1983).

The importance of invertebrates in processing large detrital particles is not limited to leaf litter, but also includes woody debris. Wood breakdown is much slower in freshwater than in terrestrial ecosystems (Harmon *et al.*, 1986) and is primarily a surface area phenomenon in freshwater habitats (Triska and Cromack, 1980). Invertebrate shredders and scrapers promote wood decomposition by scraping, gouging and tunnelling through wood (Anderson and Sedell, 1979; Dudley and Anderson, 1982), which exposes additional surfaces to microbial colonization and decomposition (Anderson and Sedell, 1979). An astonishing example of how aquatic insects can increase wood breakdown is an incidence of caddisflies causing a bridge to collapse. The collapse was a result of wood gouging by generations of hydropsychids for their retreats on the wooden bridge pilings (NTSB, 1989). In addition, feeding grooves in wood left by elmid beetles can enhance colonization by other invertebrate species (McKie and Cranston, 1998).

Invertebrate Activities that Enhance Movement of Carbon and Nutrients

Invertebrate assemblages within stream reaches influence the movement or loss of nutrients through their feeding and activity. However, this influence is difficult to estimate because of a number of problems associated with studying fine particle movement. Still, taxa from various FFGs have been found to increase particle export and thus influence the movement of energy in stream ecosystems.

The influence of the entire insect community on particle movement was addressed in studies conducted at the Coweeta Hydrologic Laboratory in western North Carolina. The application of an insecticide to a small headwater stream eliminated >90% of the aquatic insect biomass and greatly reduced secondary production (Lugthart and Wallace, 1992). This manipulation significantly reduced breakdown rates of leaf litter without significant change in microbial respiration rates of litter in treatment and reference streams (Cuffney *et al.*, 1990). In the treatment, or macroinvertebrate-reduction stream, export of FPOM was greatly reduced compared with that of nearby, untreated reference streams (Wallace *et al.*, 1982, 1991; Cuffney *et al.*, 1990). Following treatment, restoration of invertebrate shredder populations coincided with a return of leaf litter processing rates (Wallace *et al.*, 1986; Chung *et al.*, 1993) and FPOM export (Wallace *et al.*, 1991) similar to pretreatment and reference stream levels. These studies demonstrated that macroinvertebrates accounted for 25–28% of annual leaf litter processing (Cuffney *et al.*, 1990) and 56% of FPOM export (Wallace *et al.*, 1991). The 56% reduction in export during the 3-year treatment was equivalent to 161–198 kg ash-free dry mass less FPOM export to downstream reaches (Wallace *et al.*, 1991). These studies also indicate that the influence of invertebrate populations is as great on inorganic as on organic matter export (Wallace *et al.*, 1993). Lower whole-stream export of inorganic particles following pesticide treatment was attributed to at least two mechanisms: (i) the rate of particle generation by feeding activities of animals was reduced; (ii) the increased storage of leaf litter enhanced retention of particles.

We still have much to learn about transport processes of fine particles in streams. McNair *et al.* (1997) identified several major questions including: By what methods do particles become entrained in the water column? What is the length of time required for particles to reach the bottom? How far do particles travel? and What determines whether settled particles are retained or reflected? Although such questions are important to benthic organisms and ecosystem processes, the complications involved with applying the modelling efforts such as those of McNair *et al.* (1997) to heterogeneous areas of streams, which differ with respect to velocity, depth, substrate and turbulence, would appear to be overwhelming. Nevertheless, various taxa of scrapers, gatherers and filterers have been shown to enhance particle export.

Several studies have shown that grazing snails (Mulholland *et al.*, 1983; Lamberti *et al.*, 1989) and insects (Dudley, 1992) increase downstream export of FPOM from grazed substrates. Heavy grazing results in periphyton mats with closely appressed adnate forms of diatoms which are less susceptible to scouring and loss during disturbances such as large storms (Mulholland *et al.*, 1991). Experimental studies have also shown that grazing promotes nutrient turnover in periphyton communities (Steinman *et al.*, 1995). Hence, grazing may result in a consistent, prolonged release of materials to downstream reaches, in contrast with large storms that induce pulsed massive export over short time intervals. A very similar role was suggested for shredders (Wallace *et al.*, 1982; Cuffney *et al.*, 1984).

The role of gatherers in FPOM transport was implicated by Cushing *et al.* (1993). These authors found that labelled FPOM released into an Idaho stream exhibited continuous deposition and resuspension as particles moved downstream. This flux of FPOM indicated that gatherers feeding on FPOM may induce regular, diffuse downstream transmission of their food rather than local depletion (Cushing *et al.*, 1993). Other studies have also suggested that invertebrate collectors increase the downstream movement of materials. For example, in montane Puerto Rican streams, feeding activities of atyid shrimp reduce organic matter accrual on benthic substrates (Pringle *et al.*, 1993). Other invertebrate gatherers transfer fine organic matter buried in depositional areas to substratum surfaces as faeces. Larvae of *Ptychoptera* (Diptera: Ptychopteridae) possess an elongate caudal respiratory tube which allows them to feed on buried organic matter in shallow areas while maintaining contact with atmospheric air. The larvae preferentially feed in substrates that have a high percentage of fine organic matter and high microbial biomass (Wolf *et al.*, 1997). Based on gut passage times, larval life histories and AFDM of faecal pellets, Wolf *et al.* (1997) suggest that *Ptychoptera* transfer 770 g DW m^{-2} year^{-1} ($\approx$ 123 g AFDM) of buried sediments to the substrate surface as faeces. These faeces are readily available as food to other invertebrates, as well as subsequent transport to downstream areas (Wolf *et al.*, 1997).

Although filter-feeding invertebrates typically decrease losses of organic matter and nutrients within a given stream segment (Wallace and Webster, 1996), some may also increase losses. *Ametropus* spp. mayflies are unusual among filterers in that they occupy unstable, sandy-bottom areas of large rivers. *Ametropus* uses the head, mouthparts and forelegs to create a shallow pit, which initiates a unique vortex in front of the head and results in resuspension of fine organic matter as well as occasional sand grains. Some of the resuspended particles are then apparently captured by setae on the mouthparts and forelegs. Hence, *Ametropus* exploit the abundant fine organic particles entrapped in the sediments of large sandy-bottom rivers, but also increase their movement downstream (Soluk and Craig, 1988).

Invertebrate Activities that Enhance Retention

Retention of organic matter and nutrients is affected primarily by filter-feeding stream invertebrates; however, other functional groups may also be important. For example, cycling of nutrients in food webs can immobilize nutrients in stream reaches. Also, burrowing activity by invertebrates can transport surface organic matter to deeper sediments, which reduces downstream transport. In addition, structures created by some insects can serve as sites of travertine (CaCO$_3$) deposition. Nevertheless, direct removal of transported material by filter-feeders has received the bulk of attention and has been shown to have variable effects on retention depending on size of stream, abundance of filterers and taxa-specific differences in feeding.

The transformation and storage of nutrients by invertebrates can have a significant influence on nutrient export and overall efficiency of nutrient retention within streams (Merritt *et al.*, 1984), but quantifying the role of invertebrates in nutrient cycling in streams has received little study. A 6-week nutrient tracer addition ([^{15}N] NH$_4$) to a small tundra river in Alaska during the summer resulted in rapid labelling of primary producers (filamentous algae, epilithon and moss), which was then quickly incorporated into invertebrate scrapers, for example *Baetis* and chironomid larvae (Peterson *et al.*, 1997). Among filtering invertebrates, such as *Prosimulium*, the peak in label within larvae occurred further downstream than in scrapers, suggesting that these downstream filterers were relying on FPOM produced by sloughing of epilithon (Peterson *et al.*, 1997). Presumably scraper activity, including egested faeces, also contributed to the downstream increase in label found in simuliids. Those aquatic insects with long-lived larval stages (>1 year, i.e. the caddisflies *Brachycentrus* and *Rhyacophila*) retained the labelled nitrogen for at least 2 years following termination of the ^{15}N release. In fact, 1 year following the treatment these caddisflies retained a higher proportion of labelled nitrogen than the epilithon (Peterson *et al.*, 1997). It should be noted that even those animals with shorter life histories (e.g. chironomids and simuliids) emerge and transport nutrients back to the terrestrial environment, which also reduces losses of nutrients from the catchment.

Some burrowing animals increase transfer of organic matter from the stream bed surface to deeper sediments, thereby making the organic matter less amenable to downstream transport, and enhancing retention. Larvae of the European sericostomatid caddisfly, *Sericostoma personatum*, feed on surface CPOM at night and burrow into the stream bed during the day. Subsequent defecation by *S. personatum* increased subsurface sediment organic content by 75–185% compared with controls containing no sericostomatid larvae in the laboratory (Wagner, 1991). A somewhat similar phenomenon may occur with other sericostomatid larvae such as the western North American species, *Gumaga nigricula*, which transfers case-associated algae into the sediments by abrasion during burrowing (Bergey and Resh, 1994).

Although filter-feeding hydropsychid caddisflies primarily retain organic matter through their filtering (see below), their retreats and silken nets can also act as important substrates for calcium carbonate precipitation in travertine streams (Drysdale, 1999). Drysdale found that *Cheumatopsyche* larvae in a small, spring-fed stream in Australia enhanced rates of travertine deposition in three ways. First, the retreat and net act as sites for $CaCO_3$ precipitation, and become completely encrusted over time. Second, the encrusted retreats and nets that are incorporated into the travertine increase rock porosity, and increase the accumulation rate. Third, the retreats and nets increase bed surface roughness, which disrupts flow and enhances turbulence, thereby enhancing rates of CO_2 outgassing and $CaCO_3$ precipitation. Overall, these hydropsychid structures enhanced travertine deposition from about twofold in areas where background rates of precipitation were high, to about 20-fold where background rates were low (Drysdale, 1999).

Streams transport large amounts of organic matter, primarily in the form of DOM and FPOM. Most seston carried by streams consists of particles < 50 μm in diameter, and many groups including some ephemeropterans, trichopterans (Philopotamidae) and dipterans (Simuliidae and some Chironomidae) have evolved mechanisms to feed on these minute particles (Wallace and Merritt, 1980). Bivalves consume even smaller particles on average, generally ranging from ≤1 to 10 μm (Thorp and Covich, 1991). Microfilterers, such as the Philopotamidae, Simuliidae and bivalves, increase particle sizes by ingesting minute particles and egesting compacted faecal particles larger than those originally consumed. Such microfilterers perform two very important functions in streams. First, they remove fine particulate organic matter from suspension (which would otherwise pass unused through the stream segment) and second, they defecate larger particles, which are available to deposit-feeding detritivores (Wallace and Webster, 1996). Filterers were hypothesized to reduce downstream transport of suspended POM (Wallace *et al.*, 1977) and decrease spiralling distances of nutrients and organic matter (Webster and Patten, 1979; Newbold *et al.*, 1982).

Most studies of seston removal by filterers have suggested low rates of seston removal, i.e. generally well below 1% seston removal per m length of stream (Georgian and Thorp, 1992). The highest rates of seston removal by insect filter-feeders in streams have been obtained in studies that incorporated microfilterers such as Simuliidae. Morin *et al.* (1988) found that simuliid larvae ingested 0.8–1.4% of the seston per m of stream below a Quebec lake outlet; these are the highest rates of seston removal reported. The study of Morin *et al.* was made during low flows in late spring when the standing stock of black flies was high, whereas other studies, performed on an annual basis, indicated lower rates of seston removal. However, even low rates of seston removal can have a surprising impact on removal of FPOM from the water column. Recently, Wotton *et al.* (1998) studied a short length of a small Swedish stream which drained directly into the Baltic Sea, and found that 33% of the faecal pellets produced by black fly larvae were sedimented or intercepted before reaching the Baltic. Wotton *et al.* (1998) found that filter-feeder removal of seston was only 0.06% per m linear stream length. Despite this low rate of removal, Wotton *et al.* (1998) estimated that black fly faeces added 87 g C m^{-2} year^{-1} to the stream bottom using a conservative estimate of black fly larvae being in the stream for only 1 month. These faecal particles are thus available to a number of deposit-feeding collectors (Wotton *et al.*, 1998). Hall *et al.* (1996) used fluorescently labelled bacteria (FLB) and found much lower rates of removal than those reported by either Morin *et al.* (1988) or Wotton *et al.* (1998). In the headwater stream studied by Hall *et al.* (1996) only 7% of the FLB removal was by filter-feeding insects, with simuliid larvae accounting for most of the 7% removed. Instead, most FLB were removed by sedimentation. However, densities of black flies in the study of Wotton *et al.* (1998) were 52–125 times higher than those studied by Hall *et al.* (1996), which probably explains the differences in removal rates.

Rates of seston removal by filtering insects are much lower for large rivers than the above estimates for small streams (Benke and Parsons, 1990; Benke and Wallace, 1997; but see below for molluscs). In the Ogeechee River, Georgia, caddisflies removed only ≈ 0.0001% of seston per linear m of stream length; however, their removal of animal drift was about 190 times greater than seston particles such that the caddisflies were capable of removing all drifting animals in about 5 km (Benke and Wallace, 1997). Indeed, many filter-feeding macro-invertebrates consume a wide array of food types. Larger particle-feeding hydropsychids (Trichoptera) are selective towards higher quality food items such as diatoms and animal drift (Benke and Wallace, 1980, 1997; Petersen, 1985, 1989). This, and generally low rates of seston removal by hydropsychids, suggests that their major impact is on the quality and type of POM in suspension (Benke and Wallace, 1980, 1997; Petersen, 1989; Georgian and Thorp, 1992). Georgian and Thorp (1992) estimated that two *Hydropsyche* species in riffles of a small New York stream removed drifting invertebrate prey at a rate of 18% per m. The experiments by Georgian and Thorp are very relevant to ecologists studying downstream drift of invertebrates because they show that when large net-spinning caddisfly populations are present in shallow streams, *Hydropsyche* predation may reduce stream drift.

Bivalve molluscs represent another agent of seston removal by filter-feeding invertebrates. In the Ogeechee River, Georgia, Stites *et al.* (1995) used secondary production and bioenergetic efficiencies to estimate seston removal by the Asiatic clam, *Corbicula fluminea*, and found little evidence of significant seston removal. In contrast, *C. fluminea* populations in the Potomac River estuary reached extremely high densities and were estimated to filter about one-sixth of the estuary volume each day (Phelps, 1994). The author also attributed an increase in submerged aquatic vegetation to increases in water clarity and light penetration, resulting from removal of phytoplankton by *Corbicula* (Phelps, 1994).

Invasion of the zebra mussel, *Dreissena polymorpha*, into the freshwater tidal portion of the lower Hudson River resulted in massive abundance, estimated as ~ 550 billion individuals river-wide, and was considered to be the main agent behind a decline in phytoplankton (Caraco *et al.*, 1997). Summer chlorophyll concentrations decreased from 30 µg l^{-1} prior to invasion to < 5 µg l^{-1} after establishment of *Dreissena* (Caraco *et al.*, 1997). This decline in phytoplankton was followed by a strong decline (>70%) in zooplankton populations in the Hudson River, including flagellated protozoans (Pace *et al.*, 1998).

Zebra mussels are not efficient removers of sestonic bacteria, but are very effective removers of flagellated protozoans, which are major predators of bacteria. The decline in zooplankton, including flagellated protozoans, in the Hudson River resulted in an overall increase in planktonic bacteria (Findlay *et al.*, 1998). Although bacteria abundance increased, production did not, indicating that bacterial growth was limited by carbon supply following the invasion by *Dreissena* (Findlay *et al.*, 1998). In the River Spree, a lake-outflow stream in Germany, bivalve populations, including *Dreissena* and unionid mussels, were estimated to filter the entire daily discharge over a 21-km reach of the river (Welker and Walz, 1998). Furthermore, Welker and Walz (1998) observed significant downstream decreases in chlorophyll as well as in zooplankton populations, which were attributed to mussel feeding. Clearly, intense filter-feeding activity by bivalves can produce pronounced effects in food webs and has many indirect effects in rivers.

Extremely large standing stocks of filtering insects such as caddisflies and black flies can be found under certain conditions including: limited stable substrate; sufficient current velocity; and high-quality organic seston concentrations (Fremling, 1960; Parker and Voshell, 1983; Voshell, 1985; Wotton, 1987). Invariably, some of the highest levels of secondary production in the world are those of filter feeders in streams. Such high densities and production of filterers, compared with all other functional groups, are possible because filterers are using the kinetic energy of the current to exploit foods produced in other habitats and made available to them by the current (Cudney and Wallace, 1980). As a consequence, less energy is expended in search of food, and extremely high biomass per unit area can be supported (Cudney and Wallace, 1980). Thus, in habitats with high particle transport, filterers exploit the physical environment and increase particle retention.

Diversity of Roles for Predators

Predators can play numerous diverse roles at scales ranging from individuals to ecosystems, and invertebrate (and vertebrate) predators in streams are no exception. Predators can influence export and retention of energy and nutrients through their effects on the standing stocks of other functional groups (Fig. 4.2). Other mechanisms include decreasing rates of nutrient cycling by immobilizing nutrients in long-lived predator taxa versus short-lived prey. Besides direct consumption, foraging by invertebrate predators can enhance invertebrate drift and suspended FPOM, which also increases export of nutrients. Although these specific predation effects have been documented, it has often proved difficult to quantify precisely how predators influence stream processes.

Invertebrate predators can enhance retention of organic matter by retarding breakdown rates of leaf litter as well as subsequent generation of FPOM. For example, predaceous plecopterans and caddisflies significantly decreased the rate of leaf litter processing by reducing shredder populations in leaf packs (Oberndorfer *et al.*, 1984). Predator densities used by Oberndorfer *et al.* were almost 10 times those of background. However, Malmqvist (1993) used more realistic densities of predators and was not able to demonstrate a reduction in shredder densities in the presence of a predatory stonefly, *Diura*. Still, he did find that less leaf material was processed in cages with predators. In additional laboratory feeding experiments, two of three shredder species produced less FPOM when exposed to predators (Malmqvist, 1993). Finally, the study by Peterson *et al.* (1997) (see above) indicated that long-lived predator taxa (i.e. *Rhyacophila*) can retain tracer additions of nitrogen long after short-lived scraper and filterer larvae have left the stream as adults.

Invertebrate predators can also increase the rate of downstream movement of organisms and sediment. Many stream invertebrates exhibit different responses to fish and invertebrate predators, and the local impact of invertebrate

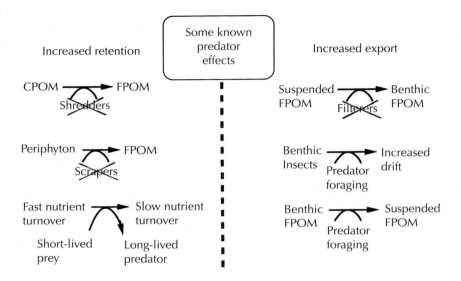

Fig. 4.2. Predators affect many lotic processes by either decreasing (left side of dashed line) or increasing (right side of dashed line) the movement of organic matter and nutrients downstream. These effects result from direct predation on various functional feeding groups or indirectly through their foraging activity. See text for details.

predation on benthic prey may exceed the impact of fish predators (Sih and Wooster, 1994). In the presence of fish, invertebrate prey often reduce movement and seek refuge in the substrate. In contrast, invertebrate predators have the ability to search in sites similar to those being used by their prey and the latter may respond by actively entering the water column and drifting downstream (Sih and Wooster, 1994; Wooster, 1994). Foraging by invertebrate predators can also influence the downstream movement of inorganic material through their physical activities. For example, foraging by a European stonefly, *Dinocras* sp., was suggested to have an erosion potential of about 200–400 kg sand m^{-2} year^{-1} at natural population densities and under favourable flow conditions (Statzner *et al.*, 1996). In addition, hungry stonefly predators tend to remove more fine sediment than fed stoneflies apparently because hungry predators forage more intensively in the interstitial spaces, thereby dislodging fine sediments (Zanetell and Peckarsky, 1996). The unusual predaceous mayfly, *Pseudiron* sp., feeds primarily on chironomid larvae inhabiting sandy substrates of large rivers. *Pseudiron* digs pits, which are enlarged by vortex currents around their head, promoting both exposure of chironomid prey and accelerating downstream movement of particles (Soluk and Craig, 1990).

Predators can also cause an increase in primary producers through a reduction in dominant primary consumers. This was well documented by

Kohler and Wiley (1997), who showed that periphyton biomass increased in several Michigan streams after the collapse of a dominant scraping caddisfly, *Glossosoma*. This collapse was caused by a parasite, a sub-category of predators, which only affected *Glossosoma*.

The results of Kohler and Wiley (1997) may also provide insight into why the overall impact of predators is difficult to detect at the community level. The microsporidian parasite acts on a specific taxon, whereas most invertebrate predators consume multiple prey taxa (Allan, 1983; Hall *et al.*, 1996). In fact, many invertebrate predators feed on prey at multiple trophic levels, including other predators (i.e. intraguild predation *sensu* Polis and Holt, 1992), which complicates conventional food chain theory (Polis, 1994; Hall *et al.*, 1996; Polis and Strong, 1996). Intraguild predation of other invertebrate predators accounts for 16–27% of all predator ingestion in a small mountain stream at Coweeta; hence, this predation is diffusely distributed among multiple taxa (Hall *et al.*, 1996).

Diffuse predation may be an important reason why the effects of predators are often difficult to show at larger scales in many streams (however, see Power, 1990; Huryn, 1998). In his review of stream food webs, Hildrew (1992) stated the problem with predators as: 'the inference is that either predation is really dynamically trivial in running waters or the experiments are unable to detect the effects through some details of scale, timing, or design'. There is certainly evidence that predators do consume a large quantity of available prey. During a 4-year litter exclusion study for an entire stream at Coweeta, total secondary production in the dominant mixed-substrate habitats declined to 22% of pre-treatment values by the 4th year, the lowest secondary production reported for streams worldwide (Benke, 1993). Litter exclusion clearly showed strong bottom-up effects extending from primary consumers to predators (Wallace *et al.*, 1997, 1999). Production by predators in the treatment stream was also strongly related to that of their prey. Based on simple bioenergetic efficiencies, the results suggest that production by predators is constrained by productivity of their prey. Thus, despite obvious bottom-up control of productivity in this system, the importance of predators (top-down) cannot be ignored as predators consume most of the benthic invertebrate production (Wallace *et al.*, 1997, 1999). Indeed, detecting such diffuse predation in this detritus-based stream would probably be impossible without multi-year studies that assessed secondary production while simultaneously curtailing the detritus food base.

Some stream studies, which have fish as top predators, have shown strong top-down effects on primary production. Such 'trophic cascades' result when feeding activities of a population at one trophic level have quantitative effects on material and energy flow through non-adjacent trophic levels (Carpenter *et al.*, 1985; Power, 1992a; Huryn, 1998). To our knowledge such trophic cascades are most pronounced in streams (or portions thereof) that have simple food chains where autochthonous production is the primary energy resource and fish are top predators. In most cases, effects were seen only in some species or for certain substrates for a short time (Power, 1990, 1992b); however, at least one study (Huryn, 1998) provides compelling evidence for strong top-down effects for an entire year. The introduction of brown trout into a New Zealand stream apparently produced sufficient top-down effects on herbivorous invertebrates to influence standing crop and increase productivity and biomass of periphyton. In contrast, in a nearby stream, the impact of native galaxias fish on herbivorous invertebrates resulted in much less pressure on herbivore species, lower periphyton biomass and productivity (Huryn, 1998). Although fish are the top consumers in these systems, we should not forget that it is through the invertebrates that the effects are propagated either up or down in food webs (Hunter and Price, 1992; Wallace and Webster, 1996).

Large-scale Physical Heterogeneity and Invertebrate Assemblages

In addition to local physical heterogeneity, there are other large-scale physical changes along stream gradients. These large-scale changes were incorporated into a general framework of stream ecosystems as the river continuum concept (RCC) (Vannote *et al.*, 1980; Minshall *et al.*, 1983). However, results from many areas around the world indicate that changes in energy inputs and biological communities proposed in the RCC are not applicable to all river systems (e.g. Statzner and Higler, 1985; Meyer, 1990; Cushing *et al.*, 1995). In the southern Appalachian Mountains, many biological attributes of streams roughly correspond to those proposed in the original RCC, although some exceptions exist (Grubaugh *et al.*, 1997). Striking differences in functional group production occur at opposite extremes of the Coweeta–Little Tennessee River continuum (Grubaugh *et al.*, 1997). For example, in the headwater stream, shredders, gatherers and predators dominate the invertebrate assemblage, whereas in the larger Little Tennessee River, 80% of the secondary production is attributable to filter-feeding taxa (Table 4.1). The most remarkable difference between sites is that the production at the river site exceeds that of the shaded, headwater stream by 20-fold.

In part, the high production at the larger Little Tennessee River site is enhanced by the presence of the aquatic macrophyte, *Podostemum ceratophyllum* (Michaux). *Podostemum* probably plays several vital roles in accounting for the high production of invertebrates, including: adding a three-dimensional structure to the substrate and providing additional habitat for invertebrates (Grubaugh *et al.*, 1997); retaining organic matter within the dense macrophyte mat and providing food to non-filtering organisms; adding surface area for colonization by epiphytic diatoms and enhancing algal availability to benthic organisms; providing refugia from foraging predators; and affording protection for invertebrates from physical disturbance such as storm flows, because *Podostemum* is usually associated with large stable substrate. Hence, it is not surprising that in eastern North America, some of the highest levels of invertebrate production in streams, especially for filterers, are associated with

Table 4.1. Habitat-weighted functional group secondary production of benthic invertebrates from two locations along the Little Tennessee River–Coweeta continuum. Catchment 53 (C 53) is a 1st order stream (Wallace *et al.*, 1997) and the Little Tennessee River is a 7th order river (Grubaugh *et al.*, 1997).

Functional feeding group	Production (g AFDM m^{-2} year^{-1})		Percentage of total production	
	C 53	Little Tennessee River	C 53	Little Tennessee River
Scrapers	<0.1	6.7	0.3	4.3
Shredders	2.8	5.3	36.4	3.4
Gatherers	2.1	12.8	26.9	8.3
Filterers	0.4	122.9	5.5	80.0
Predators	2.4	6.2	30.9	4.0
Total	7.7	153.9		

Podostemum (Parker and Voshell, 1983; Voshell, 1985; Grubaugh and Wallace, 1995; Grubaugh *et al.*, 1997).

The dissimilarities in production and community structure between upstream and downstream sites of the Little Tennessee River continuum are a result of resources being unequally distributed along the river gradient. Hall *et al.* (2000) and Rosi (1997) used the trophic basis of production method (Benke and Wallace, 1980) to estimate annual food consumption by invertebrates at both locations (Table 4.2). Note that invertebrates in the headwaters consume primarily leaf and amorphous detritus stored in the stream, whereas downstream consumption is dominated by amorphous detritus and animal tissue (Table 4.2). In the headwater stream draining a deciduous forest, there is an abundant supply of stored allochthonous detritus, and benthic organic matter exceeds that of the downstream site by more than eightfold (Table 4.3). In contrast, annual transport of organic matter per linear m of stream of the larger river site exceeds that of the headwater stream by more than 260-fold. Clearly, there are marked differences in the form of organic matter, that is, stored versus transported, available to the benthos (Table 4.3).

The functional structure of the animal community (Table 4.1) reflects this difference in organic matter supply. The high consumption required to support production at the Little Tennessee River site is a direct result of the dominance of filter-feeding hydropsychid caddisflies, which rely on the transport of organic matter and invertebrates from upstream. Obviously, the extraordinarily high consumption required to support secondary production of the Little Tennessee River could not occur without a massive subsidy of organic material from upstream areas delivered by the flow. For example, animal consumption at the study site is greater than local production (Table 4.2), which can only be sustained by input from upstream. Furthermore, the consumption required to support total secondary production in the Little Tennessee River exceeds that of the benthic organic matter standing crop, unlike at the headwater site (Table

Table 4.2. Estimated annual consumption of various food types for all macroinvertebrates to account for their secondary production in a 1st order stream (C 53) and the 7th order Little Tennessee River using the trophic basis of production methods of Benke and Wallace (1980, 1997). Data for C 53 are recalculated from July and December values from Hall *et al.* (2000) and data for Little Tennessee River from Rosi (1997).

Food type consumed	Consumption ($g \, AFDM \, m^{-2} \, year^{-1}$)		Percentage of total consumption	
	C 53	Little Tennessee River	C 53	Little Tennessee River
Algae	0.5	51.6	0.3	4.4
Leaf detritus	65.7	73.7	50.2	6.4
Wood detritus	5.3	0.0	4.1	0.0
Amorphous detritus	51.1	734.3	39.1	63.3
Fungi	1.0	0.0	0.8	0.0
Animal tissue	7.3	299.8	5.6	25.9
Total	130.9	1159.4		

4.3). This is also a consequence of the filterers in the Little Tennessee River being supported by the most available resource, FPOM in transport, which is in relatively short supply in the headwaters (Table 4.3). Although most other FFGs at the downstream site are a minor component of total production, their absolute production still surpasses that of the upstream site.

Table 4.3. Some relationships between consumption required to support macroinvertebrate production (see Table 4.2), benthic organic matter standing crops, and FPOM transport (seston) for a headwater stream (C 53) and the 7th order Little Tennessee River.

Parameter	C 53	Little Tennessee River
Consumption required for production ($kg \, AFDM \, ^{-2} \, year^{-1}$)	0.15	1.16
Benthic organic matter standing crop ($kg \, AFDM \, m^2$)	>2.65[a]	<0.32[b]
Annual FPOM transport ($kg \, AFDM \, m^{-1} \, year^{-1}$)	141[c]	37,000[d]
Organic matter standing crop: consumption	17.7	0.27
Organic matter transport: consumption	940	31,900

[a] From Wallace *et al.* (1999); does not include large woody debris.
[b] From Grubaugh *et al.* (1997); over 50% of the 0.32 kg is an aquatic macrophyte, *Podostemum ceratophyllum* (Michaux), and not readily available for invertebrate consumption.
[c] Wallace *et al.* (unpublished).
[d] Rosi (1997).

From the broader perspective of the entire basin, headwater streams tend to be shallow, easily obstructed by woody debris, have low stream power and have physical characteristics favouring retention of organic matter inputs from the surrounding forest. The invertebrate assemblage in these small, fishless streams exploits the physical environment and is dominated by shredders, gatherers and invertebrate predators (Table 4.1). Their feeding activities tend to decrease particle size of organic resources and favour downstream export of FPOM, which is more amenable to transport than larger CPOM. By contrast, downstream reaches have much higher discharge, greater stream power and less retention, which promote entrainment of organic matter. Again, the invertebrate assemblage exploits the physical characteristics of the system by a dominance of filterers (Table 4.1), which promote retention of entrained organic matter. Thus, in both the upstream and downstream areas, the invertebrate feeding assemblages have evolved to reflect the physical characteristics of the system.

Conclusions

Many of the processes that we have identified in this chapter occur at scales much smaller, i.e. within patches of substratum, than that of a stream reach. However, while small patches have received some attention among stream ecologists (Minshall, 1988; Pringle *et al.*, 1988), the more difficult problem is assessing how various patches contribute to a given stream reach (Minshall, 1988). Even within small headwater streams of the Little Tennessee River basin, localized habitats, for example moss-covered bedrock, can have an invertebrate assemblage more characteristic of larger downstream reaches (Huryn and Wallace, 1987). Furthermore, within a reach, the relative frequency of different patch types influences invertebrate assemblage structure and hence ecosystem processes within the reach. Over an entire stream continuum, there are thousands of individual patches, whose functions may vary depending on local physical attributes, for example retention or entrainment. There is also the difficult question of how materials are transferred between successive longitudinal patches (e.g. McNair *et al.*, 1997). Obviously, the size and contribution of patches with their respective invertebrate assemblages will need to be incorporated into whole-stream studies, if we are to begin to assess the contribution of benthos to processes at the ecosystem level.

These considerations raise the following questions. What spatial and temporal changes occur within and among patches? What is the relative contribution of major patches to secondary production and biotic processes within a reach? What influences the effectiveness of a patch in assemblage function and in its role in ecosystem processes? Finally, there is the broader question: How do regional differences in slope, geology, watershed area, as well as other abiotic factors, and anthropogenic activities influence patch formation and function? Clearly, it is one thing to show that individual taxa or functional groups of organisms may affect particular processes in streams; however, extrapolating these processes to entire streams will be challenging.

References

Allan, J.D. (1983) Predator–prey relationships in streams. In: Barnes, J.R. and Minshall, G.W. (eds) *Stream Ecology: Application and Testing of General Ecological Theory.* Plenum, New York, pp. 191–229.

Anderson, N.H. and Sedell, J.R. (1979) Detritus processing by macroinvertebrates in stream ecosystems. *Annual Review of Entomology* 24, 351–377.

Benke, A.C. (1993) Concepts and patterns of invertebrate production in running waters. *Verhandlungen der Internationalen Vereinigung für Theoretische und Angewandte Limnologie* 25, 15–38.

Benke, A.C. and Parsons, K.A. (1990) Modelling black fly production dynamics in blackwater streams. *Freshwater Biology* 24, 167–180.

Benke, A.C. and Wallace, J.B. (1980) Trophic basis of production among net-spinning caddisflies in a southern Appalachian stream. *Ecology* 61, 108–118.

Benke, A.C. and Wallace, J.B. (1997) Trophic basis of production among riverine caddisflies: implications for food web analysis. *Ecology* 78, 1132–1145.

Bergey, E.A. and Resh, V.H. (1994) Effects of burrowing by a stream caddisfly on case-associated algae. *Journal of the North American Benthological Society* 13, 379–390.

Brussock, P.P., Brown, A.V. and Dixon, J.C. (1985) Channel form and stream ecosystem models. *Water Resources Bulletin* 21, 859–866.

Buckingham, G.R. (1994) Biological control of weeds by insects. *Journal of the Georgia Entomological Society* (2nd Suppl.) 19, 62–78.

Caraco, N.F., Cole, J.J., Raymond, P.A., Strayer, D.L., Pace, M.L., Findlay, S.E.G. and Fischer, D.T. (1997) Zebra mussel invasion in a large, turbid river: phytoplankton response to increased grazing. *Ecology* 78, 588–602.

Carpenter, S.R., Kitchell, J.F. and Hodgson, J.R. (1985) Cascading trophic interactions and lake productivity. *BioScience* 35, 634–639.

Chung, K., Wallace, J.B. and Grubaugh, J.W. (1993) The impact of insecticide treatment on abundance, biomass, and production of litterbag fauna in a headwater stream: a study of pretreatment, treatment, and recovery. *Limnologica* 28, 93–106.

Cudney, M.D. and Wallace, J.B. (1980) Life cycles, microdistribution and production dynamics of six species of net-spinning caddisflies in a large southeastern (U.S.A.) river. *Holarctic Ecology* 3, 169–182.

Cuffney, T.F., Wallace, J.B. and Webster, J.R. (1984) Pesticide manipulation of a headwater stream ecosystem: significance for ecosystem processes. *Freshwater Invertebrate Biology* 3, 153–171.

Cuffney, T.F., Wallace, J.B. and Lugthart, G.J. (1990) Experimental evidence quantifying the role of benthic invertebrates in organic matter dynamics of headwater streams. *Freshwater Biology* 23, 281–299.

Cummins, K.W. (1973) Trophic relations of aquatic insects. *Annual Review of Entomology* 18, 183–206.

Cummins, K.W. (1986) The functional role of black flies in stream ecosystems. In: Kim, K.C. and Merritt, R.W. (eds) *Black Flies: Ecology, Population Management, and Annotated World List.* Pennsylvania State University Press, University Park, pp. 1–10.

Cummins, K.W. and Klug, M.J. (1979) Feeding ecology of stream invertebrates. *Annual Review of Ecology and Systematics* 10, 147–172.

Cushing, C.E., Minshall, G.W. and Newbold, J.D. (1993) Transport dynamics of fine particulate organic matter in two Idaho streams. *Limnology and Oceanography* 38, 1101–1115.

Cushing, C.E., Cummins, K.W. and Minshall, G.W. (eds) (1995) *River and Stream Ecosystems. Ecosystems of the World*, Vol. 22. Elsevier, Amsterdam.

Drysdale, R.N. (1999) The sedimentological significance of hydropsychid caddis-fly larvae (Order: Trichoptera) in a travertine-depositing stream: Louie Creek, Northwest Queensland, Australia. *Journal of Sedimentary Research* 69, 145–150.

Dudley, T. (1992) Beneficial effects of herbivores on stream macroalgae via epiphyte removal. *Oikos* 65, 121–127.

Dudley, T. and Anderson, N.H. (1982) A survey of invertebrates associated with wood debris in aquatic habitats. *Melanderia* 39, 1–21.

Feminella, J.W. and Hawkins, C.P. (1995) Interactions between stream herbivores and periphyton: a quantitative analysis of past experiments. *Journal of the North American Benthological Society* 14, 465–509.

Findlay, S., Pace, M.L. and Fischer, D.T. (1998) Response of heterotrophic planktonic bacteria to the zebra mussel invasion of the tidal freshwater Hudson River. *Microbial Ecology* 36, 131–140.

Fremling, C.R. (1960) Biology and possible control of nuisance caddisflies of the upper Mississippi River. *Research Bulletin of the Iowa Agriculture Experiment Station* 348, 856–879.

Frissell, C.A., Liss, W.J., Warren, C.E. and Hurley, M.D. (1986) A hierarchical framework for stream habitat classification: viewing streams in a watershed context. *Environmental Management* 10, 199–214.

Georgian, T. and Thorp, J.H. (1992) Effects of microhabitat selection on feeding rates of net-spinning caddisfly larvae. *Ecology* 73, 229–240.

Golladay, S.W., Webster, J.R. and Benfield, E.F. (1983) Factors affecting food utilization by a leaf shredding aquatic insect: leaf species and conditioning time. *Holarctic Ecology* 6, 157–162.

Gregory, S.V. (1983) Plant–herbivore interactions in stream systems. In: Barnes, J.R. and Minshall, G.W. (eds) *Stream Ecology: Application and Testing of General Ecological Theory*. Plenum, New York, pp. 157–189.

Grubaugh, J.W. and Wallace, J.B. (1995) Functional structure and production of the benthic community in a Piedmont river: 1956–1957 and 1991–1992. *Limnology and Oceanography* 40, 490–501.

Grubaugh, J.W., Wallace, J.B. and Houston, E.S. (1997) Production of benthic macro-invertebrate communities along a southern Appalachian river continuum. *Freshwater Biology* 37, 581–596.

Hall, R.O., Peredney, C.L. and Meyer, J.L. (1996) The effect of invertebrate consumption on bacterial transport in a mountain stream. *Limnology and Oceanography* 41, 1180–1187.

Hall, R.O., Wallace, J.B. and Eggert, S.L. (2000) Organic matter flow in stream food webs with reduced detrital resource base. *Ecology* (in press).

Harmon, M.E., Franklin, J.F., Swanson, F.J., Sollins, P., Gregory, S.V., Lattin, J.D., Anderson, N.H., Cline, S.P., Aumen, N.G., Sedell, J.R., Lienkaemper, G.W., Cromack, K. and Cummins, K.W. (1986) Ecology of coarse woody debris in temperate ecosystems. *Advances in Ecological Research* 15, 133–302.

Hildrew, A.G. (1992) Food webs and species interactions. In: Calow, P. and Petts, G.E.
  (eds) *The Rivers Handbook: Hydrological and Ecological Principles*, Vol. 1. Blackwell,
  Oxford, pp. 309–330.
Hunter, M.D. and Price, P.W. (1992) Playing chutes and ladders: heterogeneity and the
  relative roles of bottom-up and top-down forces in natural communities. *Ecology* 73,
  724–732.
Huryn, A.D. (1998) Ecosystem-level evidence for top-down and bottom-up control of
  production in a grassland stream system. *Oecologia* 115, 173–183.
Huryn, A.D. and Wallace, J.B. (1987) Local geomorphology as a determinant of macrofaunal production in a mountain stream. *Ecology* 68, 1932–1942.
Jacobsen, D. (1993) Trichopteran larvae as consumers of submerged angiosperms in
  running waters. *Oikos* 67, 379–383.
Jacobsen, D. and Sand-Jensen, K. (1994a) Invertebrate herbivory on the submerged
  macrophyte *Potamogeton perfoliatus* in a Danish stream. *Freshwater Biology* 31, 43–52.
Jacobsen, D. and Sand-Jensen, K. (1994b) Growth and energetics of a trichopteran larva
  feeding on fresh submerged and terrestrial plants. *Oecologia* 97, 412–418.
Jones, J.B. (1997) Benthic organic matter storage in streams: influence of detrital import
  and export, retention mechanisms, and climate. *Journal of the North American
  Benthological Society* 16, 109–119.
Kohler, S.L. and Wiley, M.J. (1997) Pathogen outbreaks reveal large-scale effects of
  competition in stream communities. *Ecology* 78, 2164–2176.
Lamberti, G.A. and Moore, J.W. (1984) Aquatic insects as primary consumers. In: Resh,
  V.H. and Rosenberg, D.M. (eds) *The Ecology of Aquatic Insects*. Praeger, New York,
  pp. 164–195.
Lamberti, G.A. and Resh, V.H. (1983) Stream periphyton and insect herbivores: an
  experimental study of grazing by a caddisfly population. *Ecology* 64, 1124–1135.
Lamberti, G.A., Gregory, S.V., Ashkenas, L.R., Steinman, A.D. and McIntire, C.D. (1989)
  Productive capacity of periphyton as a determinant of plant–animal interactions in
  streams. *Ecology* 70, 1840–1856.
Lugthart, G.J. and Wallace, J.B. (1992) Effects of disturbance on benthic functional
  structure and production in mountain streams. *Journal of the North American
  Benthological Society* 11, 138–164.
Malmqvist, B. (1993) Interactions in stream leaf packs: effects of a stonefly predator on
  detritivores and organic matter processing. *Oikos* 66, 454–462.
McDiffett, W.F. (1970) The transformation of energy by a stream detritivore, *Pteronarcys
  scotti* (Plecoptera). *Ecology* 51, 975–988.
McKie, B.G.L. and Cranston, P.S. (1998) Keystone coleopterans? Colonization by wood-
  feeding elmids of experimentally immersed woods in south-eastern Australia.
  Marine and Freshwater Research 49, 79–88.
McNair, J.N., Newbold, J.D. and Hart, D.D. (1997) Turbulent transport of suspended
  particles and dispersing benthic organisms: how long to hit bottom? *Journal of
  Theoretical Biology* 188, 29–52.
Merritt, R.W. and Cummins, K.W. (eds) (1996) *An Introduction to the Aquatic Insects of
  North America*, 2nd edn. Kendall/Hunt, Dubuque, Iowa.
Merritt, R.W., Cummins, K.W. and Burton, T.M. (1984) The role of aquatic insects in the
  processing and cycling of nutrients. In: Resh, V.H. and Rosenberg, D.M. (eds) *The
  Ecology of Aquatic Insects*. Praeger, New York, pp. 134–163.
Meyer, J.L. (1990) A blackwater perspective on riverine ecosystems. *BioScience* 40,
  643–651.

Meyer, J.L. and O'Hop, J. (1983) Leaf-shredding insects as a source of dissolved organic carbon in headwater streams. *American Midland Naturalist* 109, 175–183.

Meyer, J.L., Wallace, J.B. and Eggert, S.L. (1998) Leaf litter as a source of dissolved organic carbon in streams. *Ecosystems* 1, 240–249.

Mihuc, T.B. (1997) The functional trophic role of lotic primary consumers: generalist versus specialist strategies. *Freshwater Biology* 37, 455–462.

Minshall, G.W. (1988) Stream ecosystem theory: a global perspective. *Journal of the North American Benthological Society* 7, 263–288.

Minshall, G.W., Petersen, R.C., Cummins, K.W., Bott, T.L., Sedell, J.R., Cushing, C.E. and Vannote, R.L. (1983) Interbiome comparison of stream ecosystem dynamics. *Ecological Monographs* 53, 1–25.

Morin, A., Back, C., Chalifour, A., Boisvert, J. and Peters, R.H. (1988) Effect of black fly ingestion and assimilation on seston transport in a Quebec lake outlet. *Canadian Journal of Fisheries and Aquatic Sciences* 45, 705–714.

Mulholland, P.J., Newbold, J.D., Elwood, J.W. and Hom, C.L. (1983) The effect of grazing intensity on phosphorus spiralling in autotrophic streams. *Oecologia* 58, 358–366.

Mulholland, P.J., Steinman, A.D., Palumbo, A.V. and DeAngelis, D.L. (1991) Influence of nutrients and grazing on the response of stream periphyton communities to a scour disturbance. *Journal of the North American Benthological Society* 10, 127–142.

Naiman, R.J. and Sedell, J.R. (1979) Characterization of particulate organic matter transported by some Cascade Mountain streams. *Journal of the Fisheries Research Board of Canada* 36, 17–31.

Newbold, J.D., Elwood, J.W., O'Neill, R.V. and Van Winkle, W. (1982) Nutrient spiralling in streams: implications for nutrient limitation and invertebrate activity. *American Naturalist* 120, 628–652.

Newman, R.M. (1991) Herbivory and detritivory on freshwater macrophytes by invertebrates: a review. *Journal of the North American Benthological Society* 10, 89–114.

NTSB (1989) *Collapse of the S.R. 675 Bridge Spans over the Pocomoke River near Pocomoke City, Maryland.* PB89–916205. National Transportation Safety Board/HAR-89/04.1–80, Washington, DC.

Oberndorfer, R.Y., McArthur, J.V. and Barnes, J.R. (1984) The effect of invertebrate predators on leaf litter processing in an alpine stream. *Ecology* 65, 1325–1331.

Pace, M.L., Findlay, S.E.G. and Fischer, D. (1998) Effects of an invasive bivalve on the zooplankton community of the Hudson River. *Freshwater Biology* 39, 103–116.

Palmer, M.A., Covich, A.P., Finlay, B.J., Gibert, J., Hyde, K.D., Johnson, R.K., Kairesalo, T., Lake, S., Lovell, C.R., Naiman, R.J., Ricci, C., Sabater, F. and Strayer, D. (1997) Biodiversity and ecosystem processes in freshwater sediments. *Ambio* 26, 571–577.

Parker, C.R. and Voshell, J.R. (1983) Production of filter-feeding Trichoptera in an impounded and free flowing river. *Canadian Journal of Zoology* 61, 70–87.

Petersen, L.B.-M. (1985) Food preferences in three species of *Hydropsyche* (Trichoptera). *Verhandlungen der Internationalen Vereinigung für Theoretische und Angewandte Limnologie* 22, 3270–3274.

Petersen, L.B.-M. (1989) Resource utilization of coexisting species of Hydropsychidae (Trichoptera). *Archiv für Hydrobiologie* (Suppl.) 83, 83–119.

Petersen, R.C. and Cummins, K.W. (1974) Leaf processing in a woodland stream. *Freshwater Biology* 4, 343–368.

Peterson, B.J., Bahr, M. and Kling, G.W. (1997) A tracer investigation of nitrogen cycling in a pristine tundra river. *Canadian Journal of Fisheries and Aquatic Sciences* 54, 2361–2367.

Phelps, H.L. (1994) The Asiatic clam (*Corbicula fluminea*) invasion and system-level ecological change in the Potomac river estuary near Washington D.C. *Estuaries* 17, 614–621.

Polis, G.A. (1994) Food webs, trophic cascades and community structure. *Australian Journal of Ecology* 19, 121–136.

Polis, G.A. and Holt, R.D. (1992) Intraguild predation – the dynamics of complex trophic interactions. *Trends in Ecology and Evolution* 7, 151–154.

Polis, G.A. and Strong, D. (1996) Food web complexity and community dynamics. *American Naturalist* 147, 813–846.

Power, M.E. (1990) Effects of fish in river food webs. *Science* 250, 811–814.

Power, M.E. (1992a) Top-down and bottom-up forces in food webs: do plants have primacy? *Ecology* 73, 733–746.

Power, M.E. (1992b) Habitat heterogeneity and the functional significance of fish in river food webs. *Ecology* 73, 1675–1688.

Pringle, C.M., Naiman, R.J., Bretschko, G., Karr, J.R., Oswood, M.W., Webster, J.R., Welcomme, R.L. and Winterbourn, M.J. (1988) Patch dynamics in lotic systems: the stream as a mosaic. *Journal of the North American Benthological Society* 7, 503–524.

Pringle, C.M., Blake, G.A., Covich, A.P., Buzby, K.M. and Finley, A. (1993) Effects of omnivorous shrimp in a montane tropical stream: sediment removal, disturbance of sessile invertebrates and enhancement of understory algal biomass. *Oecologia* 93, 1–11.

Rosi, E.J. (1997) The trophic basis of production along a river continuum: temporal and spatial variability in the flow of energy in aquatic macroinvertebrate communities. Master's thesis, University of Georgia, USA.

Short, R.A. and Maslin, P.E. (1977) Processing of leaf litter by a stream detritivore: effect on nutrient availability to collectors. *Ecology* 58, 935–938.

Sih, A. and Wooster, D. (1994) Prey behavior, prey dispersal, and predator impacts on stream prey. *Ecology* 75, 1199–1207.

Soluk, D.A. and Craig, D.A. (1988) Vortex feeding from pits in the sand: a unique method of suspension feeding used by a stream invertebrate. *Limnology and Oceanography* 33, 638–645.

Soluk, D.A. and Craig, D.A. (1990) Digging with a vortex: flow manipulation facilitates prey capture by a predatory stream mayfly. *Limnology and Oceanography* 35, 1201–1206.

Statzner, B. and Higler, B. (1985) Questions and comments on the river continuum concept. *Canadian Journal of Fisheries and Aquatic Sciences* 42, 1038–1044.

Statzner, B., Fuchs, U. and Higler, L.W.G. (1996) Sand erosion by mobile predaceous stream insects: implications for ecology and hydrology. *Water Resources Research* 32, 2279–2287.

Steinman, A.D., Mulholland, P.J. and Beauchamp, J.J. (1995) Effects of biomass, light, and grazing on phosphorus cycling in stream periphyton communities. *Journal of the North American Benthological Society* 14, 371–381.

Stites, D.L., Benke, A.C. and Gillespie, D.M. (1995) Population dynamics, growth and production of the Asiatic clam, *Corbicula fluminea*, in a blackwater river. *Canadian Journal of Fisheries and Aquatic Sciences* 52, 425–437.

Thorp, J.H. and Covich, A.P. (eds) (1991) *Ecology and Classification of North American Freshwater Invertebrates.* Academic Press, San Diego.

Triska, F.J. and Cromack, K. (1980) The role of woody debris in forests and streams. In: Waring, R.H. (ed.) *Forests: Fresh Perspectives from Ecosystem Analysis.* Oregon State University Press, Corvallis, pp. 171–190.

Vannote, R.L., Minshall, G.W., Cummins, K.W., Sedell, J.R. and Cushing, C.E. (1980) The river continuum concept. *Canadian Journal of Fisheries and Aquatic Sciences* 37, 130–137.

Voshell, J.R. (1985) *Trophic Basis of Production for Macroinvertebrates in the New River below Bluestone Dam*. Department of Entomology, Virginia Polytechnic Institute and State University, Blacksburg.

Wagner, R. (1991) The influence of the diel activity pattern of the larvae of *Sericostoma personatum* (Kirby and Spence) (Trichoptera) on organic matter distribution in stream-bed sediments – a laboratory study. *Hydrobiologia* 224, 65–70.

Wallace, J.B. and Merritt, R.W. (1980) Filter-feeding ecology of aquatic insects. *Annual Review of Entomology* 25, 103–132.

Wallace, J.B. and O'Hop, J. (1985) Life on a fast pad: water-lily leaf beetle impact on water-lilies. *Ecology* 66, 1534–1544.

Wallace, J.B. and Webster, J.R. (1996) The role of macroinvertebrates in stream ecosystem function. *Annual Review of Entomology* 41, 115–139.

Wallace, J.B., Webster, J.R. and Woodall, W.R. (1977) The role of filter-feeders in flowing waters. *Archiv für Hydrobiologie* 79, 506–532.

Wallace, J.B., Webster, J.R. and Cuffney, T.F. (1982) Stream detritus dynamics: regulation by invertebrate consumers. *Oecologia* 53, 197–200.

Wallace, J.B., Vogel, D.S. and Cuffney, T.F. (1986) Recovery of a headwater stream from an insecticide-induced community disturbance. *Journal of the North American Benthological Society* 5, 115–126.

Wallace, J.B., Cuffney, T.F., Webster, J.R., Lugthart, G.J., Chung, K. and Goldowitz, B.S. (1991) A five-year study of export of fine organic particles from headwater streams: effects of season, extreme discharges, and invertebrate manipulation. *Limnology and Oceanography* 36, 670–682.

Wallace, J.B., Whiles, M.R., Webster, J.R., Cuffney, T.F., Lugthart, G.J. and Chung, K. (1993) Dynamics of inorganic particles in headwater streams: linkages with invertebrates. *Journal of the North American Benthological Society* 12, 112–125.

Wallace, J.B., Eggert, S.L., Meyer, J.L. and Webster, J.R. (1997) Multiple trophic levels of a forest stream linked to terrestrial litter inputs. *Science* 277, 102–104.

Wallace, J.B., Eggert, S.L., Meyer, J.L. and Webster, J.R. (1999) Effects of resource reduction on a detrital-based ecosystem. *Ecological Monographs* (in press).

Webster, J.R. and Meyer, J.L. (1997) Organic matter budgets for streams: a synthesis. *Journal of the North American Benthological Society* 16, 141–161.

Webster, J.R. and Patten, B.C. (1979) Effects of watershed perturbation on stream potassium and calcium dynamics. *Ecological Monographs* 49, 51–72.

Webster, J.R., Wallace, J.B. and Benfield, E.F. (1995) Organic processes in streams of the eastern United States. In: Cushing, C.E., Minshall, G.W. and Cummins, K.W. (eds) *River and Stream Ecosystems*. Elsevier, Amsterdam, pp. 103–164.

Welker, M. and Walz, N. (1998) Can mussels control plankton in rivers? – a plankto-logical approach applying a Lagrangian sampling strategy. *Limnology and Oceanography* 43, 753–762.

Wolf, B., Zwick, P. and Marxsen, J. (1997) Feeding ecology of the freshwater detritivore *Ptychoptera paludosa* (Diptera, Nematocera). *Freshwater Biology* 38, 375–386.

Wooster, D. (1994) Predator impacts on stream benthic prey. *Oecologia* 99, 7–15.

Wotton, R.S. (1987) Lake outlet blackflies – the dynamics of filter feeders at very high population densities. *Holarctic Ecology* 10, 65–72.

Wotton, R.S., Malmqvist, B., Muotka, T. and Larsson, K. (1998) Fecal pellets from a
    dense aggregation of suspension-feeders in a stream: an example of ecosystem engi-
    neering. *Limnology and Oceanography* 43, 719–725.
Zanetell, B.A. and Peckarsky, B.L. (1996) Stoneflies as ecological engineers – hungry
    predators reduce fine sediments in stream beds. *Freshwater Biology* 36, 569–577.

Webmasters in Feedback Interactions and Food Webs

Insects as Regulators of Ecosystem Development

5

T.D. Schowalter

Department of Entomology, Oregon State University, Corvallis, OR 97331–2907, USA

Introduction

One of the most important and revolutionary contributions of ecosystem research has been the concept of self-regulation (Odum, 1969; Mattson and Addy, 1975; Webster *et al.*, 1975; Schowalter, 1981). At issue is the extent to which ecosystems behave as self-regulating (cybernetic) systems. The ability of ecosystems to modify abiotic conditions, making the environment more suitable for the biota, is evident in the increased variation in temperature, relative humidity, soil structure and even precipitation when vegetation is degraded or removed (e.g. Salati, 1987; Schlesinger *et al.*, 1990; Lewis, 1998). Clearly, some ecosystems have greater ability to modify environmental conditions and resist disturbance than do others. Ecosystems with greater structure (e.g. forests) have a greater ability to shade the soil surface, modify relative humidity and abate wind within the stem zone (and thereby to buffer the biota against fire, storm damage or drought) than do ecosystems with little structure (e.g. desert or early successional ecosystems). However, such structure may require more time to redevelop following catastrophic disturbance than do simpler ecosystems.

The role of animals, and particularly insects, in the regulation of ecosystem processes has remained controversial and largely untested, because insect ecologists have rarely participated in ecosystem studies. D.A. Crossley, Jr was a pioneer in the exploration of arthropod contributions to energy and biogeochemical fluxes (Crossley and Howden, 1961; Reichle and Crossley, 1967; Crossley, 1969, 1977; Reichle *et al.*, 1969; Crossley *et al.*, 1973, 1995) and in the training of students to explore consequences of changes in arthropod populations and communities for ecosystem processes (e.g. Schowalter *et al.*, 1981a, b; Seastedt *et al.*,

1983; Seastedt and Crossley, 1984; Schowalter *et al.*, 1986; Blair and Crossley, 1988; Risley and Crossley, 1993).

Phytophagous arthropods have a widely recognized ability to modify vegetation structure and composition, often dramatically. Although traditionally viewed as destructive (a view justified for exotic species), native phytophages may adjust vegetation structure and biogeochemical fluxes to prevailing environmental conditions and thereby contribute to ecosystem stability. This paper summarizes ways in which phytophagous arthropods affect ecosystem processes and, in particular, how phytophages may regulate ecosystem development through changes in primary production and vegetation composition. Although phytophages clearly contribute to secondary production and detrital food webs, this complex topic will not be a focus of this paper.

Phytophage Effects on Ecosystem Processes

Phytophages affect ecosystem processes by consuming plant material, altering vegetation structure and composition, and initiating fluxes of energy and nutrients through secondary producers. Phytophagy is highly correlated with net primary production (Cebrián and Duarte, 1994). Phytophages typically remove 5–15% of plant biomass annually in terrestrial ecosystems (Detling, 1987; Schowalter and Lowman, 1999), but often consume a greater proportion of net primary production in ecosystems that continuously replace lost phytomass. Outbreaks of phytophagous insects can remove up to 100% of the annual production of host species. Phytophages can kill nearly all hosts during periods of chronically high intensities of phytophagy.

Outbreaks of phytophagous insects traditionally have been viewed as disturbances (together with events such as fire, storm damage and drought), having negative effects on ecosystems. In fact, phytophages function primarily in a stress-dependent and/or density-dependent manner and thereby differ fundamentally from abiotic disturbances, which tend to affect species independently of density or condition. The stress- or density-dependence of phytophages is based on: (i) their sensitivity to changes in the defensive capability of their hosts; and (ii) their increasing efficiency of discovering new hosts as host density increases, as described below.

All plants have demonstrated ability to defend themselves both physically and chemically against phytophages (e.g. Coley *et al.*, 1985; Coley, 1986). Physical defences include spines and lignified tissues, whereas chemical defences include a variety of feeding deterrents and toxins. Insect ecologists have focused on the co-evolutionary aspects of this 'chemical arms race' and largely neglected the dependence (and effects) of these interactions on ecological succession and biogeochemical cycling, although ecosystem ecologists have addressed the effects of plant lignin and other deterrent chemicals on litter decomposition (e.g. Meentemeyer, 1978; Seastedt, 1984).

Under optimal growing conditions, plants acquire sufficient energy and nutrient resources to support various metabolic processes, including production of physical and chemical defences against phytophages. These defences represent an energetic and nutritional trade-off, to the extent that the diversion of these resources to defence reduces resource availability for growth and reproduction (e.g. Coley, 1986; Herms and Mattson, 1992). Therefore, stressed plants lose the ability to allocate resources to production of energy- and nutrient-expensive defences and typically reallocate resources to maintenance of growth and/or reproduction (Coley *et al.*, 1985; Lorio, 1993; Waring and Running, 1998). Stress not only increases suitability for phytophages but often provides cues that attract phytophages (e.g. Schowalter *et al.*, 1986).

Phytophage populations grow more rapidly when the time and energy required to locate suitable hosts is reduced, that is, when suitable hosts are close together relative to the dispersal ability of the phytophage (e.g. Risch, 1981; Kareiva, 1983; Courtney, 1986; Turchin, 1988; Schowalter and Turchin, 1993). Therefore, phytophage outbreaks are most likely to occur when a dominant plant species is stressed by changing environmental conditions.

Phytophage effects on primary productivity are complex (Maschinski and Whitham, 1989). Phytophagy can stimulate, as well as suppress, primary productivity and may improve plant survival. In Chapter 6 and in Dyer *et al.* (1995) it is demonstrated that grasshopper crop and mid-gut extracts present in grasshopper regurgitants during feeding can stimulate coleoptile growth in some grass species. Knapp and Seastedt (1986) reported that accumulated standing dead vegetation limits primary productivity in grasslands. Hence, grazing typically stimulates grass productivity (McNaughton, 1979, 1993; Detling, 1987; Dyer *et al.*, 1993). Grazing can improve water and nutrient balances by reducing metabolic demands and transpiration (Webb, 1978). Parks (1993) demonstrated, in a greenhouse study, that grand fir (*Abies grandis*) seedlings defoliated by the western spruce budworm (*Choristoneura occidentalis*) had a higher survival rate during drought treatment than did non-defoliated seedlings, apparently because reduced foliar surface area reduced transpiration and improved water balance.

Primary production often peaks at low-to-moderate intensities of pruning and thinning (Fig. 5.1), supporting the grazing optimization hypothesis (McNaughton, 1979, 1993; Carpenter and Kitchell, 1984; Carpenter *et al.*, 1985; Seastedt, 1985; Detling, 1987; Paige and Whitham, 1987; Maschinski and Whitham, 1989; Williamson *et al.*, 1989; Dyer *et al.*, 1993, 1995; Lovett and Tobiessen, 1993). The widespread ability of plants to compensate for tissue loss to phytophages may depend on the availability of resources (Coley *et al.*, 1985; Trumble *et al.*, 1993) from abiotic or biotic pools, and is most conspicuous in ecosystems characterized by relatively constant productivity, for example aquatic, grassland and tropical ecosystems (Cebrián and Duarte, 1994). However, compensation may occur over long time periods in some (especially temperate forest) ecosystems. Wickman (1980) and Alfaro and Shepherd (1991) reported that increased diameter growth of defoliated trees persisted for

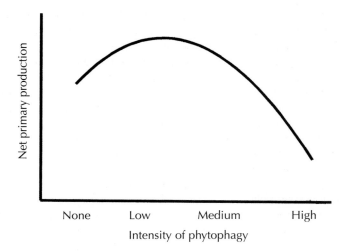

Fig. 5.1. Relationship between intensity of phytophagy and net primary production. Net primary production often peaks at low to moderate intensities of phytophagy, supporting the grazing optimization hypothesis.

decades after severe defoliation in conifer forests. Romme *et al.* (1986) reported that annual wood production following mountain pine beetle (*Dendroctonus ponderosae*) outbreaks equalled or exceeded pre-outbreak levels within 10 years.

Differential intensities of phytophagy among plant species in the ecosystem affect vegetation dynamics. Phytophages frequently determine the patterns of plant species occurrence across spatial and temporal gradients. Louda *et al.* (1990) and Louda and Rodman (1996) experimentally documented restriction of host plant geographic or habitat ranges by phytophages. Torres (1992) reported a sequential rise to dominance of herbaceous plant communities in Puerto Rico during the months following Hurricane Hugo. As the plant species characterizing each successive sere became dominant, the concentrated resource promoted outbreaks of host-specific phytophages that suppressed these species and facilitated their replacement by other plant species that, in turn, induced outbreaks of their associated phytophages. Boring *et al.* (1988) found that the black locust borer (*Megacyllene robiniae*) facilitates the replacement of black locust (*Robinia pseudo-acacia*) by later successional hardwoods in southeastern USA, and Schowalter *et al.* (1981b) reported that southern pine beetle (*Dendroctonus frontalis*) accelerates replacement of pines by hardwoods in southern USA. As plant species are replaced by others that use water, carbon and nutrient resources differently, changes occur in soil pH, moisture and nutrient levels, and rates and seasonal patterns of fluxes (Boring *et al.*, 1988).

Phytophages also affect ecosystem processes that are related to vegetation structure and/or the intensity and distribution of phytophagy among plant species. The vegetation canopy intercepts and filters light, precipitation and wind, thereby shading and cooling the soil surface, reducing water droplet impact and erosion and channelling water back into the atmosphere, and reducing windspeed and acquiring nutrients from the airstream (e.g. Lewis, 1998). Phytophages increase vegetation porosity by reducing foliage surface area or plant density. This increases penetration of light, precipitation and wind to the understorey and soil surface (e.g. Schowalter *et al.*, 1991). Increased light penetration can increase the photosynthetic efficiency of remaining foliage (Webb, 1978). At the same time, reduced foliage surface area or plant density reduce transpiration and metabolic demand for water and nutrients.

Phytophages affect biogeochemical cycling through changes in turnover of plant material. Some consumed nutrients are concentrated in phytophage tissues, enhancing their food value for predators or scavengers (Seastedt and Tate, 1981), but this effect on nutrient flux is relatively unimportant at the ecosystem level (Schowalter and Crossley, 1983). The most important effects of phytophages on nutrient fluxes appear to be through leaching of nutrients from damaged plant surfaces (Kimmins, 1972; Seastedt *et al.*, 1983; Schowalter *et al.*, 1991; Lovett *et al.*, 1996), stimulation of forest floor processes through altered rates, seasonality and quality of litter inputs (Zlotin and Khodashova, 1980; Hollinger, 1986; Grier and Vogt, 1990; Lovett and Ruesink, 1995), and alteration of vegetation structure and consequent changes in patterns of nutrient use (Ritchie *et al.*, 1998). Hollinger (1986) reported that during an outbreak of the California oak moth, *Phryganidia californica*, fluxes of nitrogen and phosphorus to the ground more than doubled, and faeces and insect remains accounted for 60–70% of the total nitrogen and phosphorus fluxes. Ritchie *et al.* (1998) found that herbivory tended to reduce the abundance of plant species with N-rich tissues in an oak savanna in the north central USA. Grier and Vogt (1990) and Lovett and Ruesink (1995) found that a major effect of honeydew or folivore faeces may be carbohydrate-fuelled, microbial immobilization of nitrogen.

Schowalter *et al.* (1981a), Schowalter (1995) and Schowalter and Ganio (1999) reported that catastrophic disturbance (clearcutting and storm damage) in temperate and tropical forests resulted in a consistent shift in dominance from folivorous to sap-sucking phytophages. This shift is accompanied by changes in biogeochemical processes. In undisturbed forest, nutrient flux occurs primarily as canopy throughfall, foliage fragments and folivore faeces (sufficient to maintain low levels of microbial activity and immobilization of limited amounts of nitrogen), whereas in disturbed forest, flux occurs primarily as honeydew (Schowalter *et al.*, 1981a) which provides abundant labile carbohydrates to fuel increased microbial immobilization of nitrogen (Grier and Vogt, 1990). The slow fluxes of nutrients resulting from folivory and sap-sucking represent alternative mechanisms for controlling fluxes of nitrogen and other nutrients within and from the ecosystem. Therefore, the complementary functions conferred by species diversity and species replacement during succession can be viewed as a means to stabilize primary production and resource availability under changing conditions.

Phytophages as Regulators

Phytophages could affect ecosystem properties, without necessarily providing regulation. However, phytophages have been shown to respond to changes in vegetation density or physiological condition in ways that provide both positive and negative feedbacks, depending on the direction of deviation in primary production from nominal levels. The extent to which phytophages regulate ecosystem development is controversial but worthy of consideration. Our approach to managing ecosystem resources depends on our perspective of ecosystems as random assemblages of interacting species or as self-regulating entities.

Objections to the concept of self-regulation of ecosystem properties are based primarily on three arguments. Early objections were based on the dependence of self-regulation on group (species assemblages) selection. However, kin selection, co-evolution of mutualistic interactions, and other forms of cooperation or density-dependent regulation among species have been widely demonstrated (Berryman *et al.*, 1987). In fact, selection should favour inter-actions that stabilize ecosystem parameters, thereby maintaining conditions favourable to the co-evolving species. More recent objections are based on the apparent lack of centralized mechanisms for communicating changes in eco-system properties and triggering responses that return the ecosystem to nominal conditions and on the apparent instability of ecosystems subject to catastrophic disturbances (e.g. Engleberg and Boyarsky, 1979). However, communication mechanisms can be identified (e.g. volatile chemicals used to assess and locate resources) and even individual organisms vary in their homeostatic ability. Ultimately, the ability of an organism or an ecosystem to persist depends on its ability to acquire resources at a rate sufficient to balance losses. Furthermore, although the degree of deviation following disturbances may seem extreme from a human perspective, many ecosystems have persisted over evolutionary and geologic time.

Ecologists have not reached a consensus on a definition of ecosystem stability or the criteria to measure it. For example, does stability require constant species composition or constant net primary production? How quickly must the ecosystem return to nominal conditions to be considered qualitatively stable? Clearly, frequently disturbed ecosystems maintained by frequent disturb-ances at early successional stages (e.g. grasslands) recover more quickly than do more complex but less frequently disturbed ecosystems that require longer time periods to recover (e.g. forests). Finally, what spatial scale is most appro-priate for evaluating stability? Most studies have focused on the multiple successional pathways followed by relatively small patches (e.g. Pickett and White, 1985), but the dynamic patch pattern at the landscape scale may represent stable proportions of various successional stages and community types composing the larger-scale ecosystem (Schowalter *et al.*, 1981b).

Net primary production is a parameter subject to regulation through density- or stress-dependent feedback, as described above. Net primary production modifies a number of other ecosystem parameters, including ecosystem structure, climate, and energy and nutrient storage and fluxes. Net primary production largely controls rates of energy, water and nutrient use by ecosystems. Imbalances occur when resource demand exceeds resource availability, and potentially threaten the entire community. Therefore, natural selection should favour interactions among the co-evolved members of the community that tend to stabilize net primary production and resource use.

Accumulating evidence indicates that various mechanisms in natural ecosystems interact to maintain nominal ecosystem properties and to return ecosystem properties to nominal levels following deviation. The dominant organisms are adapted to survive disturbances or environmental changes that recur regularly with respect to generation time. In other words, adaptation to prevailing conditions (evolution) constitutes a coarse adjustment that reduces ecosystem deviation from nominal conditions, as a result of disturbances or other regular environmental changes (e.g. seasonal cycles), and that contributes to persistence of species composition. For example, many grassland and pine forest species are adapted to survive low intensity fires and drought (e.g. underground rhizomes and insulating bark, respectively) which characterize the ecosystems that they dominate. Clearly, anthropogenic changes in ecosystem conditions, especially loss or introduction of species, will disrupt regulatory functions.

All ecosystems are subject to periodic catastrophic disturbances that devastate the biota and to long-term environmental changes that lead to further adaptation (speciation). These processes lead to species replacement. Ecosystem diversity provides for species replacement and continuity of ecosystem functions, such as protection of the soil surface and conservation of limiting nutrients. Rapid colonization by early successional plant species maintains ground coverage, water flux and net primary production during the period before later successional species (longer-lived with slower reproductive rates) recover dominance, but some early or mid-successional stages are capable of inhibiting further succession (MacMahon, 1981; Boring *et al.*, 1988).

Self-regulation requires mechanisms for sensing and communicating deviation in system operation to feedback mechanisms that return system operation to prescribed levels, thereby minimizing deviation in system operation. Phytophagous insects, in particular, possess several important properties of cybernetic regulators.

First, insects are small and have high reproductive rates and short life spans. These attributes make them highly responsive to environmental changes, as demonstrated by the capacity of insect populations to increase (or decrease) by several orders of magnitude within short periods of time. Populations of larger, less fecund, or longer-lived organisms respond more slowly to environmental changes. For example, individual trees and vertebrates may persist for long periods under adverse conditions, but be unable to reproduce and ensure population continuity (e.g. Temple, 1977; Franklin *et al.*, 1992). Furthermore, insect biomass is generally small and inexpensive (in terms of energy and nutrients) to maintain, but can be amplified quickly and dramatically.

Second, insects detect airborne chemical signals that can indicate changes in resource quality (e.g. host physiological condition) and quantity (density). Volatile chemicals produced by various organisms represent a mechanism for communicating ecosystem status and eliciting feedback responses. Behavioural flexibility of animal populations, including insects, permits a direct functional response (e.g. reorientation, improved efficiency through learning) followed by a numerical (reproductive) response to changing availability of resources.

Third, the diversity of insect species represents a comparable diversity of ecological strategies. Insects account for 70–80% of all species in ecosystems where relatively complete inventories of species are available. Each species represents a unique combination of tolerance ranges for various environmental conditions and of ecological functions. This ensures continuity of ecological functions as species intolerant of changing environmental conditions are replaced by tolerant species (Table 5.1; e.g. Schowalter and Ganio, 1999).

Finally, phytophagous insects respond to changes in net primary production in a stress- or density-dependent manner and increase or decrease rates of primary production and nutrient fluxes to restore balance (Mattson and Addy, 1975; Schowalter, 1981) (Fig. 5.2), as well as alter the direction of succession as environmental conditions change (Davidson, 1993). Davidson (1993) compiled data showing that phytophages can advance or retard (or reverse) succession, depending on the extent to which phytophagy is concentrated on earlier or later successional species. Phytophages may be critical to successional advance from persistent (inhibitive) seres (e.g. MacMahon, 1981; Schowalter,

Table 5.1. Complementary changes in phytophagous arthropod abundances within functional groups as a result of canopy-opening disturbance in temperate and tropical forest canopies. Temperate data are for partially harvested and undisturbed old-growth Douglas-fir at the H.J. Andrews Experimental Forest Long Term Ecological Research Site in western Oregon during 1992 (Schowalter, 1995), and tropical data are for tabonuco in hurricane-generated gaps and intact forest at the Luquillo Experimental Forest Long-term Ecological Research Site in Puerto Rico during 1991–1995 (Schowalter and Ganio, 1999).

	Andrews Forest		Luquillo Forest	
	Disturbed (no. kg^{-1} foliage)	Undisturbed (no. kg^{-1} foliage)	Disturbed (no. kg^{-1} foliage)	Undisturbed (no. kg^{-1} foliage)
Folivores	5	8	65	96
Lepidoptera	4	5	15	35
Hymenoptera	1	3	0	0
Coleoptera	0	0	3	23
Sap-suckers	500	170	220	170
Homoptera 1	29	4	14	5
Homoptera 2	8	20	17	26

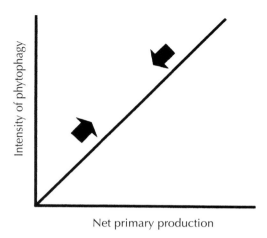

Fig. 5.2. Stimulation of primary production from low levels and suppression of primary production from high levels by phytophages (see Fig. 5.1) stabilizes primary production.

1981). Changes in penetration of light, water and nutrients to the soil surface, as a result of canopy opening by phytophages, can stimulate nutrient mobilization or immobilization and alter water and nutrient balances (e.g. Parks, 1993; Lovett and Ruesink, 1995).

Outbreaks of phytophagous insects are most likely to occur under two inter-related conditions, both of which represent responses to departure from nominal ecosystem conditions, often as a result of anthropogenic factors (Schowalter, 1985; Schowalter and Lowman, 1999). First, adverse environmental conditions, such as inadequate water or nutrient availability, changing climate and atmospheric pollution, cause changes in plant physiological conditions that can increase suitability for phytophages. High intensities of phytophagy under these conditions tend to restore balances between water or nutrient uptake and loss, or, in extreme cases, reduce biomass of the most stressed plants, regardless of their abundance, and promote replacement by more tolerant plants (e.g. Ritchie *et al.*, 1998; Schowalter and Lowman, 1999). Second, high densities of particular plant species, as a result of artificial planting or of inhibitive successional stages, improve host availability for associated phytophages. High intensities of phytophagy represent a major mechanism for disrupting such plant dominance structure and accelerating succession. Torres (1992) found that phytophagous insects were instrumental in suppressing dominant plant species and facilitating their replacement through several seres of plant succession in Puerto Rico following Hurricane Hugo. The locust borer, *Megacyllene robiniae*, is instrumental in facilitating the replacement of early successional, closed-canopy black locust forest by later successional mixed hardwood forest in the south-eastern USA (Boring *et al.*, 1988). The following example illustrates the tailoring of vegetation composition, including reversal of chronic anthropogenic changes, as a result of phytophage responses to host density and stress (see Schowalter *et al.*, 1981b; Hagle and Schmitz, 1993).

Succession from pioneer pine forest to late successional fir forest in western North America can be advanced or reversed, depending primarily on moisture availability and condition of the dominant vegetation. Under conditions of adequate moisture typical of riparian corridors and high elevations, mountain pine beetle advances succession by facilitating the replacement of shade-intolerant host pines by shade tolerant (but fire-intolerant) understorey firs. The more arid conditions prevailing at lower elevations historically ensured a high frequency of low intensity fires that maintained an open pine-dominated woodland. Fire suppression during the past century has led to increased densities of understorey firs which have become increasingly stressed by competition for limited soil water. Inevitable drought, which occurred during the past decade, exacerbates moisture limitation and increases the vulnerability of the firs to several defoliators and bark beetles (Fig. 5.3). The resulting insect-induced fir mortality functions in the absence of fire to maintain the pine sere. However, extensive fir mortality also increases the likelihood of catastrophic fires which, historically, were rare in these forests. Such fires result in eventual re-establishment of the pine forest.

Insects have been blamed for the health problems of these forests and have repeatedly been targets of suppression efforts. Accumulating evidence indicates that outbreaks of native species are triggered by increasing vegetation density and/or stress and function to reduce water and nutrient demands in ecosystems where these resources are limiting. Therefore, changes in intensity of phytophagy by native species appear to regulate vegetation density, biomass and/or species composition at levels that are sustainable under prevailing environmental conditions and that maintain characteristic temperature, moisture and nutrient conditions necessary for associated species.

To what extent can insects contribute to stability and 'health' of various ecosystems? Phytophagous insects are capable of regulating primary production by stimulating production from low rates and suppressing production from high rates, as described above (Figs 5.2 and 5.3). Because primary production controls many other ecosystem variables, stable net primary production could contribute to overall ecosystem stability. Few studies have addressed the long-term effects of phytophages on variability in ecosystem parameters. Romme *et al.* (1986) suggested that bark beetles increased variability in production of pine forests over a 10-year period, relative to uninfested forest. However, increased variability in some ecosystem parameters may be offset by reduced competition or other potentially destructive changes over the longer term.

Clearly, long-term studies will be necessary to evaluate the contribution of phytophages to ecosystem stability. Manipulation of phytophage abundance is often difficult, but is necessary to evaluate phytophage effects on variation in ecosystem parameters. Our management of ecosystem resources, and in particular our approach to managing phytophagous insects, requires that we

Fig. 5.3. Phytophage modification of succession in central Sierran mixed conifer ecosystems during 1998. Understorey white fir (*Abies concolor*), the late successional dominant, is increasingly stressed by competition for water in this arid forest type. An outbreak of the Douglas-fir tussock moth (*Orgyia pseudotsugata*) has completely defoliated the white fir, contributing to persistence of the earlier, more drought-tolerant, pine-dominated sere in the absence of fire.

understand the extent to which phytophages affect variation in climate, primary production, nutrient availability or other conditions that affect ecosystem stability.

Conclusions

Phytophagous insects clearly affect vegetation structure and composition through density-dependent and stress-dependent feedback on plant growth and survival. Phytophage-induced changes in vegetation structure and composition in turn affect primary productivity, biogeochemical cycling, successional processes and even microclimatic conditions in ways that appear to reduce variability in ecosystem conditions and facilitate recovery following disturbances. Insects are ideal regulators because of their small, easily amplified biomass, sensitivity to ecosystem conditions communicated through volatile chemicals carried on the airstream, and diversity of niches and effects on primary production and other ecosystem processes. If phytophages respond to changes in host productivity and condition in ways that stabilize primary production and ecosystem development in a cybernetic manner, then 'pest' suppression may be counterproductive in natural ecosystems.

References

Alfaro, R.I. and Shepherd, R.F. (1991) Tree-ring growth of interior Douglas-fir after one year's defoliation by Douglas-fir tussock moth. *Forest Science* 37, 959–964.

Berryman, A.A., Stenseth, N.C. and Isaev, A.S. (1987) Natural regulation of herbivorous forest insect populations. *Oecologia* 71, 174–184.

Blair, J.M. and Crossley, D.A., Jr (1988) Litter decomposition, nutrient dynamics and litter microarthropods in a southern Appalachian hardwood forest 8 years following clearcutting. *Journal of Applied Ecology* 35, 683–698.

Boring, L.R., Swank, W.T. and Monk, C.D. (1988) Dynamics of early successional forest structure and processes in the Coweeta Basin. In: Swank, W.T. and Crossley, D.A., Jr (eds) *Forest Hydrology and Ecology at Coweeta*. Springer-Verlag, New York, pp. 161–179.

Carpenter, S.R. and Kitchell, J.F. (1984) Plankton community structure and limnetic primary production. *American Naturalist* 124, 159–172.

Carpenter, S.R., Kitchell, J.F. and Hodgson, J.R. (1985) Cascading trophic interactions and lake productivity. *BioScience* 35, 634–639.

Cebrián, J. and Duarte, C.M. (1994) The dependence of herbivory on growth rate in natural plant communities. *Functional Ecology* 8, 518–525.

Coley, P.D. (1986) Costs and benefits of defense by tannins in a neotropical tree. *Oecologia* 70, 238–241.

Coley, P.D., Bryant, J.P. and Chapin, F.S., III (1985) Resource availability and plant antiherbivore defense. *Science* 230, 895–899.

Courtney, S.P. (1986) The ecology of pierid butterflies: dynamics and interactions. *Advances in Ecological Research* 15, 51–131.

Crossley, D.A., Jr (1969) Comparative movement of [106]Ru, ^{60}Co, and ^{137}Cs in arthropod food chains. In: Nelson, D.J. and Evans, F.C. (eds) *Symposium on Radioecology*, USAEC CONF-670503. US Atomic Energy Commission, Washington, DC, pp. 687–695.

Crossley, D.A., Jr (1977) The roles of terrestrial saprophagous arthropods in forest soils: current status of concepts. In: Mattson, W.J. (ed.) *The Role of Arthropods in Forest Ecosystems*. Springer-Verlag, New York, pp. 49–56.

Crossley, D.A., Jr and Howden, H.F. (1961) Insect–vegetation relationships in an area contaminated by radioactive wastes. *Ecology* 42, 302–317.

Crossley, D.A., Jr, Coulson, R.N. and Gist, C.S. (1973) Trophic level effects on species diversity in arthropod communities of forest canopies. *Environmental Entomology* 2, 1097–1100.

Crossley, D.A., Jr, Blood, E.R., Hendrix, P.F. and Seastedt, T.R. (1995) Turnover of cobalt-60 by earthworms (*Eisenia foetida*) (Lumbricidae, Oligochaeta). *Applied Soil Ecology* 2, 71–75.

Davidson, D.W. (1993) The effects of herbivory and granivory on terrestrial plant succession. *Oikos* 68, 23–35.

Detling, J.K. (1987) Grass response to herbivory. In: Capinera, J.L. (ed.) *Integrated Pest Management on Rangeland: a Shortgrass Prairie Perspective*. Westview Press, Boulder, Colorado, pp. 56–68.

Dyer, M.I., Turner, C.L. and Seastedt, T.R. (1993) Herbivory and its consequences. *Ecological Applications* 3, 10–16.

Dyer, M.I., Moon, A.M., Brown, M.R. and Crossley, D.A., Jr (1995) Grasshopper crop and midgut extract effects on plants: an example of reward feedback. *Proceedings of the National Academy of Sciences USA* 92, 5475–5478.

Engleberg, J. and Boyarsky, L.L. (1979) The noncybernetic nature of ecosystems. *American Naturalist* 114, 317–324.

Franklin, J.F., Swanson, F.J., Harmon, M.E., Perry, D.A., Spies, T.A., Dale, V.H., McKee, A., Ferrell, W.K., Means, J.E., Gregory, S.V., Lattin, J.D., Schowalter, T.D. and Larsen, D. (1992) Effects of global climatic change on forests in northwestern North America. In: Peters, R.L. and Lovejoy, T.E. (eds) *Global Warming and Biological Diversity*. Yale University Press, New Haven, Connecticut, pp. 244–257.

Grier, C.C. and Vogt, D.J. (1990) Effects of aphid honeydew on soil nitrogen availability and net primary production in an *Alnus rubra* plantation in western Washington. *Oikos* 57, 114–118.

Hagle, S. and Schmitz, R. (1993) Managing root disease and bark beetles. In: Schowalter, T.D. and Filip, G.M. (eds) *Beetle–Pathogen Interactions in Conifer Forests*. Academic Press, London, pp. 209–228.

Herms, D.A. and Mattson, W.J. (1992) The dilemma of plants: to grow or defend. *Quarterly Review of Biology* 67, 283–335.

Hollinger, D.Y. (1986) Herbivory and the cycling of nitrogen and phosphorus in isolated California oak trees. *Oecologia* 70, 291–297.

Kareiva, P. (1983) Influence of vegetation texture on herbivore populations: resource concentration and herbivore movement. In: Denno, R.F. and McClure, M.S. (eds) *Variable Plants and Herbivores in Natural and Managed Systems*. Academic Press, New York, pp. 259–289.

Kimmins, J.P. (1972) Relative contributions of leaching, litterfall, and defoliation by *Neodiprion sertifer* (Hymenoptera) to the removal of cesium-134 from red pine. *Oikos* 23, 226–234.

Knapp, A.K. and Seastedt, T.R. (1986) Detritus accumulation limits productivity of tallgrass prairie. *BioScience* 36, 662–668.

Lewis, T. (1998) The effect of deforestation on ground surface temperatures. *Global and Planetary Change* 18, 1–13.

Lorio, P.L., Jr (1993) Environmental stress and whole-tree physiology. In: Schowalter, T.D. and Filip, G.M. (eds) *Beetle–Pathogen Interactions in Conifer Forests*. Academic Press, London, pp. 81–101.

Louda, S.M. and Rodman, J.E. (1996) Insect herbivory as a major factor in the shade distribution of a native crucifer (*Cardamine cordifolia* A. Gray, bittercress). *Journal of Ecology* 84, 229–237.

Louda, S.M., Keeler, K.H. and Holt, R.D. (1990) Herbivore influences on plant performance and competitive interactions. In: Grace, J.B. and Tilman, D. (eds) *Perspectives on Plant Competition*. Academic Press, San Diego, pp. 413–444.

Lovett, G.M. and Ruesink, A.E. (1995) Carbon and nitrogen mineralization from decomposing gypsy moth frass. *Oecologia* 104, 133–138.

Lovett, G. and Tobiessen, P. (1993) Carbon and nitrogen assimilation in red oaks (*Quercus rubra* L.) subject to defoliation and nitrogen stress. *Tree Physiology* 12, 259–269.

Lovett, G.M., Nolan, S.S., Driscoll, C.T. and Fahey, T.J. (1996) Factors regulating throughfall flux in a New Hampshire forested landscape. *Canadian Journal of Forest Research* 26, 2134–2144.

MacMahon, J.A. (1981) Successional processes: comparisons among biomes with special reference to probable roles of and influences on animals. In: West, D.C., Shugart, H.H. and Botkin, D.B. (eds) *Forest Succession: Concepts and Application*. Springer-Verlag, New York, pp. 277–304.

Maschinski, J. and Whitham, T.G. (1989) The continuum of plant responses to herbivory: the influence of plant association, nutrient availability, and timing. *American Naturalist* 134, 1–19.

Mattson, W.J. and Addy, N.D. (1975) Phytophagous insects as regulators of forest primary production. *Science* 190, 515–522.

McNaughton, S.J. (1979) Grazing as an optimization process: grass-ungulate relationships in the Serengeti. *American Naturalist* 113, 691–703.

McNaughton, S.J. (1993) Grasses and grazers, science and management. *Ecological Applications* 3, 17–20.

Meentemeyer, V. (1978) Macroclimate and lignin control of litter decomposition rates. *Ecology* 59, 465–472.

Odum, E.P. (1969) The strategy of ecosystem development. *Science* 164, 262–270.

Paige, K.N. and Whitham, T.G. (1987) Overcompensation in response to mammalian herbivory: the advantage of being eaten. *American Naturalist* 129, 407–416.

Parks, C.G. (1993) The influence of induced host moisture stress on the growth and development of western spruce budworm and *Armillaria ostoyae* on grand fir seedlings. PhD dissertation, Oregon State University, Corvallis, USA.

Pickett, S.T.A. and White, P.S. (eds) (1985) *Ecology of Natural Disturbance and Patch Dynamics*. Academic Press, New York.

Reichle, D.E. and Crossley, D.A., Jr (1967) Investigation on heterotrophic productivity in forest insect communities. In: Petrusewicz, K. (ed.) *Secondary Productivity of Terrestrial Ecosystems: Principles and Methods*. Państwowe Wydawnictwo Naukowe, Warszawa, Poland, pp. 563–587.

Reichle, D.E., Shanks, M.H. and Crossley, D.A., Jr (1969) Calcium, potassium, and sodium content of forest floor arthropods. *Annals of the Entomological Society of America* 62, 57–62.

Risch, S.J. (1981) Insect herbivore abundance in tropical monocultures and polycultures: an experimental test of two hypotheses. *Ecology* 62, 1325–1340.

Risley, L.S. and Crossley, D.A., Jr (1993) Contribution of herbivore-caused greenfall to litterfall nitrogen flux in several southern Appalachian forested watersheds. *American Midland Naturalist* 129, 67–74.

Ritchie, M.E., Tilman, D. and Knops, J.M.H. (1998) Herbivore effects on plant and nitrogen dynamics in oak savanna. *Ecology* 79, 165–177.

Romme, W.H., Knight, D.H. and Yavitt, J.B. (1986) Mountain pine beetle outbreaks in the Rocky Mountains: regulators of primary productivity? *American Naturalist* 127, 484–494.

Salati, E. (1987) The forest and the hydrologic cycle. In: Dickinson, R.E. (ed.) *The Geophysiology of Amazonia: Vegetation and Climate Interactions.* John Wiley & Sons, New York, pp. 273–296.

Schlesinger, W.H., Reynolds, J.F., Cunningham, G.L., Huenneke, L.F., Jarrell, W.M., Virginia, R.A. and Whitford, W.G. (1990) Biological feedbacks in global desertification. *Science* 247, 1043–1048.

Schowalter, T.D. (1981) Insect herbivore relationship to the state of the host plant: biotic regulation of ecosystem nutrient cycling through ecological succession. *Oikos* 37, 126–130.

Schowalter, T.D. (1985) Adaptations of insects to disturbance. In: Pickett, S.T.A. and White, P.S. (eds) *Ecology of Natural Disturbance and Patch Dynamics.* Academic Press, New York, pp. 235–252.

Schowalter, T.D. (1995) Canopy arthropod communities in relation to forest age and alternative harvest practices in western Oregon. *Forest Ecology and Management* 78, 115–125.

Schowalter, T.D. and Crossley, D.A., Jr (1983) Forest canopy arthropods as sodium, potassium, magnesium and calcium pools in forests. *Forest Ecology and Management* 7, 143–148.

Schowalter, T.D. and Ganio, L.M. (1999) Invertebrate communities in a tropical rain forest canopy in Puerto Rico following Hurricane Hugo. *Ecological Entomology* 24, 191–201.

Schowalter, T.D. and Lowman, M.D. (1999) Forest herbivory by insects. In: Walker, L.R. (ed.) *Ecosystems of the World: Ecosystems of Disturbed Ground.* Elsevier, Amsterdam, pp. 269–285.

Schowalter, T.D. and Turchin, P. (1993) Southern pine beetle infestation development: interaction between pine and hardwood basal areas. *Forest Science* 39, 201–210.

Schowalter, T.D., Webb, J.W. and Crossley, D.A., Jr (1981a) Community structure and nutrient content of canopy arthropods in clearcut and uncut forest ecosystems. *Ecology* 62, 1010–1019.

Schowalter, T.D., Coulson, R.N. and Crossley, D.A., Jr (1981b) Role of southern pine beetle and fire in maintenance of structure and function of the southeastern coniferous forest. *Environmental Entomology* 10, 821–825.

Schowalter, T.D., Hargrove, W.W. and Crossley, D.A., Jr (1986) Herbivory in forested ecosystems. *Annual Review of Entomology* 31, 177–196.

Schowalter, T.D., Sabin, T.E., Stafford, S.G. and Sexton, J.M. (1991) Phytophage effects on primary production, nutrient turnover, and litter decomposition of young Douglas-fir in western Oregon. *Forest Ecology and Management* 42, 229–243.

Seastedt, T.R. (1984) The role of microarthropods in decomposition and mineralization processes. *Annual Review of Entomology* 29, 25–46.

Seastedt, T.R. (1985) Maximization of primary and secondary productivity by grazers. *American Naturalist* 126, 559–564.

Seastedt, T.R. and Crossley, D.A., Jr (1984) The influence of arthropods on ecosystems. *BioScience* 34, 157–161.

Seastedt, T.R. and Tate, C.M. (1981) Decomposition rates and nutrient contents of arthropod remains in forest litter. *Ecology* 62, 13–19.

Seastedt, T.R., Crossley, D.A., Jr and Hargrove, W.W. (1983) The effects of low-level consumption by canopy arthropods on the growth and nutrient dynamics of black locust and red maple trees in the southern Appalachians. *Ecology* 64, 1040–1048.

Temple, S.A. (1977) Plant-animal mutualism: coevolution with dodo leads to near extinction of plant. *Science* 197, 885–886.

Torres, J.A. (1992) Lepidoptera outbreaks in response to successional changes after the passage of Hurricane Hugo in Puerto Rico. *Journal of Tropical Ecology* 8, 285–298.

Trumble, J.T., Kolodny-Hirsch, D.M. and Ting, I.P. (1993) Plant compensation for arthropod herbivory. *Annual Review of Entomology* 38, 93–119.

Turchin, P. (1988) The effect of host-plant density on the numbers of Mexican bean beetles, *Epilachna varivestis. American Midland Naturalist* 119, 15–20.

Waring, R.H. and Running, S.W. (1998) *Forest Ecosystems: Analysis at Multiple Scales*, 2nd edn. Academic Press, San Diego.

Webb, W.L. (1978) Effects of defoliation and tree energetics. In: Brookes, M.H., Stark, R.W. and Campbell, R.W. (eds) *The Douglas-fir Tussock Moth: a Synthesis*, USDA Forest Service Technical Bulletin 1585. USDA Forest Service, Washington, DC, pp. 77–81.

Webster, J.R., Waide, J.B. and Patten, B.C. (1975) Nutrient recycling and the stability of ecosystems. In: Howell, F.G., Gentry, J.B. and Smith, M.H. (eds) *Mineral Cycling in Southeastern Ecosystems*, CONF-740513. USDOE Energy Research and Development Administration, Washington, DC, pp. 1–27.

Wickman, B.E. (1980) Increased growth of white fir after a Douglas-fir tussock moth outbreak. *Journal of Forestry* 78, 31–33.

Williamson, S.C., Detling, J.K., Dodd, J.L. and Dyer, M.I. (1989) Experimental evaluation of the grazing optimization hypothesis. *Journal of Range Management* 42, 149–152.

Zlotin, R.I. and Khodashova, K.S. (1980) *The Role of Animals in Biological Cycling of Forest-steppe Ecosystems* (trans. French, N.R.). Dowden, Hutchinson & Ross, Stroudsburg, Pennsylvania.

Herbivores, Biochemical Messengers and Plants: Aspects of Intertrophic Transduction

6

M.I. Dyer

Institute of Ecology, University of Georgia, Athens, GA 30602–2202, USA

Introduction

In the last quarter of the 20th century the role of herbivores in an ecological context, at least as defined by many biologists throughout several preceding decades, has taken on new dimensions. Rather than viewing consumers as simple conduits of energy and nutrients through all of Earth's terrestrial and aquatic ecosystems, or as pests competing for resources that otherwise humans would harvest and utilize, now we realize that herbivores have a much larger impact than previously realized in their ecological associations, and a more important story to tell in terms of fundamental biology and evolution. Curiously, the new aspects, which have come from a large variety of studies on how plants and herbivores interact, result from earnest debate between workers in two disciplines in biology, each with differing overall objectives: (i) neo-Darwinists dedicated to understanding the evolutionary framework involving herbivorous animals and their host plants, principally through employing a competitive-interactions paradigm; and (ii) ecosystem- and community-level scientists striving to understand processes that govern large, complex systems.

The two divergent disciplines have developed their ideas and scientific bases more or less simultaneously, and, for much of the time, without much apparent inter-communication. This competitive nature contributing to the separation has become more apparent in the 1990s, partially perhaps because of increasing competition for shrinking support for research, but largely because of convergence on topics involved with environmental problems throughout the world. Such convergence has pitched the two disciplines with contrasting objectives into sometimes contentious, invidious debate at meetings and in the herbivory literature. Those systems around which such controversies have focused include: (i) tropical and temperate forests; (ii) grasslands; (iii) temporal wetlands; (iv) rivers and estuaries; (v) agroecosystems; and (vi) possibly urban landscapes.

While the tenor of the debate has subsided somewhat, rhetorical arguments between disciples of the two branches of biology noted earlier still erupt sporadically (for an example of one of the most recent, see Levin, 1993). None the less, these arguments about how to view plant–herbivore associations have resulted in healthy discussions that point towards new dimensions of what formerly many have considered a fixed or stable view of herbivory *per se* (for the most recent synthesis see Karban and Baldwin, 1997). In this paper I focus on some of the contrasting viewpoints and, from my own work, present what I think may formulate some new dimensions.

What Constitutes an Interaction between a Plant and a Herbivore?

Historically, most workers have considered herbivory as a linear transfer of food-stuffs from a live plant to an animal dependent upon the plant for its survival (Fig. 6.1A). Such a simple fact as this becomes quickly apparent to anyone who has spent even the smallest amount of time watching any given animal feed on a plant. Herbivores (both invertebrates and vertebrates), for a host of reasons, have developed many specialized mechanisms to avail themselves of this food supply, such as: (i) mouth parts for cutting and chewing; (ii) stylets for probing and sucking plant juices; (iii) burrowing directly into plant stems, roots or leaves; and (iv) multiple-compartment digestive systems. Plants, according to a vast literature, have developed ways to ward off invading herbivores through: (i) developing external barriers, such as thorns and tough, thick or waxy integuments; (ii) a broad array of toxicants that (a) preclude digestion, (b) arrest development, or (c) kill the invader outright; (iii) a variety of tactics to avoid predation from herbivores whereby, on average, they escape in time and space; and (iv) an ability to outgrow the herbivory effect (see Herms and Mattson, 1992, review). These adaptations derive from positive feedbacks imparted by the herbivore (Fig. 6.1B).

At the same time, herbivores developed their strategies which allowed their populations to prosper, or at least not to go extinct in the face of plant defences. This general viewpoint has elicited a so-called 'arms race' between plants and their herbivores (Karban and Baldwin, 1997), a designation that has lasted for several decades and appears in many general articles and text books, but has become one with more and more questions raised about its aptness. The set of emergent properties that most workers can agree upon centres on the idea of 'inducibility' of a variety of plant and herbivore responses that arise from this unique association.

I will not recite a lengthy presentation of differing viewpoints about plant–herbivore relationships from the standpoint of defensive interactions.

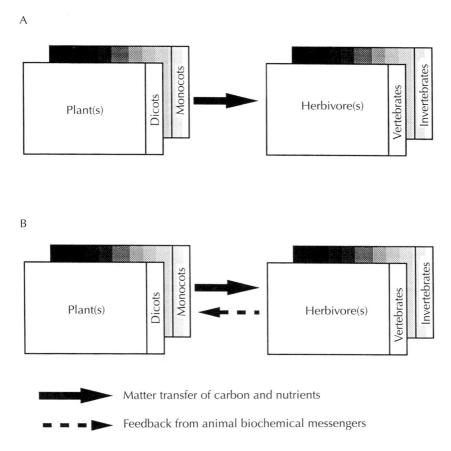

Fig. 6.1. Two theoretical representations of herbivory. (A) Studies that focus on material transfers from a plant to an animal, simply regarding the association as animals feeding upon plants. In most cases this model represents theoretical and functional food webs in any given biotic community. (B) Studies that incorporate a feedback to the plant add the concept of interconnectivity to the system where each of the two associates can affect the state of the other in a variety of ways.

Instead I refer the reader to several recent excellent books and reviews, mostly about how plants defend themselves against predation (see Gilbert and Raven, 1975; Wallace and Mansell, 1976; Crawley, 1983; Denno and McClure, 1983; Spencer, 1988; Herms and Mattson, 1992; Bernays and Chapman, 1994; Polis and Winemiller, 1996; and Karban and Baldwin, 1997).

Nearly 40 years ago Hairston *et al.* (1960) set a paradigm for many ecologists who have argued vociferously for top-down predator-based controls as the means for setting homeostasis in ecosystems, commonly referred to as the 'green world' hypothesis (a position restated in Hairston and Hairston,

1993). In this approach trophic cascades become the central controllers of ecosystems. As the hypothesis states, in a simple system consisting of predators, herbivores and primary producers, the predators will regulate the herbivores, thus releasing the primary producers from top-down regulation and hence, in theory, the world remains green!

But, in the past decade, work using food web theory suggests that top-down regulation alone does not account for determining ecosystem function and homeostatic mechanisms (Polis, 1994; Polis and Winemiller, 1996). Rather, Polis (1994) argues that other interactive mechanisms help to explain better the abundances and patterns of species and their biomass across terrestrial landscapes, and probably in aquatic systems as well. Lastly, nowhere in the 'green world' hypothesis does there exist the notion of positive feedbacks between plants and herbivores that bear a special relationship in helping to maintain Earth's highly varied ecosystems.

What Constitutes Herbivory?

To most the above question might seem trivial. However, when one starts to examine a diverse literature, the answer becomes much less so! To those who investigate evolutionary questions, herbivory takes on a slightly different aspect compared with those who focus on ecosystem- or landscape-level questions. Why should such a seemingly straightforward topic create such differences of opinion? Part of the answer comes from asking: who poses the question?

One can define herbivory from an autoecological viewpoint – the process of consumption of a plant or its parts by an animal (Fig. 6.1A) – or a synecological paradigm – combined plant–animal interactions that arise as a function of consumption of plants or their parts by an animal which affect both the herbivore and its plant host (Fig. 6.1B). Clearly the two viewpoints share elements in the definition, but they also differ markedly.

Furthermore the general literature seems to regard a case of *nominal herbivory*, which implies some sort of long-term stasis one encounters in any given system, that, without exception, shows a level greater than zero. Consequently we can define nominal herbivory from the autoecological approach as the rate of consumption of a plant or its parts by an animal when the plant population remains in steady state, or from the synecological approach as the identity and level of interactions involved in the biotic community under steady-state conditions that arise as a function of consumption by animals.

Now I come to perhaps the most important topic, that of the null hypotheses from which the two schools of ecology must operate. First the autoecological null hypothesis for herbivory: *the process in which animals consume plant tissue where responses of individual plants eaten show the same productivity measures as those in individuals not eaten.* Then the synecological null hypothesis for herbivory: *a process in all biotic communities which has as its basis the consumption of plants or plant tissues whereby productivity in patches eaten by animals shows the same productivity as that in patches not eaten.* By using these distinctions we can see that the two ecological disciplines may, more often than not, use differing metrics by which they measure the impacts of herbivory, and that they may come upon differing interpretations in their results. Brown and Allen (1989) give a clear picture of how such scalar problems may affect the interpretation of the influences of herbivores in nature.

One of the most useful ways to regard herbivory lies in its role as an inducer of a variety of important biological processes (Karban and Baldwin, 1997). By taking the basic elements of Fig. 6.1B, I propose the following relationships, which, on a functional basis, can cover much of the extant literature dealing with plant–animal interactions (Fig. 6.2). The remarkable thing about plant–animal interactions is the induction of biological actions that occur in each of the two partners but do not occur under any circumstances until the herbivore actually starts to feed on the plant host. These inducible relationships may have more to do with the plant than the animal, yet each partner takes away from herbivory a new state of being. Such actions range from receiving food, on the part of the herbivore, to the production of semiochemicals and replacement tissues in the plant (Herms and Mattson, 1992; Karban and Baldwin, 1997). In short, the responses resulting from this interacting pair represent a large amount of cellular and biochemical activity which can only have come from long-standing coevolutionary associations ascribed to symbionts.

Some of the fundamental activities listed in Fig. 6.2 include: (i) exogenous induction of cell division and elongation; and (ii) the use of endogenous and exogenous signals as biochemical messengers which the plant utilizes for defence from further herbivory. The outcome of these induction responses can not only alter the plant physiology and development itself, but on a collective basis, which one measures as a function of herbivory incidence, also results in altered structure, productivity and yields in the plants fed upon by the herbivore. Further, these processes, once set in motion, may quickly reorganize plant population dynamics and community structure, which, throughout a growing season, can in turn affect larger, scalar ecological units, such as ecosystem dynamics, and one could even consider landscape impacts in various plant–animal associations. The key to this entire scenario lies in the fact that, on average, in terrestrial ecosystems at least, one cannot think of any set of plant species on Earth that avoids at least some degree of herbivory. Even more importantly, some ecosystems, such as those dominated by graminoids, operate under the certainty of herbivore association and stress.

Therefore, in the long run we must consider the question of plant susceptibility to herbivory as moot. Thus, the real question in herbivory and ecosystem studies becomes: given an unspecified (but real and significant) probability that an animal will feed on any given plant, how will it respond to the loss of its tissue on an internal physiological basis, and how well will it fare with its immediate neighbours (conspecific or not) that may or may not have had equivalent

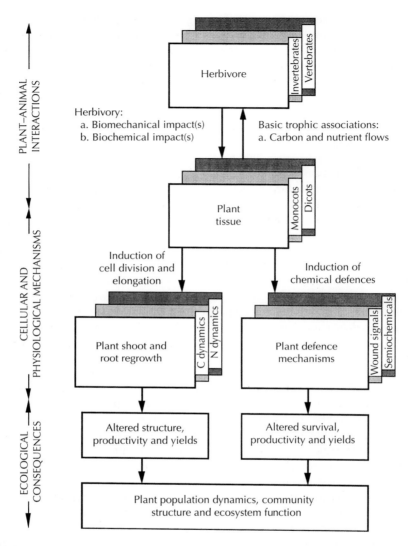

Fig. 6.2. This theoretical framework places scale of association into plant–animal associations via the need to study differing concepts to answer differing questions that one can pose at each level. The plant–herbivore association, following Fig. 6.1B, suggests two lineages involving induction impacts: one, considered more traditional in terms of numbers of reported studies in the literature, involves negative feedback induction of chemical defences which can alter the population structure of the community in question; and the second involves a positive feedback through a signal from the herbivore to the plant in which the plant reorganizes its growth and development within the community, which one measures in terms of productivity. An uninvestigated potential exists in that plants subjected to varying levels of herbivory might invoke both types of inducible responses.

tissue removal? Such are the questions that lie before a student interested in understanding the role of herbivory in terms of evolution and ecosystem function. For the remainder of the chapter I will address one small, but important new element in this complex biological equation, that of specific inducer feedback from the herbivore to the plant in ecosystems dominated by graminoids.

The Role of Biochemical Messengers in Plant–Herbivore Relationships

As I pointed out above, how might a plant respond to the initial stress of herbivory? Two ways come to mind: (i) from signals endogenous to the plant whereby the wounding process sets in motion biochemical messages that within some organ a herbivory event has occurred, a message that the plant can transmit throughout the rest of the plant; and (ii) from an exogenous signal delivered by the herbivore that tends to reset plant physiological thresholds needed for either defence or growth responses. A great deal about the former exists in the literature, but we know of only a few examples of the latter.

Karban and Baldwin (1997, pp. 27–28) give a thorough synopsis of endogenous post-herbivory (or pathogenic invasion) responses from the standpoint of induced plant defence. They list six criteria that one must apply to determining any putative endogenous systemic plant signalling system arising from a herbivore wound: (i) an identity of the inducer; (ii) rapid messenger generation at the wound site; (iii) a known ability for the plant to elicit a meaningful response; (iv) a dispersal mechanism that transmits the message throughout the plant; (v) a known and reasonable time lag consistent with the induction response; and (vi) a dose–response induction consistent with stress measured in the plant. Endogenous signalling systems identified thus far include: (i) oligosaccharides (most commonly associated with plant pathogens (Ryan, 1987)); (ii) a polypeptide (identified as systemin) involved in the wound response (Pearce *et al.*, 1991; McGurl *et al.*, 1992); (iii) salicylic acid in response to pathogen invasion (Raskin, 1992); (iv) ethylene circulation throughout the plant in response to wounding (Rhoades, 1985); (v) abscisic acid in response to wounding (Peña Cortés *et al.*, 1989); (vi) production of jasmonic acid and its derivatives in response to wounding (Staswick, 1992); and (vii) electrical signals sent throughout the plant as a function of the wounding response (Wildon *et al.*, 1992).

Many of these foregoing examples involve the induced production of proteinase inhibitors (PIs, a secondary plant metabolite) by the plant which inhibit proteolytic digestive enzymes in the herbivore (most commonly an insect herbivore). Thus, once set in motion by the primary herbivore, the presence of the PI ingested during subsequent herbivory episodes severely limits the potential for subsequent herbivores invading the plant to digest their meals, and thus causes reduced growth and development, if not outright starvation. Additional examples include the production of phytoalexins and other compounds that interfere with the herbivore's developmental physiology. In general this group of inducins acts to reduce further plant stress once the signal resulting from the initial attack has sounded its biological alarm.

A second induction response involves the exogenous introduction of an animal signal – essentially an animal biochemical messenger system – that: (i) induces a plant defence sequence, or (ii) induces and regulates fundamental growth and development processes. For perhaps the best example of the first type of messenger system, work in J.H. Tumlinson's USDA Gainesville, Florida, laboratories (Turlings and Tumlinson, 1992) has resulted in the description of a fatty acid (named volicitin), produced in the digestive system of lepidopteran caterpillars and present in oral secretions, that elicits production of volatile compounds capable of free-air detection by a parasitic wasp (Alborn *et al.*, 1997). The volatile signals guide a female wasp to a caterpillar attacking a plant, whereupon she deposits her eggs on the invading insect. After the eggs hatch, the wasp larvae burrow into the caterpillar and destroy it.

Dyer and co-workers (Dyer and Bokhari, 1976; Dyer *et al.*, 1982, 1995) have suggested a second animal biochemical messenger system. This putative system involves the deposition of an exogenous messenger during the feeding process that induces the plant to reorganize its fundamental growth and development processes. Dyer (1980) reported the first such messenger found, a highly conserved mitogenic peptide, epidermal growth factor (EGF). The postulated plant responses to animal messengers involve cell division (Kato *et al.*, 1993) and elongation (Moon *et al.*, 1994; Dyer *et al.*, 1995) (together constituting plant growth), increased photosynthesis and translocation of labile photosynthate throughout the plant (Dyer *et al.*, 1991, 1993a; Holland, 1996; Holland *et al.*, 1997; Dyer, unpublished), and possibly interactions with plant growth hormones that tend to reset the plant's phenological development following herbivory (Dyer *et al.*, 1995). Herms and Mattson (1992) suggest that this type of response may provide the plant with an ability to outgrow the herbivory stress, but give little support for this hypothesis from the physiological literature. Thus, most of the evidence for this positive feedback response exists mainly in sporadic publications throughout the ecological literature going back several decades.

The Exogenous Animal Biochemical Messenger: Mostly Chimerical or Master Controller?

The hypothesis that I posed in the early 1980s (Dyer, 1980; Dyer *et al.*, 1982) that animal biochemical messengers deposited during herbivory might act as major controllers of plant function, has posed an enigma for the scientific community. Indeed, the most severe critics of the herbivore positive feedback and coevolutionary ideas postulated by myself and McNaughton (1979) have termed them 'red herrings' (Belsky *et al.*, 1993). In the face of such criticism, both Dyer and McNaughton have defended their stance (Dyer *et al.*, 1993b;

McNaughton, 1993) and have gained adherence in terms of evolutionary and ecological function (de Mazancourt *et al.*, 1998; de Mazancourt *et al.*, 1999, unpublished observations), as well as from a plant physiology standpoint (Kato *et al.*, 1995). Their work may even point towards unique interactions in molecular biology (Kato *et al.*, 1995; Dyer and Hollenberg, unpublished).

What exactly does the animal biochemical messenger idea entail? First of all, many ecologists have encountered the functional aspects of grazer–plant interactions leading to non-linear relationships in field studies where grazed systems often out-produce ungrazed systems, and furthermore show an optimization of plant productivity at low to moderate herbivory levels (McNaughton, 1979; Hilbert *et al.*, 1982; Dyer *et al.*, 1993b), a response that strongly resembles hormesis (Stebbing, 1982; Bailer and Oris, 1998; Stebbing, 1998)[1]. Both Dyer and McNaughton have considered that herbivory contains two discrete, yet interacting processes: (i) biomechanical-induced stress to plants during feeding; and (ii) biochemical signalling of herbivory through several potential messenger systems as a function of the deposition of oral regurgitants originating from the digestive system of the herbivore. However, in these cases, instead of acting as a messenger that activates solely a defensive strategy to avoid further herbivory, the putative messenger system induces new levels of growth and development within the plant. This induced plant activity may come from the biomechanical stresses, but both Dyer and co-workers (Dyer, 1980; Moon *et al.*, 1994; Dyer *et al.*, 1995) and McNaughton (1985) have shown that the biochemical signal may act alone. In reality, the two plant stressing mechanisms probably act together in a synergistic manner to amplify the apparent herbivory signal.

The apparent action of an animal biochemical signal provokes several questions about its evolutionary and ecological roles beyond the scope of this chapter. Notwithstanding these questions, I will review some of the evidence that suggests the influence of the animal biochemical messengers on plants, and relate recent work that hints where this endeavour may lead us.

Background on animal biochemical messengers

My own recognition of potentials for the role of an animal induction in plant–herbivore interactions arose during the late 1960s and early 1970s as I sought to understand agricultural–avian ecology in the western Lake Erie basin (south-western Ontario and northern Ohio). My field data covering several years of work (essentially 1964 to 1971) showed time and again that icterid bird flocks feeding in fields of ripening field maize did not always result in crop reduction. In fact, on average, such apparent predation regularly increased yields to the farmer (Dyer, 1973, 1975). A search of the literature gave hints that others had observed similar results with bird populations (Dawson, 1970).

Early experiments (Dyer, 1976) suggested that both the mechanical aspect of damaging a maize ear and the introduction of raw saliva could affect both maize grain biomass and total protein content. At about the same time workers at Texas A & M University suggested that thiamine in cattle saliva could affect the productivity potential of grasses in the southern great plains (Reardon *et al.*, 1974). Following this work Dyer and Bokhari (1976) suggested that grasshoppers feeding on blue grama plants held in laboratory microcosms could stimulate aboveground plant production and apparent root and/or microbial respiration in their belowground systems.

Further field and laboratory experiments with raw saliva from a variety of vertebrates during the late 1970s yielded positive results (Dyer, unpublished data) and Dr Carlos Bonilla, then at Colorado State University, Fort Collins, suggested that I consider testing various known cell growth regulators and biochemical messengers in animal saliva. In 1979 and 1980 I found evidence that epidermal growth factor (EGF), a peptide in all known vertebrate salivary systems, could regulate growth and development in sorghum (*Sorghum bicolor* var. Rio) seedlings (Dyer, 1980) (Fig. 6.3) and when applied to either field maize (*Zea mays*) leaves or injected directly into the growing maize ears, could increase ear length (Fig. 6.4A) and associated biomass yield 5–10% above controls (Fig. 6.4B).

McNaughton (1985) substantiated the earlier Reardon *et al.* (1974) work on thiamine. No further work emerged on the exogenous biochemical messenger hypothesis until a team in Japan (Kato *et al.*, 1991, 1993, 1995) repeated Dyer's earlier work on EGF and amplified studies of its effects on maize coleoptile growth (Fig. 6.5A), maize mesocotyl growth (Fig. 6.5B), and maize root growth

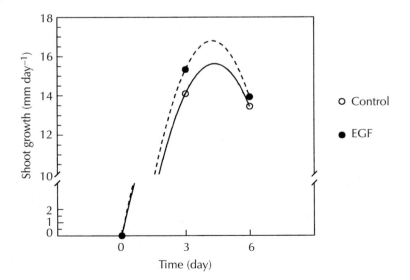

Fig. 6.3. Results redrawn from Dyer (1980) giving the first indication that plants, here *Sorghum bicolor* var. Rio, can recognize and respond to epidermal growth factor (EGF), an animal polypeptide mitogen and growth regulator.

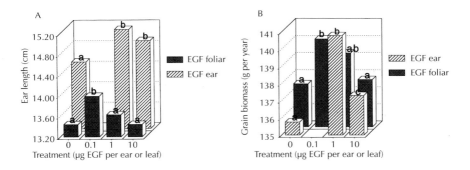

Fig. 6.4. Results from field experiments conducted at Beltsville Agricultural Research Center in 1979 and 1980 (M.I. Dyer, unpublished data) which demonstrate that maize (*Zea mays*) can recognize EGF and respond by altering ear length (A) and grain yield (B). Two experiments, consisting of: (i) applying EGF in varying concentrations at the three- to five-leaf stage of growth; and (ii) injecting varying concentrations of EGF directly into the growing maize ear during the milk stage of development, show non-linear, optimal responses.

(Fig. 6.5C). They also demonstrated that dicotyledonous plants also recognized and responded to EGF, obtaining a significant increase in growth rate of carrot (*Daucus carota*) cells *in vitro* (Fig. 6.6A) and root primordia formation in adzuki bean (*Vigna angularis*) epicotyl cuttings (Fig. 6.6B). I have presented the total experience known for EGF effects on plants in Table 6.1.

I noted above that virtually all vertebrates produce EGF (Hollenberg, 1979; Hollenberg, personal communication) and, with respect to potential ecological connections, secrete sufficient quantities in all body fluids (ng ml^{-1}), particularly in saliva and urine (Carpenter and Wahl, 1990), necessary to induce the plant responses cited. EGF, since its discovery over three decades ago (Cohen, 1962), has become one of the most researched compounds in molecular and cellular biology and biochemistry. Workers in these fields now publish over 900 papers on EGF annually (Carpenter and Wahl, 1990).

Savage and Cohen (1972) described the EGF 53 amino acid sequence in a single-chained molecule, including locations of its three disulphide bonds (molecular size = 6.1 kDa). Functionally EGF and EGF-like members in this family of small peptides act as strong mitogens, but with a wide range of controls over other cellular and biochemical processes (Hollenberg, 1979; Carpenter and Wahl, 1990). EGF and the EGF-like peptide family show homologies in their sequence identity, ranging from 20% among all members of the peptide growth factor family, to 70% between humans and laboratory mice, to as high as 90% among other EGF subgroups (Carpenter and Wahl, 1990). This overlap, known as sequence conservation, suggests that EGF represents one of the oldest, most conserved and bioactive molecules known. Even though no one has studied plants to determine receptor homologies for EGF or EGF-like peptides as reported for

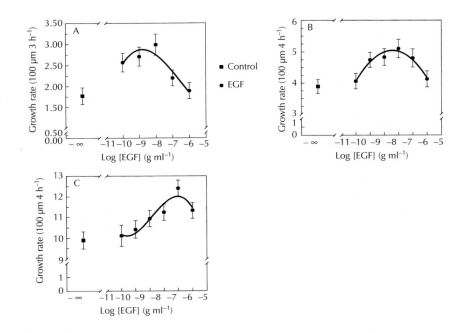

Fig. 6.5. Results redrawn from laboratory experiments conducted by Kato *et al.* (1991, 1995) show that plants recognize EGF and respond through altered growth responses correlated with the concentration of the growth regulator. (A) Shows effects on maize coleoptiles, (B) shows growth responses of maize mesocotyls, and (C) demonstrates the effects on maize root growth.

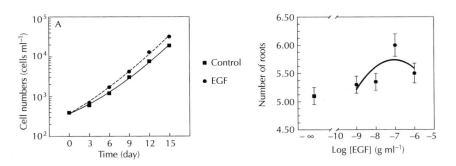

Fig. 6.6. Results redrawn from Kato *et al.* (1993) laboratory experiments demonstrating that dicots as well as monocots respond to EGF treatments. (A) Shows higher growth rates of carrot cells treated with EGF compared with untreated when held in suspension over a 15-day period, and (B) shows the effects of EGF on adzuki bean adventitious root primordia as a function of EGF concentration.

Table 6.1. The effects of epidermal growth factor (EGF) on plant growth and development in monocots and dicots. All references reported bell-shaped curve growth responses; dashes indicate no information available. In addition, Moon *et al.* (1994) showed that EGF interacts with indole-3-acetic acid (IAA) to give a synergistic reaction in maize, oats and sorghum. This finding suggests that plants recognize the exogenous animal biochemical messengers, such as EGF, through their own systems regulated by phytohormones.

		Plant response				
Cell division	Leaf physiology	Whole-plant development	Coleoptile and/or shoot growth	Root growth	Epicotyl/ mesocotyl growth	Ear growth, C and N in grain yield
Monocots:						
—	Maize, Dyer (unpublished ^{11}C studies)	Maize, Dyer (unpublished ^{11}C and field studies)	Sorghum, Dyer (1980); Moon *et al.* (1994); McNaughton (personal communication) Maize, Kato *et al.* (1991); Moon *et al.* (1994) Oats, Moon *et al.* (1994)	Maize, Kato *et al.* (1995)	Maize, Kato *et al.* (1995)	Maize, Dyer (1980)
Dicots: Carrots, Kato *et al.* (1993)	—	—	—	Adzuki bean, Kato *et al.* (1993) Mustard, Dyer *et al.* (1982)	—	—

animals (Carpenter, 1987), somehow plants recognize the exogenous introduction of this messenger. Thus I simply note that there exists a high probability that
EGF or EGF-like compounds produced in high levels by salivary glands and kidneys
in vertebrates – precisely the organs one would consider for the potential of any
signal transduction from herbivores to plants – act as an intertrophic messenger
system. As I will show, it is very likely that a parallel system exists for invertebrates.

From these observations, several questions arise, dominated by asking: in
the light of an apparent major interaction between plants and herbivores today,
what, if any, role did EGF and EGF-like peptides have in the evolution of our
modern fauna? How might herbivores and plants interact in such a way as to
employ the transferral of EGF or its peptide family as intertrophic messengers?
How then do plants recognize such a biochemical signal? Do they indeed have
the cellular mechanisms, such as receptors and biochemical pathways involved
with transcriptional attributes of EGF and EGF-like compounds, to function as
plant growth regulators as they do in animal systems (see Hollenberg, 1979;
Carpenter, 1987; Carpenter and Wahl, 1990)? The pursuit of answers to any
but a few of these questions lies outside the framework of this paper, but I will
return briefly to this topic in the closing comments.

Evidence for herbivore biochemical messengers in the grasshopper model: the midgut

In 1990 Dyer and Hollenberg (unpublished data) conducted experiments that
suggest invertebrate herbivores – specifically several species of grasshoppers – may
have EGF-like peptides in their alimentary systems. To date no one has continued
these investigations to determine more about the EGF-like peptides. However,
recent work in the Department of Entomology, University of Georgia, demonstrates that highly conserved peptides may exist widely between invertebrates and
vertebrates (Huang *et al.*, 1998). In this instance two peptides isolated from
endocrine cells within the midgut of corn earworm larvae (*Helicoverpa zea*) show
a close relationship to pancreatic polypeptides isolated previously from lower
vertebrates (Huang *et al.*, 1998). Furthermore, some evidence exists that these
same small peptides occur in *Romalea* midgut (M.R. Brown, J.W. Crim and M.I.
Dyer, unpublished data). Thus, a large and unknown set of proteins and peptides
may show a high degree of conservancy between invertebrates and vertebrates,
which we can expect to show major regulatory functions in both phyla.

In recent work conducted in my laboratory we have isolated a midgut
grasshopper fraction (GHF) (using the lubber grasshopper, *Romalea guttata* as an
experimental model) that mimics much of the EGF function cited for plants (Dyer
et al., 1995; Dyer, unpublished data). Complementary field work conducted on
above- to belowground translocation of ^{14}C tagged photosynthates in maize as a
function of *Romalea* feeding intensity shows a very rapid inducible pattern with
an intensity directly proportional to the numbers of feeding insects (Holland *et
al.*, 1997; Kisselle, 1998). Thus far, no one has conducted experiments that

would link the laboratory findings and the field study results. None the less, new work following the protocol shown in Fig. 6.7 attempting to identify a putative biochemical messenger in the midgut and/or salivary glands of the *Romalea* model suggests that *Romalea* has the capacity to signal a plant of herbivory.

Dyer *et al.* (1995) reported that midgut fractions from *Romalea* induce increased growth in a standardized coleoptile bioassay. New work confirms that this fraction, referred to as GHF$_{20}$,[2] thus far isolated by two different acid extraction procedures followed by solid phase extraction (SPE), interacts with the plant auxin IAA (indole-3-acetic acid) and gives a non-linear or possible hormetic response in our coleoptile bioassay (see Dyer *et al.*, 1995 for methods).[3]

In the next step using a semi-preparative HPLC column with absorbance set at 280 nm, a software package estimated that the GHF$_{20}$ sample yielded 27 specific major and minor absorbance peaks over a 60 min fractionation period (Fig. 6.8). After dividing the HPLC peaks into three different samples, each covering roughly one-third of the fractionation period, only residues from the 'B' fraction[4] with a dominant single peak at 32 min (Fig. 6.8) indicated significant levels of bioactivity (Fig. 6.9).

In the next step (analysis of the residues that include the 'B' fraction from an analytical column at an absorbance of 215 nm) we found three major peaks, but because we had insufficient residues to conduct the bioassay we have not yet had an opportunity to test each peak. However, from the work with the semi-preparative column we know that the absorbance peak nearest to the 'A' sample (see Fig. 6.8) with limited retention has never shown any bioactivity, while the other two with prominent peaks contain residues that have shown significant bioactivity in the past. Now the task becomes to obtain sufficient material to conduct bioassays of the bioactive residues, and to continue with identification of any active compound(s).

Further evidence of biochemical messengers: the grasshopper salivary glands

At the time we dissected the midgut of each *Romalea*, we also removed the salivary glands[5] and segregated them by date. To date, we have taken the extraction and bioassay steps through the SPE phase. Here we also found that the 20% MECN/0.1% TFA fraction showed significant bioactivity and, equally important, fractions obtained from concentrations >20% MECN showed no effect on coleoptile elongation in the bioassays.

Discussion

Throughout this chapter I have presented the theme that herbivores do a lot more in ecosystems than just eat plants, although that alone has made them an important focus for major investigations in biology, biochemistry, cellular and

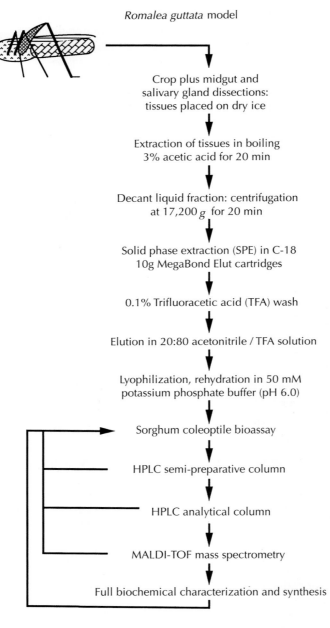

Romalea guttata model

Crop plus midgut and
salivary gland dissections:
tissues placed on dry ice

Extraction of tissues in boiling
3% acetic acid for 20 min

Decant liquid fraction: centrifugation
at 17,200 g for 20 min

Solid phase extraction (SPE) in C-18
10g MegaBond Elut cartridges

0.1% Trifluoracetic acid (TFA) wash

Elution in 20:80 acetonitrile / TFA solution

Lyophilization, rehydration in 50 mM
potassium phosphate buffer (pH 6.0)

Sorghum coleoptile bioassay

HPLC semi-preparative column

HPLC analytical column

MALDI-TOF mass spectrometry

Full biochemical characterization and synthesis

Fig. 6.7. Protocol for collecting specific midgut and salivary gland plant growth factors from *Romalea guttata*, a grasshopper model intended as a representative of insect herbivores/omnivores. As each extraction goes through the sequence it passes on to the next step only when it shows altered growth responses in a standardized sorghum coleoptile bioassay described by Dyer *et al.* (1995).

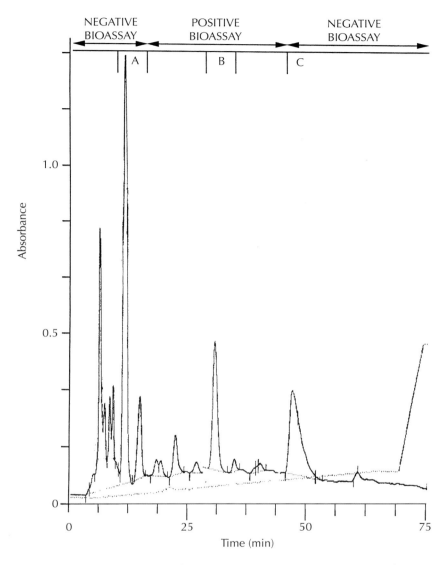

Fig. 6.8. Representative HPLC chromatogram using a semi-preparative column (see Fig. 6.7) and a gradient starting with 20% acetonitrile (MECN) with absorbance set at 280 nm. Initially a residue from the solid phase extraction (SPE) step using an extraction medium consisting of 20% MECN in an 80% trifluoroacetic acid ([TFA] = 0.1% TFA in HPLC grade water) showed: (i) no bioactivity for materials collected in the first 18 min, (ii) significant bioactivity for materials collected from 20 to 36 min, and (iii) no bioactivity for any residues collected after 36 min. Subsequent collection from residues from the three peaks in the 'A', 'B' and 'C' designations in the graph show bioactivity only in peak 'B'. No information exists currently about potential bioactivity in residues collected near the active 'B' peak.

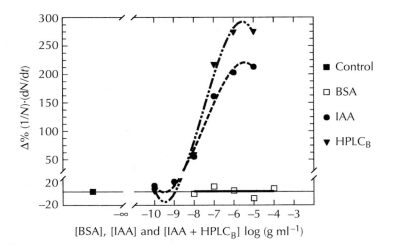

Fig. 6.9. Results from the bioassay of the 'B' peak residue obtained from *Romalea* midgut GHF_{20} fractions as shown in Fig. 6.8 expressed in terms of specific growth rates (SGR) over a 24 h incubation period. Relative to the IAA controls, the *Romalea* midgut HPLC residue resulted in a significant ($P = 0.0087$) increase in SGR, with peak response in the bioassay 275% above untreated and bovine serum albumin (BSA) controls at levels of 10^{-6} g ml^{-1}, and $\geq 23\%$ above IAA controls at levels from 10^{-5} to 10^{-7} g ml^{-1} of the *Romalea* midgut isolate.

molecular biology, and ecology in this century. To me it seems ironic that perhaps the major impact of herbivores has remained so unrecognized and certainly unheralded. Indeed, in a recent book review in *Science*, Deane Bowers (Bowers, 1998) called attention to the most recent, and perhaps most comprehensive book on herbivory (Karban and Baldwin, 1997) by picking up the theme that Charles Darwin, whom everyone enjoys citing as the progenitor of their specific study with evolutionary overtones, missed the induction responses of plants following herbivory. However, Karban and Baldwin (1997) and Bowers (1998) focus only partially on the herbivore–plant associations presented in Fig. 6.1B, and mention nothing about the potentials for plant regrowth dynamics on one side of the picture featured in Fig. 6.2. While presenting a sound picture of the endogenous herbivore signal following feeding, they have either ignored or missed what I continue to argue represents the even greater potential involving an exogenous signalling hypothesis in plant–animal associations.

Herbivory: first or secondary principles?

As I noted during the course of the paper, in the past 20 years I have proposed at various times that herbivores, by passing a signal to the plant during feeding, can reorganize a plant's entire physiological machinery, and thus can alter significantly its growth and development. Furthermore, since we understand that herbivory has spatial and temporal distributions in practically all plant systems on Earth, it, by definition, has a larger and more prominent function in scalar studies ranging from the individual to the landscape. By inducing a non-linear and non-monotonic response in plants (e.g. a hypothetical hormetic response) as a function of feeding intensity, perhaps we can then consider the process of herbivory as one of first principles in system function. Few, if any, other biotic processes control as many elements in primary production as herbivory: (i) it (herbivory) can regulate much of the leaf area that plants use to intercept energy from the sun (Dyer *et al.*, 1998); (ii) it passes a signal that the plant can recognize and utilize to regain lost photoactive tissue, even to the point that graminoids tend to produce more than they lost; (iii) the signal in essence resets the phenological clock in the plant that allows the regrowth of tissue; (iv) the signal helps the plant hide away important foodstuffs by altering translocation and associated C and nutrient sinks; (v) the signal helps the plant in ways as yet unknown to reorganize its ability to cycle and recycle important nutrients, particularly N; and (vi) if we can put sufficient stock in a few studies, the signal may provide the basis for natural selection of populations which may operate either better or worse as a function of the induction signal intensity and subsequent processes that emanate from the signal.

Ecological function of exogenous biochemical messengers: further interpretations

The ecological and biomedical evidence accumulated thus far suggests that EGF or a biochemical messenger of EGF-like qualities may exist in all vertebrate herbivores. In addition we should expect that additional messengers also exist in vertebrate salivary systems, such as the case for thiamine (Reardon *et al.*, 1974; McNaughton, 1985). And, we should not expect all plants to respond to exogenous messengers in the same way as the few cited in Table 6.1 respond. Indeed, one of Sam McNaughton's undergraduate students recently undertook a study to examine the effects of EGF on several graminoids from the Serengeti and obtained mixed results, even though he found the same results with sorghum seedlings that Dyer (1980) reported (S.J. McNaughton, personal communication). Lastly, I urge caution in interpreting findings that compare vertebrate and invertebrate systems. Even though I argued for highly conserved systems earlier, we also should expect that a large suite of exogenous messenger systems may exist, some conserved between the two phyla and some not.

Exogenous herbivore messengers in plants: an evolutionary framework

For EGF and EGF-like peptides to perform as plant induction messengers as I have hypothesized, we must consider how this association may have arisen.

Several potential paths may exist, but the most simple approach that I can perceive has to do with the puzzling case that comes to us from the biomedical sciences for several messenger peptides. Because the biomedical community can find the small regulatory peptides, such as EGF, nerve growth factor, insulin and literally hundreds of others, in a lineage of organisms that have evolved more recently than ancient prokaryotes, they reason the origin of many of these compounds occurred at the dawn of biological life on Earth. Thus, instead of asking questions about how natural selection pressures may have picked or chosen various traits on the basis of what bioactive peptide an organism may have found advantageous, or have needed to survive, the central question may become: how did the existence of the messenger help to guide the survival of the organisms that employed this specialized biochemical molecule as a messenger? In the first sense one would expect the answers to how herbivores and plants have co-developed (perhaps co-evolved) to come from the neo-Darwinist paradigm based in *competitive interactions*. However, if the highly conserved EGF or EGF-like molecule(s) play a major role in modern biology and ecology of plant–herbivore relationships, then we have to try to understand the role these molecules played in the origination of co-evolution of plants and herbivores, and that basis might come from a *symbiotic association*. That is to say, *if EGF or EGF-like messengers function in plant–herbivore interrelationships as I postulate, then when the first animal took the first bite out of the first plant, the so-called 'exogenous messenger system' already existed*. Thus, perhaps herbivory, as we understand it now, arose as a function of biochemical pre-adaptation which came about from the molecular–cellular processes that today we call transduction in or among single cell organisms, or any cell in any plant or animal tissue. While this might at first seem a radical proposal, I argue that if herbivores and plants exchange information through messengers such as EGF or EGF-like peptides, the intertrophic mechanisms may possess homologies with those well known for the cell. In the case of EGF, this would constitute a remarkable case: the peptide would serve both as the intertrophic messenger and the cellular messenger that results in, as I have pointed out several times, major alteration of plant basic growth and development patterns.

Where to go now?

I have laid out my reasoning for the existence of an exogenous animal messenger system governing plant growth and development, but admittedly have yet to achieve a complete picture. Many of the Karban and Baldwin (1997) criteria mentioned above apply to studies of exogenous messengers as well as endogenous messengers. Alborn *et al.* (1997) have satisfied the criteria for identifying an exogenous messenger involved with plant defence, but in view of my arguments for messenger systems pertinent to fundamental plant growth and development, the information base falls short. Indeed, at this time I can satisfy only parts of the six basic criteria, as presented in the following.

1. *Messenger identity.* EGF and at least one unknown set of HPLC fractions from the *Romalea* model may act as exogenous animal messengers controlling plant growth and development, but, with respect to the protocol in Fig. 6.7, may also give an incomplete picture.

2. *Endogenous plant messenger transfer at wound site.* Evidence exists to suggest that the plant recognizes a signal specific to a herbivore (Dyer and Bokhari, 1976; Dyer, 1980; Dyer *et al.*, 1991; Holland, 1996; Holland *et al.*, 1997; Kisselle, 1998; Dyer, unpublished data), but no one has specifically searched for an endogenous plant messenger system at the point of damage. Such messengers may follow the biochemical and molecular models already described, but they also might follow completely different pathways.

3. *Plant response.* Ecological literature, as I have summarized briefly, arguably gives ample evidence for a growth and development response *vis-à-vis* the herbivore optimization model, and thus perhaps an example of hormesis in a community or ecosystem context.

4. *Message dispersal throughout the plant.* Evidence for this criterion, much as for number 2, relies more on ecological and, to some degree, on plant physiological studies that show, for instance, that above- to belowground transference occurs, but no studies exist to examine the processes from the standpoint of cellular or molecular mechanisms.

5. *Lag time.* The physiological studies conducted thus far suggest that a rapid response system (within minutes to hours) exists in plants (Dyer *et al.*, 1991; Dyer, unpublished data). However, again no one has studied specific mechanisms for their inherent reaction times.

6. *Messenger concentrations.* Estimated concentrations of EGF exist for two media (saliva and urine, Carpenter and Wahl, 1990) which may function as the transfer agent to the plant, but no one has reported the amounts actually entering the plant at the site of the wound. I have conducted experiments using ^{14}C tagged midgut contents present in oral regurgitants, but have not yet completed the study.

In this chapter I have tried to lay out the framework for viewing plant–herbivore interrelationships involving principles of C and nutrient transfer from both holistic and reductionist arguments which I believe have major implications for evolutionary and management viewpoints. For meaningful advance on the topic, we must examine what has emerged from the standpoint of ecological through modern cellular and molecular approaches. However, to fully appreciate such advances, we must then return to ecological principles for the final tests of veracity.

Finally, in a call to the future, if we obtain information further validating the idea that biochemical messengers in herbivores feeding above ground signal the plant utilizing a cascade of messengers that induce post-feeding differential growth and development, I then recommend we focus attention on belowground herbivores. The conference from which this book is derived highlighted several strong candidates for such new research, particularly the major invertebrate species that have direct contact with plant roots.

Acknowledgements

Over the past several years many people have helped me either with field work or laboratory studies. I greatly appreciate their input. Many colleagues helped me deal with complex molecular and biochemical techniques, all of which have come to exist since my days as a graduate student; indeed, these patient and knowledgeable people have helped me to continue my work as a student! Foremost, I thank Professor Morley Hollenberg, University of Calgary, Alberta (an authority on animal peptides with biomedical importance) and Professor E.P. Odum, University of Georgia (a world authority on ecosystem behaviour) for helping me bridge the enormous gap I created for myself by even thinking of tackling a programme that spans molecules to ecosystems. Many others have contributed to the most recent studies, but in particular (and in alphabetical order): Mark Brown, Sherry Farley, Mark Hunter, Brian Leigh, Sam McNaughton, Angela Moon-Avila, Cheryl Pearson, Bernard Renaux, Justin Swartz and Linda Wallace. The support for this work has come from the National Science Foundation Ecological and Evolutionary Physiology Program Grant IBN-9630347 through the University of Georgia Research Foundation.

I leave my last comments to thank the honouree of this conference, Professor D.A. Crossley, Jr, for uncounted hours of discussion, arguments and good fellowship. Years ago he stepped forward, following a particularly severe medical crisis in my life, and made me understand a great many things one cannot divine for oneself. So, I hope that Dac finds this paper, one that the organizers, Drs Dave Coleman and Paul Hendrix invited me to present in honour of his retirement after a long and profitable career, a suitable dedication on my part. But now, after having taken a short detour via a small emotional outpouring, I hasten back to the real world to say: *'Dac, it's still your turn to buy!'*

Notes

1. Hormesis, an old concept that has recently regained attention, carries an increasingly important distinction among many working with toxicology and in aquatic systems. Its meaning, taken from the Greek root, *hormē*, indicating assault or rapid motion, as in impulse, pertains to the non-linear stimulating effect of sub-inhibitory concentrations of any substance, often a toxin, on any organism (Dorland, 1957). Bailer and Oris (1998) define hormesis as 'a dose response relationship that is stimulatory at low doses, but is inhibitory at higher doses'. For the purposes of plant–animal inter-actions, that definition translates to an initial induction signal from the herbivore to the plant that first stimulates plant growth once the interaction phase passes through a threshold, followed then by a maximum response, and ends in an inhibitory phase; e.g. the herbivore optimization curve (HOC) described by McNaughton (1979) and Hilbert *et al.* (1982). Statisticians refer to this response as the 'ß curve' (Stebbing, 1998). By definition, for the non-linear plant response to grazing a 'growth hormesis zone' develops where compensatory growth exceeds that of ungrazed control plants as a function of grazing intensity.

2. The designation comes from our discovery of a grasshopper factor (GHF) that SPE isolates with 20% acetonitrile (MECN) when combined with a 0.1% trifluoroacetic acid (TFA) solution. We have also tested grasshopper midgut residues derived from other concentrations of acetonitrile (e.g. those designated GHF_{40}, GHF_{60} and GHF_{80}) in our coleoptile bioassay. While those fractions contain apparent peptides, not one has shown any effect on coleoptile elongation.

3. We have replaced the somewhat complex and tedious cold extraction method utilized by Dyer *et al.* (1995) with a more efficient method designed to isolate small molecular weight peptides. We now take the grasshopper tissues directly from a $-80°C$ freezer and place them in boiling 3% acetic acid for 20 min, then decant, centrifuge and conduct the SPE fractionation using C-18 MegaBond Elut cartridges and an extraction medium consisting of 80:20, 0.1% trifluoroacetic acid and acetonitrile (Fig. 6.7).

4. Typically for each crude extraction we use approximately a 1:10 mass to volume ratio between tissue and acetic acid extractant. In such extractions we select frozen tissues, which to date have consisted of both male and female samples, for a specific date of collection and treat these as a unit in our extraction, SPE and HPLC series. For example, in one case we extracted 152 midguts (74% males, 26% females) which contained a wet weight of 42.3 g in 500 ml of 3% hot acetic acid. The residue yield for the peaks in sample 'B', which consisted of a yellowish-white powder consistent with the appearance of a protein or peptide, amounted to a mass of < 1 mg. For each bioassay experiment we require 0.22 mg.

5. Adult and late instar *Romalea* have a pair of prominent salivary glands that resemble grape clusters located on either side of the anterior part of the crop. At the time we dissected the grasshopper we opened up the ventrum with scissors and first removed the salivary glands and placed them on a small sheet of aluminium foil resting on dry ice to quick-freeze the tissue. We extracted the salivary glands using only the hot 3% acetic acid method and used the SPE techniques described for the midgut fractionation.

References

Alborn, H.T., Turlings, T.C.J., Jones, T.H., Stenhagen, G., Loughrin, J.H. and Tumlinson, J.H. (1997) An elicitor of plant volatiles from beet armyworm oral secretion. *Science* 276, 945–949.

Bailer, A.J. and Oris, J.T. (1998) Incorporating hormesis in the routine testing of hazards. *Belle Newsletter* 6(3), 2–5.

Belsky, A.J., Carson, W.P., Jensen, C.L. and Fox, G.A. (1993) Overcompensation by plants: herbivore optimization or red herring? *Evolutionary Ecology* 7, 109–121.

Bernays, E.A. and Chapman, R.F. (1994) *Host-plant Selection by Phytophagous Insects.* Chapman and Hall, New York.

Bowers, D. (1998) Not noticed by Darwin? *Science* 280, 1543.

Brown, B.J. and Allen, T.F.H. (1989) The importance of scale in evaluating herbivory inputs. *Oikos* 54, 189–194.

Carpenter, G. (1987) Receptors for epidermal growth factor and other polypeptide mitogens. *Annual Review of Biochemistry* 56, 881–914.

Carpenter, G. and Wahl, M.I. (1990) The epidermal growth factor family. In: Sporn, M.B. and Roberts, A.B. (eds) *Peptide Growth Factors and Their Receptors I.* Springer-Verlag, Berlin, pp. 69–171.

Cohen, S. (1962) Isolation of a mouse submaxillary gland protein accelerating incisor eruption and eyelid opening in the new-born animal. *Journal of Biological Chemistry* 237, 1555–1562.

Crawley, M.J. (1983) *Herbivory: the Dynamics of Animal–Plant Interactions*. University of California Press, Berkeley.

Dawson, D.G. (1970) Estimation of grain loss due to sparrows (*Passer domesticus*) in New Zealand. *New Zealand Journal of Agricultural Research* 13, 681–688.

Denno, R.F. and McClure, R.F. (eds) (1983) *Variable Plants and Herbivores in Natural and Managed Systems*. Academic Press, New York.

Dorland's Illustrated Medical Dictionary (1957) 23rd edn. W.B. Saunders, Philadelphia.

Dyer, M.I. (1973) Plant–animal interactions: the effects of red-winged blackbirds on corn growth. In: Cones, H.R., Jr and Jackson, W.B. (eds) *Proceedings of Sixth Bird Control Seminar*. Bowling Green State University, Bowling Green, Ohio, pp. 229–241.

Dyer, M.I. (1975) The effects of red-winged blackbirds (*Agelaius phoeniceus*) on biomass production of corn grains (*Zea mays* L.). *Journal of Applied Ecology* 12, 719–726.

Dyer, M.I. (1976) Plant–animal interactions: simulation of bird damage on corn ears. In: Jackson, W.B. (ed.) *Proceedings of Seventh Bird Control Seminar*. Bowling Green State University, Bowling Green, Ohio, pp. 173–179.

Dyer, M.I. (1980) Mammalian epidermal growth factor stimulates sorghum seedling growth. *Proceedings of the National Academy of Sciences USA* 77, 4826–4837.

Dyer, M.I. and Bokhari, U.G. (1976) Plant–animal interactions: studies of the effects of grasshopper grazing on blue grama grass. *Ecology* 57, 762–772.

Dyer, M.I., Detling, J.K., Coleman, D.C. and Hilbert, D.W. (1982) The role of herbivory in grasslands. In: Estes, J.R., Tyrl, R.J. and Brunken, J.N. (eds) *Grasses and Grasslands*. University of Oklahoma Press, Norman, pp. 255–295.

Dyer, M.I., Acra, M.A., Wang, G.M., Coleman, D.C., Freckman, D.W., McNaughton, S.J. and Strain, B.R. (1991) Source-sink carbon relations in two *Panicum coloratum* ecotypes in response to herbivory. *Ecology* 72, 1472–1483.

Dyer, M.I., Coleman, D.C., Freckman, D.W. and McNaughton, S.J. (1993a) Measuring heterotrophic-induced source-sink relationships in *Panicum coloratum* with the ^{11}C technology. *Ecological Applications* 3, 654–665.

Dyer, M.I., Turner, C.L. and Seastedt, T.R. (1993b) Herbivory and its consequences. *Ecological Applications* 3, 10–16.

Dyer, M.I., Moon, A.M., Brown, M.R. and Crossley, D.A., Jr (1995) Grasshopper crop and midgut extract effects on plants: an example of reward feedback. *Proceedings of the National Academy of Sciences USA* 92, 5475–5478.

Dyer, M.I., Turner, C.L. and Seastedt, T.R. (1998) Biotic interactivity between grazers and plants: relationships contributing to atmospheric boundary layer dynamics. *Journal of the Atmospheric Sciences* 55, 1247–1259.

Gilbert, L.E. and Raven, P.H. (eds) (1975) *Coevolution of Animals and Plants*. University of Texas Press, Austin, Texas.

Hairston, N.G., Jr and Hairston, N.G., Sr (1993) Cause–effect relationships in energy flow, trophic structure, and interspecific interactions. *The American Naturalist* 142, 379–411.

Hairston, N.G., Smith, F. and Slobodkin, L. (1960) Community structure, population control and competition. *The American Naturalist* 94, 421–425.

Herms, D.A. and Mattson, W.J. (1992) The dilemma of plants: to grow or defend. *Quarterly Review of Biology* 67, 283–335.

Hilbert, D.W., Swift, D.M., Detling, J.K. and Dyer, M.I. (1982) Relative growth rates and the grazing optimization hypothesis. *Oecologia* (Berlin) 51, 14–18.

Holland, J.N. (1996) Effects of above-ground herbivory on soil microbial biomass in conventional and no-tillage agroecosystems. *Applied Soil Ecology* 2, 275–279.

Holland, J.N., Cheng, W. and Crossley, D.A., Jr (1997) Herbivore-induced changes in plant carbon allocation: assessment of below-ground C fluxes using carbon-14. *Oecologia* 107, 87–94.

Hollenberg, M.D. (1979) Epidermal growth factor-urogastrone, a polypeptide acquiring hormonal status. *Vitamins and Hormones* 37, 69–110.

Huang, Y., Brown, M.R., Lee, T.D. and Crim, J.W. (1998) RF-amide peptides isolated from the midgut of the corn earworm, *Helicoverpa zea*, resemble pancreatic polypeptide. *Insect Biochemistry and Molecular Biology* 28, 345–356.

Karban, R. and Baldwin, I.T. (1997) *Induced Responses to Herbivory*. University of Chicago Press, Chicago.

Kato, R., Itagaki, K., Uchida, K., Shinomura, T.H. and Harada, Y. (1991) Effects of an epidermal growth factor on the growth of *Zea* coleoptiles. *Plant Cell Physiology* 32, 917–919.

Kato, R., Nagayama, K., Suzuki, T., Uchida, K., Shimomura, T.H. and Harada, Y. (1993) Promotion of plant cell division by an epidermal growth factor. *Plant Cell Physiology* 34, 789–793.

Kato, R., Shimoyama, T., Suzuki, T., Uchida, K., Shimomura, T.H. and Harada, Y. (1995) Effects of an epidermal growth factor on growth of *Zea* primary roots and mesocotyls. *Plant and Cell Physiology* 36, 197–199.

Kisselle, K.W. (1998) Microbial use of root-derived carbon and litter-derived carbon in conventional and no tillage soils. PhD dissertation, University of Georgia, Athens, USA.

Levin, S.A. (1993) Grazing theory and rangeland management. *Ecological Applications* 3, 1.

McGurl, B., Pearce, G., Orozco-Cardenas, M. and Ryan, C.A. (1992) Structure, expression, and antisense inhibition of the systemin precursor gene. *Science* 255, 1570–1573.

McNaughton, S.J. (1979) Grazing as an optimization process: grass-ungulate relationships in the Serengeti. *The American Naturalist* 113, 691–703.

McNaughton, S.J. (1985) Interactive regulation of grass yield and chemical properties by defoliation, a salivary chemical, and inorganic nutrition. *Oecologia* (Berlin) 65, 478–486.

McNaughton, S.J. (1993) Grasses and grazers, science and management. *Ecological Applications* 3, 17–20.

de Mazancourt, C., Loreau, M. and Abbadie, L. (1998) Grazing optimization and nutrient cycling: when do herbivores enhance plant production? *Ecology* 79, 2242–2252.

de Mazancourt, C., Loreau, M. and Abbadie, L. (1999) Grazing optimization and nutrient cycling: potential impact of large herbivore in a savanna ecosystem. *Ecological Applications* 9, 784–797.

Moon, A.M., Dyer, M.I., Brown, M.R. and Crossley, D.A., Jr (1994) Epidermal growth factor interacts with indole-3-acetic acid and promotes coleoptile growth. *Plant Cell Physiology* 35, 1173–1177.

Pearce, G., Strydom, D., Johnson, S. and Ryan, C.A. (1991) A polypeptide from tomato leaves induces wound-inducible proteinase inhibitor proteins. *Science* 253, 895–898.

Peña Cortés, H., Sánchez-Serrano, J.J., Mertens, R., Willmitzer, L. and Prat, S. (1989) Abscisic acid is involved in the wound-induced expression of the proteinase inhibitor

II gene in potato and tomato. *Proceedings of the National Academy of Sciences USA* 86, 9851–9855.

Polis, G.A. (1994) Food webs, trophic cascades and community structure. *Australian Journal of Ecology* 19, 121–136.

Polis, G.A. and Winemiller, K.O. (eds) (1996) *Food Webs: Integration of Patterns and Dynamics*. Chapman and Hall, New York.

Raskin, I. (1992) Role of salicylic acid in plants. *Annual Review of Plant Physiology and Plant Molecular Biology* 43, 439–463.

Reardon, P.O., Leinweber, C.L. and Merrill, L.B. (1974) Response of sideoats grama to animal saliva and thiamine. *Journal of Range Management* 27, 400–401.

Rhoades, D.F. (1985) Offensive-defensive interactions between herbivores and plants: their relevance in herbivore dynamics and ecological theory. *The American Naturalist* 125, 205–238.

Ryan, C.A. (1987) Oligosaccharide signaling in plants. *Annual Review of Cell Biology* 3, 295–317.

Savage, C.R., Jr and Cohen, S. (1972) Epidermal growth factor and a new derivative. Rapid isolation procedures and biological and chemical characteristics. *Journal of Biological Chemistry* 247, 7609–7611.

Spencer, K. (ed.) (1988) *Chemical Mediation of Coevolution*. Academic Press, San Diego.

Staswick, P.E. (1992) Jasmonate, genes, and fragrant signals. *Plant Physiology* 99, 804–807.

Stebbing, A.R.D. (1982) Hormesis – the stimulation of growth by low levels of inhibitors. *The Science of the Total Environment* 22, 213–234.

Stebbing, A.R.D. (1998) A theory for growth hormesis. *Mutation Research* 403, 249–258.

Turlings, T.C.J. and Tumlinson, J.H. (1992) Systemic release of chemical signals by herbivore-injured corn. *Proceedings of the National Academy of Sciences USA* 89, 8399–8402.

Wallace, J.W. and Mansell, R.J. (eds) (1976) *Biochemical Interaction Between Plants and Insects*. Plenum Press, New York.

Wildon, D.C., Thain, J.F., Minchin, P.E.H., Grubb, I.R., Reilly, A.J., Skipper, Y.D., Doherty, H.M., O'Donnell, P.J. and Bowles, D.J. (1992) Electrical signaling and systemic proteinase inhibitor induction in the wounded plant. *Nature* 360, 62–65.

Soil Invertebrate Controls and Microbial Interactions in Nutrient and Organic Matter Dynamics in Natural and Agroecosystems

7

C.A. Edwards

Soil Ecology Program, Botany and Zoology Building, The Ohio State University, Columbus, OH 43210, USA

Introduction

It is generally accepted that soil-inhabiting invertebrates and microorganisms are very important in facilitating the breakdown of organic matter. However, research into the finer details of the diverse interactions between various groups of invertebrates and microorganisms in the decomposition process has developed mainly over the last 25 years. Such research was accelerated in the 1960s continuing into the 1970s by the International Biological Program, which focused on primary and secondary production. There were not many detailed reviews on invertebrate–microbial interactions until this topic was the theme of a symposium on the 'Biological Interactions in Soil' held at the Ohio State University in Columbus (Edwards *et al.*, 1988). There are further discussions of interactions between soil organisms in *Fundamentals of Soil Ecology* (Coleman and Crossley, 1996) and in some more general review-type publications (Whitford *et al.*, 1982; Hunt *et al.*, 1987; Martin, 1987; Moore *et al.*, 1988; Andren *et al.*, 1990; Siepel, 1994; Edwards *et al.*, 1995a; van Straalen and Butovsky, 1998). Butovsky and van Straalen (1998) and Pokarzehvskii *et al.* (1998) discussed the effects of pollution on soil food webs.

The total populations and biomass of microorganisms and invertebrates in agroecosystems are enormous. Indeed, they total more in overall biomass (Table 7.1) than the vertebrates that live above ground. It is clear that the total biomass of the microorganisms in soil greatly exceeds that of the invertebrates, except for earthworms. In spite of these differences, there is increasing evidence that the

Table 7.1. Numbers and biomass of soil invertebrates and microorganisms.

	no. m^{-2}	no. g^{-1}	kg ha^{-1}
Microflora			
Bacteria	10^{13}–10^{14}	10^8–10^9	400–5000
Actinomycetes	10^{12}–10^{13}	10^7–10^8	400–5000
Fungi	10^{10}–10^{11}	10^5–10^6	1000–15,000
Algae	10^9–10^{10}	10^4–10^6	10–500
Invertebrates			
Protozoa	10^9–10^{10}	10^4–10^5	20–200
Nematoda	10^6–10^7	10–10^2	10–150
Acarina	10^3–10^6	1–10	5–150
Collembola	10^3–10^6	1–10	5–150
Earthworms	10–10^3		100–15,000
Others	10^2–10^4		10–100

importance of the role of invertebrates, in controlling the pattern and rates of organic matter decomposition processes, greatly exceeds their numbers and biomass relative to that of microorganisms, in soils and litter (Moore *et al.*, 1988).

The interactions between microorganisms and the different functional groups of invertebrates are extremely complex. During the last 25 years, there have been many attempts at linking and understanding these interactions better, through development and construction of complex food web models which allow energy and material flows to be estimated. However, we are a long way from being able to make a full trophic level analysis of organisms in any soil ecosystem. Moreover, identifying specific interactions between different groups of soil invertebrates and microorganisms is made more difficult because of our lack of detailed knowledge of the food and feeding habits of many soil-inhabiting invertebrates (Walter, 1988). Additionally, although invertebrates can be separated conveniently into the functional groups such as plant feeders, predators, carnivores and detritivores (saprophages, fungivores, bacteriovores), such a classification can be confusing because particular groups of invertebrates may fall into several of these categories or even be omnivores. Moreover, even a single species can change its food and feeding habits in response to food availability and environmental stress.

Another obstacle to constructing and understanding soil food webs is that, particularly among microorganisms, there has been extensive discussion of functional redundancy between different species. This assumes that if one or more species or groups of organisms are absent or eliminated from a functional ecological niche, one or more other species or groups can take their place. This greatly complicates the identification of consistent food web linkages, particularly if species from different major groups have similar functions.

Finally, there is the complex issue of scale. Most of the food webs that have been proposed in the literature limit their discussions to the microscale, i.e.

defining interactions between soil microorganisms (protozoa, algae, actinomycetes, bacteria and fungi) and soil-inhabiting invertebrates (nematodes, enchytraeids and microarthropods, particularly oribatid and prostigmatid mites and springtails). There have been hardly any attempts to scale these basic food webs into full trophic level chains addressing ecosystem processes. The question of whether microscale studies can provide data to facilitate a full understanding of ecosystem function is discussed elsewhere in this volume (Chapter 1).

Soil Food Webs in Natural and Agroecosystems

The idea of decomposition food webs was addressed initially in forest and other natural ecosystems. One of the earliest formulations of a soil food web, representing the decomposer community of litter and soil, was presented by Edwards *et al.* (1970) (Fig. 7.1). These authors considered that it was very difficult to assign specific groups of organisms into positions in a food web, although some organisms could be placed into appropriate trophic levels or components equivalent to compartments in their model. Under their concept, a particular compartment could be a donor or recipient of material from many other compartments. Models constructed in this manner have had conceptual value in describing flows and complexity, but are limited functionally to calculating energy or material flows only at fixed points in time, since it is difficult to make accommodation for the changing food habits of species of soil organisms during different life history stages or environmental pressures. The energy approach is a good way of addressing the predator–prey relationships, because of a strong correlation between energy content and protein content in animals. However, most other authors have based these issues on the basis of nutrient flows (Polis, 1994). For instance, a shortage of microbial protein may be compensated for by invertebrates turning to carnivory instead of to other foods.

One of the earlier food web models that defined the interactions between more specific groups of soil organisms was that designed by Moore *et al.* (1988), modified from Hunt *et al.* (1987), which addressed the soil detrital food web in a shortgrass steppe or prairie (Fig. 7.2). These workers considered that the detrital system was too complex to study at the level of individuals or species populations and they lumped taxonomic groups into units or compartments that function similarly. For instance, this involved separating nematodes into fungivorous, bacteriophagous and predatory groups, and separating mites into two different groups of fungal feeders (slow-growing cryptostigmatid mites and faster-growing species) and predaceous species. The model food web that they designed for a shortgrass prairie was developed from measurements in the field of the abundance and dynamics of different groups of soil organisms. These data enabled hypotheses to be set up and tested subsequently in the laboratory in more detail, to confirm the validity of the model. This type of food web has been devised and used by several other workers to develop rather more detailed and specific food webs.

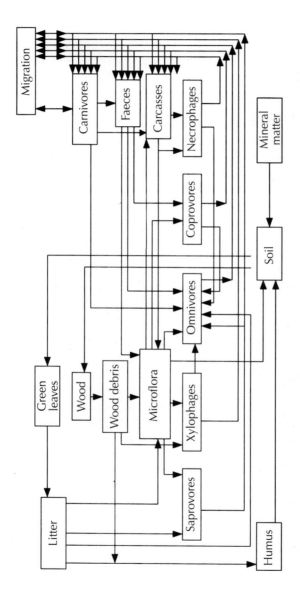

Fig. 7.1. Energy flow and nutrient cycles through soil populations. Exchange along foods chains (modified from Edwards *et al.*, 1970).

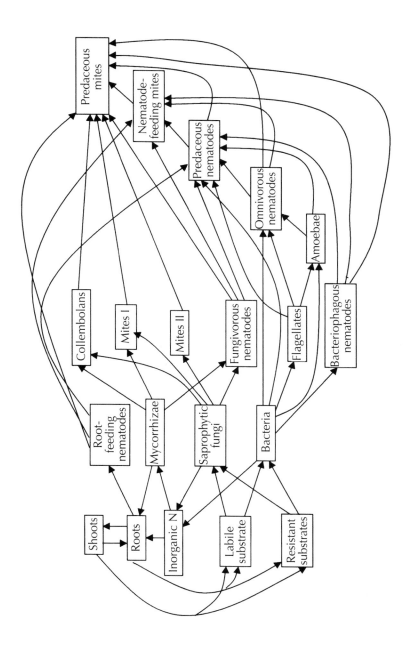

Fig. 7.2. Diagrammatic representation of a detrital food web in shortgrass prairie (Hunt *et al.*, 1987).

A model food web for a mull soil, which involved several trophic levels including aboveground vertebrates, was that of Schaefer (1995), but this food web was very generalized and difficult to validate. Another simplified and generalized food web, also involving trophic implications, was published by Bottomley (1998). This model placed microarthropods, nematodes and earthworms in a central position in the decomposition process, but did not take or develop any argument for their key roles in decomposition any further.

Two groups of soil biologists in Sweden and The Netherlands have discussed food webs in agroecosystems. A Swedish integrated soil ecology project entitled 'Ecology of Arable Land' was initiated in 1979 and continued until 1988. This investigated carbon and nitrogen flows in farm cropping systems with different inputs and sources of C and N. The processes of nitrogen deposition, leaching, nitrification, denitrification, primary production and changes in population of soil-inhabiting invertebrates and microorganisms were studied in detail and associated with rates of decomposition using litter bags and other research tools. The studies on soil organisms focused on assessing the effects of cropping systems on soil communities but also on nutrient flows through these communities (Persson and Rosswall, 1983; Andren *et al.*, 1990). Based on these studies, Andren *et al.* (1988) presented a tentative food web of decomposing barley straw in their agroecosystem (Fig. 7.3). In spite of a numerous and diversified decomposer community, exhibiting clear successional trends in litter bags, they were able to present a simple decomposition model, with temperature and moisture as driving forces, which provided an excellent fit with measured total biomass nitrogen and water-soluble materials in the decomposing litter. They considered that although some of the organisms involved may decrease or increase in abundance and, as a result, some of the pathways may change, the rates of mass loss of litter may not change significantly in a particular agroecosystem. However, they emphasized that such food web models may be misleading, since many soil-inhabiting invertebrates exhibit a lower food specificity than indicated and opportunistic feeding may be more general than conventionally accepted, for example astigmatid mites which are considered fungivorous have also been shown to be voracious nematode predators (Walter *et al.*, 1986).

In 1985, the Dutch Programme on Soil Ecology of Arable Farming Systems in Wageningen started a multidisciplinary research programme on the effects of management practices on the biotic and abiotic properties of arable soils (Brussaard *et al.*, 1990). They compared two management practices: conventional (inorganic fertilizer) and integrated (organic fertilizer, reduced tillage, reduced use of pesticides). A food web concept was developed based on the results from the Lovinkhoeve experimental farm (De Ruiter *et al.*, 1994; Vreeken-Buijs, 1998) (Fig. 7.4). Brussaard *et al.* (1990) distinguished seven different microarthropod functional groups in the Dutch agroecosystem.

1. Omnivorous Collembola, which feed on fungi, algae and organic matter.
2. Cryptostigmatid mites, which feed on fungi and organic matter.

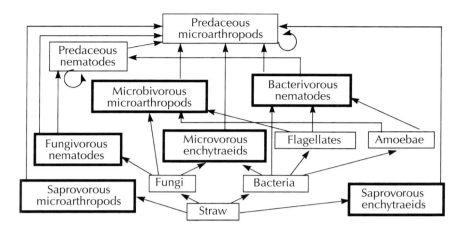

Fig. 7.3. A simplified food web based on decomposing barley straw in agroecosystems (modified from Andren *et al.*, 1988).

3. Omnivorous non-cryptostigmatid mites, which feed on fungi, organic matter and possibly nematodes.
4. Bacterivorous mites, which feed on bacteria only.
5. Nematophagous mites, which are specialized predators of nematodes.
6. Predatory Collembola, which take various kinds of prey.
7. Predatory mites, which feed on various kinds of prey, including predatory Collembola.

Their hypothesis was that the integrated management system would result in a higher organic matter content of the soil, which would in turn favour the microarthropods, which as a result might increase rates of nitrogen mineralization. They designed a relatively complex food web which subdivided major groups (Fig. 7.4), and subsequently made a comparison between the Lovinkhoeve food webs and the food webs studied in Sweden (Lagerlöf and Andren, 1988; Andren *et al.*, 1990) and Georgia, USA (Hendrix *et al.*, 1986) with special emphasis on the microarthropod functional groups. The Dutch researchers concluded from combined evidence that, although the total biomass of microarthropod functional groups was small, they make a significant contribution of more than 30% to overall soil nitrogen mineralization (Verhoeff and Brussaard, 1990). In further studies in this programme, Vreeken-Buijs (1998) distinguished major contributions of various groups of mites and springtails to processes in the soils. She concluded that the positive effects of microarthropods on decomposition were often balanced by negative effects, and concluded that protozoa and nematodes exerted a stronger controlling influence on decomposition than did microarthropods.

In studies on agroecosystems at Horseshoe Bend, Athens, USA, which compared conventionally tilled agroecosystems with no-till systems, Hendrix *et*

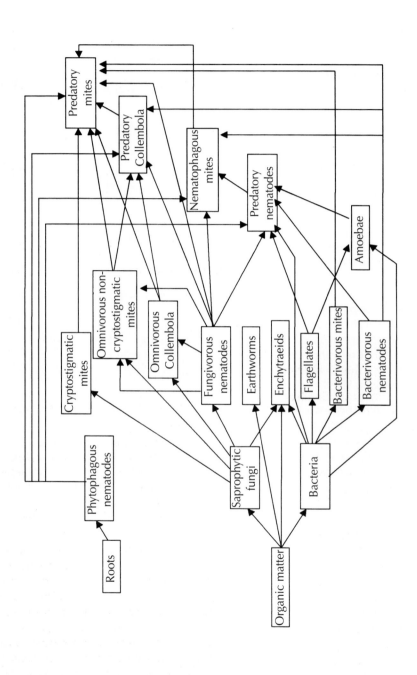

Fig. 7.4. The proposed food web at the Lovinkhoeve Experimental Farm in the Netherlands (modified from De Ruiter *et al.*, 1994).

al. (1986) constructed conceptual model food webs to trace nutrient transfers
(Fig. 7.5a and b). They concluded that although a comparable model could be
constructed for both tillage systems with similar relationships, the relative
importance of different groups of soil-inhabiting invertebrates varied greatly in
response to changes in inputs. They considered that conventional tillage
systems were bacterially dominated, with protozoa, enchytraeid worms and
bacterivorous nematodes playing key roles (Fig. 7.5a), whereas no-till systems
were fungally dominated with key roles being played by earthworms, fungivorous microarthropods, particularly springtails, and fungivorous nematodes
(Fig. 7.5b). This is a clear demonstration of how the manipulation of a single
major input can completely change the main nutrient pathways and roles of
invertebrates and microorganisms in an agroecosystem.

Thus, although there appear to be a number of common patterns in the
various broader food web designs that have been postulated (Chapter 8), we are
still a considerable distance from a detailed understanding of the broader interactions between microorganisms and invertebrates in food webs or how these
influence the decomposition process.

Key Invertebrate Controls in Natural and Agroecosystems

From the preceding discussion and the overall soil ecological literature, there is
an increasing body of evidence that various groups of soil-inhabiting invertebrates can play key roles in the decomposition process that greatly exceed their
biomass relative to that of microorganisms in agroecosystems. Space does not
allow the presentation of a comprehensive review of all of the evidence for such
controlling mechanisms among all the diverse groups of invertebrates. Instead,
in this chapter I propose to review only five groups of soil-inhabiting invertebrates, in terms of their possible roles as key invertebrate controls in the decomposition process. The groups that I have selected for such further discussion are
nematodes, earthworms, microarthropods, termites and ants, since there seems
to be more clear evidence in the literature about the role of these invertebrates,
as control organisms in food webs, than for other groups.

Nematodes (Nematoda)

Nematodes are major components of most soil ecosystems, both in terms of
numbers and potential functional effects. They have more diverse roles in soil
ecosystems than almost any other group of soil invertebrates, and various
different functional groups can be identified readily from the structure of the
mouthparts (Parmelee *et al.*, 1994). Clearly, nematodes are major components
of the food webs already discussed, acting as either bacteriovores, fungivores,
omnivores or predators (Hunt *et al.*, 1987; Andren *et al.*, 1988; De Ruiter *et al.*,
1994; Schaefer, 1995; Bottomley, 1998).

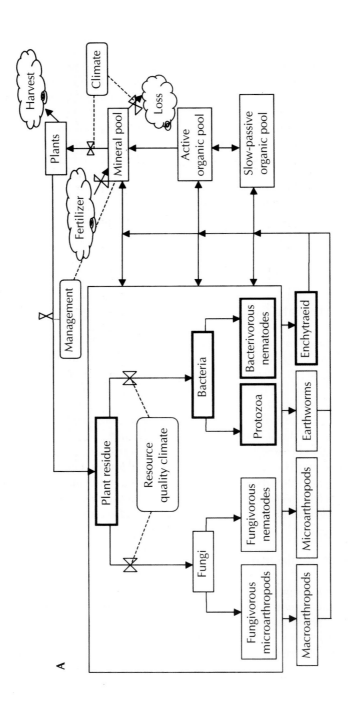

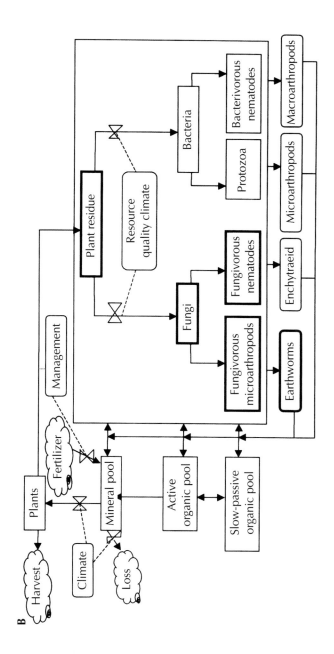

Fig. 7.5. Conceptual model of nutrient fluxes in (A) conventional tillage and (B) no-tillage acroecosystem (Hendrix *et al.*, 1986). Boxes, nutrient storages; clouds, nutrient sources or sinks; arrows, nutrient transfer pathways. Valve symbols indicate that nutrient transfers are influenced by factors connected with dotted lines.

It has been suggested (Whitford *et al.*, 1982) that free-living nematodes, especially bacterivorous species, make up the bulk of the desert soil fauna. In desert ecosystems, the organic matter is aggregated mainly under shrubs, in depressions, or associated with the roots of ephemeral plants, which die and leave roots that decompose. Using field studies and applying selective chemical inhibitors, such as broad-spectrum nematicides and fungicides, Santos and Whitford (1981) studied the relationships between various groups of soil organisms, during the various stages of decomposition, and found considerable interactions and competition between microarthropods, nematodes and microorganisms. For instance, exclusion of tydeid mites from a system resulted in increases in cephalobid nematode populations and decreases in bacterial populations. Their data suggested that predatory mites maintained grazing populations of nematodes at fairly low levels, which increased microbial activity and the overall rates of decomposition. They also found, at other desert sites, that inhibition of predatory mites resulted in increases of several orders of magnitude in populations of those species of nematodes that graze on bacteria and fungi.

Based on these results they constructed a model food web for desert soils (Fig. 7.6). They hypothesized that the faecal material from the mites provided the energy base for fungi and bacteria which were under the indirect control of free-living nematodes. These interactions are clearly dependent upon rainfall and soil moisture and the decomposition is a 'pulse' phenomenon triggered by rainfall, clearly under the control of predatory mites.

Earthworms (Oligochaeta)

Earthworms have been quoted as playing a key role in soil ecosystems by many authors including Edwards *et al.* (1995b), Edwards and Bohlen (1996) and

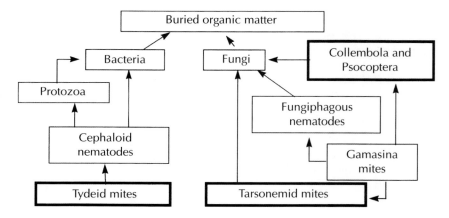

Fig. 7.6. Trophic relationships among soil biota in buried litter (modified from Santos and Whitford, 1981).

Edwards (1998). There is clear evidence that they can accelerate organic matter decomposition and nutrient release in agroecosystems (Blair *et al.*, 1995). They exert a controlling activity through their strong interactions with microorganisms in the decomposition process (Brown, 1995). These interactions affect not only organic matter breakdown and nutrient release, but also the dispersal of microorganisms, particularly vesicular arbuscular mycorrhizae and plant pathogens, and also the transmission of nematodes through soil (Edwards and Fletcher, 1988). Additionally, through their interactions with microorganisms, phytohormone-like plant growth regulators (PGRs) and other plant growth-influencing materials including free enzymes and humic materials that affect plants may be produced (Fig. 7.7) (Krishnamoorthy and Vajranabhaiah, 1986).

The interactions between microorganisms and organic matter in the guts of earthworms are complex and poorly understood (Blair *et al.*, 1995), but clearly the fragmentation of organic matter increases the surface area of organic matter available for microbial colonization and the casts are much more microbially active than the organic matter and soil consumed. There have been relatively few detailed studies of the role of microorganisms in the diet of earthworms. Edwards and Fletcher (1988) and Morgan (1988) reared earthworms axenically on pure cultures of bacteria, fungi and protozoa and identified a number of these microorganisms on which earthworms could grow. They concluded that fungi were the most important microorganisms in the diet of earthworms (Table 7.2), but that a number of different kinds of microorganisms were needed for earthworms to grow satisfactorily to maturity.

In long-term studies of the role of earthworms in agroecosystems (Parmelee *et al.*, personal communication) in which earthworm populations were

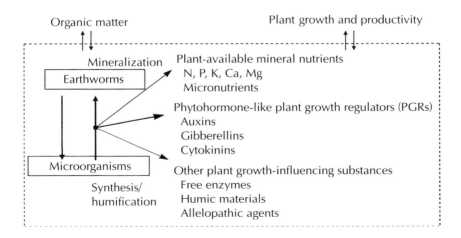

Fig. 7.7. Plant growth-influencing materials produced from interactions between earthworms and microorganisms.

Table 7.2. Role of microorganisms in the diet of earthworms (from Edwards and Fletcher, 1988).

Microbial group	Importance in diet
Bacteria	Minor
Algae	Moderate
Protozoa	Important but less than fungi
Fungi	Major
Nematodes	Uncertain
Mixtures	Essential

maintained at indigenous densities, increased (by inoculation) or eliminated as far as possible (by electroshocking), there were major effects of earthworms on microbial biomass (Fig. 7.8). When large populations of earthworms were present, microbial biomass decreased significantly (Fig. 7.8A). Blair *et al.* (1995) hypothesized that earthworm feeding resulted in a smaller, but perhaps more active, microbial biomass.

Micro- and mesoarthropods

Mites (Acarina) and springtails (Collembola) are found in very large numbers in most agricultural and natural soils (Table 7.1). They are diverse groups including phytovores, bacteriovores, fungivores, predators and omnivores; moreover, they may change food preferences under food stress. Due to their small size and low biomass, relative to that of microorganisms and macroinvertebrates, such as earthworms, termites and ants, their importance in decomposition is often discounted. However, in many of the food webs illustrated in this chapter, many different groups of microarthropods could be seen to occupy key positions in food chains. In exclusion experiments, using chemicals to selectively exclude groups of organisms, and in litter bag studies, it is becoming increasingly obvious that microarthropods can play key roles in natural and agricultural ecosystems.

For instance, Hunt *et al.* (1987) considered that springtails and different groups of mites played key roles in the detrital food web of a shortgrass prairie (Fig. 7.2). Andren *et al.* (1988) concluded that microbivorous and saprovorous microarthropods played key roles in the food web of decomposing barley straw (Fig. 7.3). De Ruiter *et al.* (1994) defined cryptostigmatid, bacterivorous, omnivorous and predatory mites and predatory and omnivorous Collembola as playing key roles in the food web in their agricultural soil (Fig. 7.4). Hendrix *et al.* (1986) placed fungivorous microarthropods, especially Collembola, in a key role in no-tillage agroecosystems, although they were much less important in conventionally tilled systems (Fig. 7.5). Whitford *et al.* (1982) and Santos and Whitford (1981) concluded that tydeid and tarsonemid mites had important roles in controlling decomposition in desert ecosystems (Fig. 7.6). From these

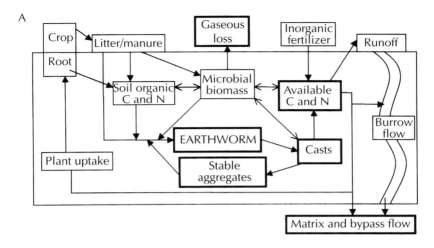

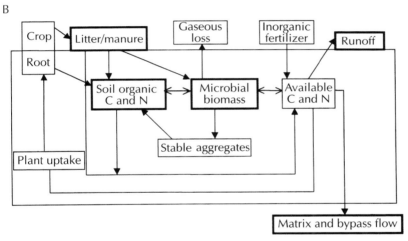

Fig. 7.8. Changes in microbial biomass and available carbon and nitrogen in the presence (A) and absence (B) of earthworms (Bohlen, Edwards and Parmelee, unpublished).

data and other research, especially work in the International Program, it was concluded that there is good evidence that microarthropods can play key roles in influencing the activities of microorganisms in decomposition.

Termites

Termites are social insects living in complex nests with a wide range of form and function. Termites consume organic matter and soil and use their faeces for constructing parts of their complex nest systems. They do not incorporate organic matter into the soil matrix in ways similar to earthworms; instead, they consume large quantities of organic matter and break it down almost completely. They can act as detritivores or become pests in agroecosystems because they often consume or destroy crop plants. Some groups of termites, such as the Macrotermitinae, deposit all their faeces in fungus gardens. The fungi in these gardens belong to the *Basidiomycetes* in the genus *Termitomyces* with a number of species. Particular species of the termites appear to be associated with particular species of the fungus, although occasionally more than one fungal species may be present (Martin, 1987). This is a truly symbiotic relationship where the fungi grow on fungus combs, which are highly convoluted sponge-like structures, the surfaces of which are covered with a growth of mycelium and numerous white spheres or nodules or synnemata (aggregations of conidiophores and conidia). The termites feed on these fungal combs and it has been shown that they are an essential component of the termites' diet, that is, this is an obligate relationship.

Price (1988) also considered that termites play a key controlling role in decomposition and he listed a range of ways in which they contribute to organic matter turnover, including the effects of fungivorous nematodes on fungi which break down toxins that would kill protozoa in the termite gut.

Ants

Attine ants of the subfamily Mymecimae, Tribe Allini, have a fascinating obligate interactive relationship with symbiotic fungi (Martin, 1987). These ants include some of the more abundant and familiar leaf-cutting ants in tropical areas of South and Central America. All of these ants cultivate fungi on plant substrates brought into the nest by foraging workers. Fresh or decaying plant material is most commonly used to set up these fungal gardens. The fungal gardens contain fragile, sponge-like structures composed of small pieces of plant material held together by dense mycelial growth. The microbial complex is usually a single filamentous fungus associated with yeasts and bacteria. A typical fungal garden is hemispherical with a diameter 15–30 cm and a single nest may contain several hundred such gardens. The fungi are a major source of nutrition for the ants, particularly the swollen tips (gongylidia), which unite into clusters (staphylae). These organs are extremely rich in nutrients, but although these are the main source of nutrients for the ants, the workers probably also ingest sap from the plants. It has been suggested that the fungal gardens produce enzymes that assist the ants in digesting this sap (Martin, 1987). Clearly, the ant is an essential control to the function of the fungus in these ecosystems.

Conclusions

The examples presented here demonstrate clearly that the interactions between soil microorganisms and invertebrates are very complex and variable with ecosystem, space and time. There is relatively sparse evidence of a common structure to food webs in soil ecosystems, although as more food webs are proposed, some common links and interactions are beginning to emerge. There is compelling and consistent evidence from exclusion experiments using litter bags or chemicals that various groups of soil-inhabiting invertebrates play key control roles in the decomposition process, particularly where relatively recalcitrant plant litter is involved. Without the fragmentation by soil-inhabiting invertebrates, the decomposition process slows down significantly. The groups of invertebrates we have selected as probable key controls in the decomposition process (nematodes, earthworms, microarthropods, termites and ants) can all be demonstrated clearly to exert such a key role in decomposition.

There is little doubt that other groups of invertebrates such as woodlice, millipedes and centipedes can also exert control. There is an urgent need for much further research into the structure and modelling of soil food webs.

References

Andren, O., Paustian, K. and Rosswall, T. (1988) Soil biotic interactions in the functioning of agroecosystems. *Agriculture, Ecosystems and Environment* 24, 57–67.

Andren, O., Lindberg, T., Bostran, U., Clarholm, M., Hansson, A-C., Johansson, G., Lagerlöf, J., Paustian, K., Persson, J. and Petterson, R. (1990) Ecology of arable land: organisms, carbon and nitrogen cycling. *Ecological Bulletin* 40, 85–126.

Blair, J.M., Parmelee, R.W. and Lavelle, P. (1995) Influences of earthworms on biogeochemistry. In: Hendrix, P.F. (ed.) *Earthworm Ecology and Biogeography in North America*. Lewis Publishers, Boca Raton, Florida, pp. 127–158.

Bottomley, P.J. (1998) Microbial ecology. In: Sylvia, D.M., Fuhrmann, J.J., Hartel, P.G. and Zuberer, D.A. (eds) *Principles and Applications of Soil Microbiology*. Prentice-Hall, Englewood Cliffs, New Jersey, pp. 149–167.

Brown, G.G. (1995) How do earthworms affect microfloral and faunal community diversity? *Plant and Soil* 170, 209–231.

Brussaard, L., Van Veen, J.A., Kooistra, M.J. and Lebbink, G. (1990) The Dutch Programme on Soil Ecology of Arable Farming Systems. I. Objectives, approach and some preliminary results. *Ecological Bulletin* 39, 35–40.

Butovsky, R.O. and van Straalen, N.M. (eds) (1998) *Pollution-induced Changes in Soil Invertebrate Food Webs*. Vrije Universitet, Amsterdam.

Coleman, D.C. and Crossley, D.A., Jr (1996) *Fundamentals of Soil Ecology*. Academic Press, San Diego.

De Ruiter, P.C., Bloem, J., Bouwman, L.A., Didden, W.A.M., Hoenderboan, G., Lebbink, H.J., Marinissen, J.C.Y., De Vos, J.A., Vreeken-Buijs, M.J., Zwart, K.B. and Brussaard, L. (1994) Simulation of dynamics of nitrogen mineralization in the below-ground food webs of two arable farming systems. *Agriculture, Ecosystems and Environment* 51, 199–208.

Edwards, C.A. (ed.) (1998) *Earthworm Ecology*. CRC Press/Lewis Publishers, Boca Raton, Florida.

Edwards, C.A. and Bohlen, P.J. (1996) *Earthworm Biology and Ecology*, 3rd edn. Chapman & Hall, London.

Edwards, C.A. and Fletcher, K.E. (1988) Interactions between earthworms and micro-organisms in organic matter breakdown. *Agriculture, Ecosystems and Environment* 24, 235–247.

Edwards, C.A., Reichle, D.E. and Crossley, D.A., Jr (1970) The role of soil invertebrates. In: Reichle, D.E. (ed.) *Turnover of Organic Matter and Nutrients in Ecological Studies: Analyses and Synthesis*, Vol. 1. Springer-Verlag, Berlin, pp. 147–172.

Edwards, C.A., Stinner, B.R., Stinner, D.H. and Rabatin, S. (1988) *Biological Interactions in Soil*. Elsevier, Amsterdam.

Edwards, C.A., Abe, T. and Striganova, B.R. (1995a) *Structure and Function of Soil Communities*. Kyoto University Press, Kyoto, Japan.

Edwards, C.A., Bohlen, P.J., Linden, D.R. and Subler, S. (1995b) Earthworms in agro-ecosystems. In: Hendrix, P.F. (ed.) *Earthworm Ecology and Biogeography in North America*. Lewis Publishers, Boca Raton, Florida, pp. 185–214.

Hendrix, P.F., Parmelee, R.W., Crossley, D.A., Jr, Coleman, D.C., Odum, E.P. and Groffman, P.M. (1986) Detritus food webs in conventional and no-tillage agroecosystems. *BioScience* 36, 374–380.

Hunt, H.W., Coleman, D.C., Ingham, E.R., Elliott, E.T., Moore, J.C., Rose, S.L., Reid, C.P.P. and Morley, C.R. (1987) The detrital food web in a shortgrass prairie. *Biology and Fertility of Soils* 3, 57–68.

Krishnamoorthy, R.V. and Vajranabhaiah, S.N. (1986) Biological activity of earthworm casts: an assessment of plant growth promoters in the casts. *Proceedings of the India Academy of Science (Animal Science)* 95(3), 341–351.

Lagerlöf, J. and Andren, O. (1988) Abundance and activity of soil mites (Acarina) in four cropping systems. *Pedobiologia* 32, 129–145.

Martin, M.M. (1987) Invertebrate–microbial interactions. In: *Explorations in Chemical Ecology*. Comstock Publ., Cornell University Press, Ithaca, New York.

Moore, J.C., Walter, D.E. and Hunt, J.W. (1988) Arthropod regulation of micro- and mesobiota in below-ground detrital food webs. *Annual Review of Entomology* 33, 419–439.

Morgan, M.H. (1988) The role of microorganisms in nutrition of *Eisenia foetida*. In: Edwards, C.A. and Neuhauser, E.H. (eds) *Earthworms in Waste and Environmental Management*. SPB Academic Publishing, The Hague, pp. 71–82.

Parmelee, R.W., Bohlen, P.J. and Edwards, C.A. (1994) Analysis of nematode trophic structure in agroecosystems. In: Collins, H.P., Robertson, G.P. and Klug, M.G. (eds) *Significance and Regulation of Biodiversity. Plant and Soil*. Kluwer Academic, Dordrecht, pp. 203–207.

Persson, J. and Rosswall, T. (1983) Opportunities for research in agricultural ecosystems – Sweden. In: Lowrance, R.R., Todd, R.L., Asmussen, L.E. and Leonard, R.A. (eds) *Nutrient Cycling in Agricultural Ecosystems*. Special Publication No. 23, University of Georgia, Athens, pp. 61–71.

Pokarzehvskii, A.D., van Straalen, N.M., Butovsky, R.O., Verhoef, S.C. and Filimonova, Z.V. (1998) The use of detrital food webs to predict ecotoxicological effects of heavy metals. In: Butovksy, R.O. and van Straalen, N.M. (eds) *Pollution-Induced Changes in Soil Invertebrate Food Webs*. Vrije Universiteit, Amsterdam, pp. 1–9.

Polis, G.A. (1994) Food webs, trophic cascades and community structure. *Australian Journal of Ecology* 19, 121–136.

Price, P. (1988) An overview of organismal interactions in evolutionary and ecological time. *Agriculture, Ecosystems and Environment* 24, 369–377.

Santos, P.F. and Whitford, W.G. (1981) The effects of microarthropods in litter decomposition in a Chihuahuan desert ecosystem. *Ecology* 62, 654–663.

Siepel, H. (1994) Structure and function of soil microarthropod communities. PhD thesis, Institute of Forestry and Nature Research (IBN_DLO), The Netherlands.

van Straalen, N.M. and Butovsky, D.A. (1998) *Bioindicator Systems for Soil Pollution*, NATO ASI Series. Kluwer Academic Publishers, Dordrecht.

Verhoeff, H.A. and Brussaard, L. (1990) Decomposition and nitrogen mineralization in natural and agroecosystems; the contribution of soil animals. *Biogeochemistry* 11, 175–211.

Vreeken-Buijs, M. (1998) Ecology of microarthropods in arable soil. PhD thesis, Landbouw Universitet, Wageningen, The Netherlands.

Walter, D.E. (1988) Nematophagy by soil arthropods from the Shortgrass Steppe, Chihuahuan Desert and Rocky Mountains of the Central United States. *Agriculture, Ecosystems and Environment* 24, 307–316.

Walter, D.E., Hudgens, R.A. and Freckman, D.W. (1986) Consumption of nematodes by fungivorous mites, *Tyrophagus* sp. (Acarina: Astigmata: Acaridae). *Oecologia* (Berlin) 70, 357–361.

Whitford, W.G., Freckman, D.W., Santos, P.F., Elkins, N.Z. and Parker, L.W. (1982) The role of nematodes in decomposition in desert ecosystems. In: Freckman, D.W. (ed.) *Nematodes in Soil Ecosystems*. University of Texas Press, Austin, Texas, pp. 98–115.

8

Invertebrates in Detrital Food Webs along Gradients of Productivity

J.C. Moore[1] and P.C. de Ruiter[2]

[1]*Department of Biological Sciences, University of Northern Colorado, Greeley, CO 80639, USA;* [2]*Department of Environmental Studies, University of Utrecht, 3508 TC Utrecht, The Netherlands*

Introduction

Hutchinson (1959) and Hairston *et al.* (1960) exposed the Janus-faced nature of the forces that organize communities. On the one hand, thermodynamics cannot be ignored. The amount of available energy after each predator–prey interaction diminishes and limits both diversity and trophic structure. On the other hand, trophic structure influences the placement of available energy in a community. Consumers are regulated by their predators, creating a cascade of regulation and biomass. The extensive publications on latitudinal gradients in diversity with productivity (Rosenzweig, 1995), and the validation of cascades of biomass in the aquatics literature (Carpenter *et al.*, 1985, 1987) are triumphs of modern ecology.

Oksanen *et al.* (1981) extended these ideas to include whole trophic levels. From their work in arctic communities, they developed hypotheses of how the community structure of primary producer-based communities is influenced along steep productivity gradients. Communities develop with increased productivity as proposed by Hutchinson (1959), but the regulation that consumers have on lower trophic levels (Hairston *et al.*, 1960) is dependent on their position in the community relative to the top predator.

A separate line of studies has shown regularities in trophic structure that are tied to productivity, diversity and stability. Here, the diversity of production cannot be ignored. Communities are not random collections of species. Instead, communities are organized into compartments/assemblages of species based on different energy inputs (types and rates) in a way that is important to their stability (May, 1972; Moore and Hunt, 1988; de Ruiter *et al.*, 1995). The compartments are structured similarly to the entire community, yet respond to disturbances in a quasi-independent manner (Moore and de Ruiter, 1997). Moreover, communities with higher diversity of production (more compartments) support more consumers (Moore and Hunt, 1988) and, at least for certain grasslands, communities with a higher diversity of producers are more productive and less prone to unstable oscillations following disturbances (Naeem *et al.*, 1994; Tilman *et al.*, 1996).

As appealing as these concepts may be, they have been criticized in part because of the interpretation of ecological theory within a narrow and linear range of productivity as discussed below. First, there is empirical evidence that undermines these concepts. For example, trophic cascades of biomass are rare in terrestrial systems (Strong, 1992), and comparisons of food webs worldwide have not yielded the predicted relationship that ecosystems with higher levels of primary productivity would possess longer food chains (Pimm and Lawton, 1977; Briand and Cohen, 1987). Second, experiments conducted at different points along the productivity gradient or with organisms that process energy at different rates yield different results. These results, which seem to contradict the concept of regularity in trophic structure, are actually demonstrations of the non-linearity of the relationship between trophic structure and productivity. For example, assumptions about the role of predators in trophic structure have typically been tested on a part of the productivity gradient that is not sufficiently steep to confirm or refute the thermodynamic argument that food chains would increase in length with increased productivity (Moore *et al.*, 1993). Third, models designed to represent the dynamics of primary producer-based systems of plants/algae, herbivore and predators, do not consider the role of detritus (defined here as any form of non-living organic matter – faeces, carcasses, hair, plant leaf litter, etc.) in systems, even though most of the productivity is not consumed by herbivores or predators (Moore and Hunt, 1988; Moore *et al.*, 1988; Polis *et al.*, 1989; Polis, 1994; Polis and Strong, 1996). Lastly, it is typically assumed that the mechanisms operating to shape communities operate linearly along a productivity gradient with little regard to the diversity of production, even though models of simple and complex systems suggest otherwise (Rosenzweig and MacArthur, 1963; Rosenzweig, 1971; May, 1973, 1976; Moore *et al.*, 1993). In short, theory, experimentation and observation are out of synchrony with one another.

In this chapter we explore possible interactions between productivity, dynamic stability and community structure along real and model gradients of productivity for soil and sediment food webs. We will proceed by constructing models that explicitly include aspects of ecological energetics. The feasibility (likelihood of all species maintaining positive steady states at a given level of productivity), distribution of biomass at each trophic level (trophic structure), stability (local stability) and recovery times (return-time) of these models will be assessed. Next, we will compare the feasibility, trophic structures, stability and recovery times of models that represent assemblages of organisms that differ in their energetic efficiencies and reproductive rates (e.g. endotherms versus ectotherms, or slow versus fast growers). Finally, we present empirical evidence that supports our proposition that communities exhibit a continuum of trophic structures that is governed by productivity.

Primary Producer and Detritus-based Models

We developed two classes of models designed to study the effects of minor perturbations on the local stability and resilience of simple systems (Moore *et al.*, 1993; de Ruiter *et al.*, 1995). The first class of models are based solely on primary production (May, 1973), while a second class of models describe food chains that are based on inputs of dead organic material that we define as detritus (Fig. 8.1A). Primary producers are modelled as:

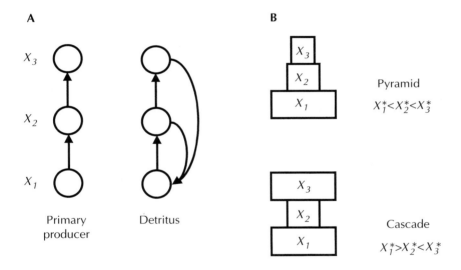

A

X_3

X_2

X_1

Primary producer   Detritus

B

X_3

X_2

X_1

Pyramid

$X_1^* < X_2^* < X_3^*$

X_3

X_2

X_1

Cascade

$X_1^* > X_2^* < X_3^*$

Fig. 8.1. (A) Differential equations were developed to describe the dynamics of species within primary producer-based (left) and detritus-based (right) food chains of length 2, 3 and 4. Length has been defined here as the number of species (circles) within a chain. Other authorities may define length as the number of transfers (arrows) between basal resources and top predators within the chain. Primary producers are typically higher plants and algae but include any autotroph (e.g. chemosynthetic bacteria). Detritus includes all non-living organic material. The arrows feeding back to detritus represent corpses, unconsumed kill and faeces. Detritus also enters from outside the food chain (not depicted in diagram). (B) A *pyramid of biomass* and a *cascade of biomass* defined in terms of the distribution of biomass that accumulates at steady state by trophic position. The boxes representing the biomass at each trophic position are not drawn to scale.

$$\frac{dX_1}{dt} = r_1 X_1 - \sum_{i,j=2}^{k} c_{1j} X_1 X_j \tag{1}$$

where X_1 and X_j represent the population densities of primary producers and herbivores, respectively, expressed as biomass (e.g. units of g C m^{-2}), r_1 is the specific growth rate (time^{-1}), and c_{1j} the consumption coefficient of herbivores on the primary producers with units of (g C m^{-2}) time^{-1}.

Detritus was modelled as entering a system from an external source and then cycling internally within the system (Moore *et al.*, 1993):

$$\frac{dX_d}{dt} = R_d + \sum_{i,j=1}^{k} (1 - a_i) c_{ij} X_i X_j + \sum_{i=1}^{k} d_i X_i - \sum_{i,j=1}^{k} c_{ij} X_i X_j \tag{2}$$

where R_d represents the allochthonous input rate of detritus, for example plant leaf litter and debris. Detritus also cycles autochthonously as metabolic wastes (egested material) and unconsumed kill (leavings and carcasses), $(1 - a_{ij}) c_{ij} X_i X_j$, and non-predatory death (dead bodies) of consumers, $d_i X_i$.

The dynamics of a consumer (detritivores, herbivores and predators), X_i, are similar within primary producer and detritus food chains:

$$\frac{dX_i}{dt} = -d_i X_i + \sum_{i,j=2}^{k} a_i p_i c_{ij} X_i X_j. \tag{3}$$

Consumers die at a specific rate, d_i (time^{-1}), and increase as a function of prey consumed, $a_i p_i c_{ij} X_i X_j$, their assimilation efficiency a_i (%), and their production efficiency p_i (%).

Assessing the trophic structure of communities

The number of trophic levels and the distribution of biomass within the trophic levels are two aspects of trophic structure. The number of trophic levels is governed by the availability of energy. We assessed the number of trophic levels by first determining whether the system was feasible. A system is feasible if the steady-state values of all $X_i^* > 0$. The feasibility of a system must be considered to reconcile mathematical possibilities and biological reality (Roberts, 1974). For example, the equations presented above can generate systems whereby the steady-state densities of one or more of the species is less than zero. This would not be a problem from a mathematical standpoint. However, such a system makes no sense from a biological standpoint (i.e. how can a community possess negative population densities?).

The distribution of biomass within trophic levels results in trophic structures that range from a pyramid of biomass at one extreme to a cascade of biomass at the other. The pyramid of biomass is the familiar trophic structure whereby the largest pool is located at the base of the community, followed by successively smaller pools at the next higher trophic levels (Fig. 8.1A). The cascade of biomass is a trophic structure at steady state characterized by alternating levels of biomass beginning with the first trophic level and ending with the upper trophic level. For example, systems with two and three trophic levels would possess a cascade of biomass if $X_1^* < X_2^*$ and $X_1^* > X_2^* < X_3^*$, respectively (Fig. 8.1B).

Transitions in trophic structure

Communities will undergo transitions in trophic structure along a gradient of productivity. First, the food chain has to be feasible in the sense that there is sufficient productivity within the ecosystem so that all species maintain positive steady-state densities, i.e. all $X_n^* > 0$. For the primary producer-based food chain, this condition is met if $r_1 = r_{F2}$:

$$r_{F2} > \frac{d_2}{a_2 p_2} \frac{c_{11}}{c_{12}}. \tag{4}$$

The threshold at r_{F2} is equal to $|\alpha_{11}|$, indicating a direct relationship to the degree of intraspecific competition on the part of the producer. The detritus-based food chain is feasible over the entire range of productivity provided there is sufficient productivity to support a single organism (see Equation 4).

Second, given that the two-species food chains are feasible, for a cascade of biomass to occur, the equilibrium density of the species at the upper trophic level must exceed that of the lower trophic position, i.e. $X_2^* > X_1^*$. For the primary producer-based food chain this occurs when $r_1 = r_{C2}$:

$$r_{C2} > \frac{d_2}{a_2 p_2}\left(1 + \frac{c_{11}}{c_{12}}\right) \tag{5}$$

and, for the detritus-based food chain when $R_d = R_{C2}$:

$$r_{C2} > \frac{d_2}{a_2 b_2 c_{12}}\left(\frac{d_2}{p_2} - d_2\right). \tag{6}$$

For both types of food chains, the level of productivity at which the cascade of biomass occurs ($X_2^* > X_1^*$) exceeds the level of productivity required to ensure that the chains are feasible. This implies that once a food chain has been established, it would initially take on the appearance of a pyramid of biomass ($X_1^* > X_2^*$). As productivity increases, the pyramid would give way to a cascade of biomass ($X_1^* < X_2^*$). The difference between the thresholds of productivity where the food chains become feasible and where the food chains exhibit a cascade ($r_{C2} - r_{F2}$) is $d_2/(a_2 p_2)$, or $|\alpha_{12}|$. Therefore, the interaction strength of a predator on its prey governs the range of productivity wherein the trophic structure appears as a pyramid of biomass, and the point at which it switches to a cascade of biomass.

If we extend this argument to three species and beyond, we get a glimpse of why trophic cascades are so rare. We restrict the formal presentation of the argument to the three-species primary producer food chain. The steady-state densities are:

$$X_1^* = \frac{r_1 a_3 p_3 c_{23} - d_3 c_{11}}{c_{11} a_3 p_3 c_{23}} \tag{7a}$$

$$X_2^* = \frac{d_3}{a_3 p_3 c_{23}} \tag{7b}$$

$$X_3^* = \frac{r_1 c_{12} a_2 p_2 c_{23} a_3 p_3 - c_{11} d_2 c_{23} a_3 p_3 - c_{11}^2 a_2 p_2 d_3}{c_{11} c_{23}^2 a_3 p_3} \ . \tag{7c}$$

The level of productivity where the three-species food chain is feasible ($r_1 = r_{F3}$) is:

$$r_{F3} = r_{C2} - \frac{d_2}{a_2 p_2} - \frac{c_{11}^2 d_3}{c_{12} c_{23} a_3 p_3}. \tag{8}$$

The level of productivity where the three-species food chain exhibits a cascade of biomass (r_{C3} when $X_3^* > X_2^*$) is:

$$r_{C3} = r_{F3} + \frac{d_3}{a_3 p_3 a_2 p_2}. \tag{9}$$

The level of productivity where the three-species food chain first exhibits a cascade is greater than the level of productivity where the chain becomes feasible. Furthermore, the point at which the three species becomes feasible is near the point where the two-species food chain cascades, i.e. $r_{F2} < r_{F3} < r_{C2} < r_{C3}$. If we extend this reasoning to four levels and beyond, the differences between the productivity that ensures feasibility of four levels and the rate that initiates cascades for each trophic position precludes the widespread existence of trophic cascades of biomass (Fig. 8.2).

Assessing the dynamic stability of communities

Stability has been defined in several different ways. We have adopted the concepts of local stability and resilience. A community is locally stable if all species return to their steadystate following a minor disturbance. We start with a system of differential equations that describes the trophic interactions (feeding relationships) among species as presented above. The steady-state abundances of all species, X_i^*, are determined by setting the differential equations to zero and solving for all X_i. If the system is feasible (all $X_i^* > 0$), we assess the local stability of the system by evaluating the eigenvalues of the

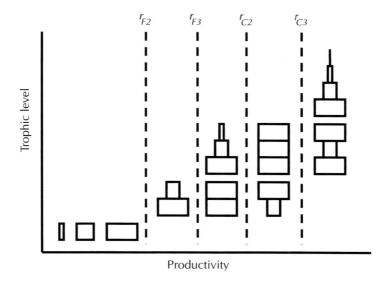

Fig. 8.2. The proposed relationship of trophic structures with increased productivity (NPP or rates of detritus input). The vertical dashed lines represent thresholds in trophic structures that occur at points along the productivity gradient. Note that for the first three trophic positions, the occurrence of cascades of biomass for systems with 2 and 3 trophic positions (r_{C2} and r_{C3}) are further up the gradient than the points at which there is sufficient energy to support 2 and 3 levels (r_{F2} and r_{F3}). The positioning of these thresholds along the productivity gradient is a function of the energetic efficiencies of the organisms that are present. The boxes representing the biomass at each trophic position are not drawn to scale.

Jacobian matrix for a set of equations that describe deviations from the steady-state values. We define x_I as a small deviation in the abundance of any species i, $x_i = X_i^* - X_i$, resulting in the following set of equations:

$$\frac{dx_i}{dt} = \sum_{i=1}^{k} \alpha_{ij} x_j \tag{10}$$

where,

$$\alpha_{ij} = \left[\frac{\partial\left(\frac{dx_i}{dt}\right)}{\partial(dx_j)} \right]^* . \tag{11}$$

Stability is assessed in the neighbourhood or 'local' vicinity of the steady-state abundances. The system of k species is locally stable if and only if each of the eigenvalues, λ_i, of the $k \times k$ Jacobian matrix (with α_{ij} as its elements) has a negative real part, i.e. λ_i is either a negative real number or is a complex number with a negative real part (see May, 1973 and Jefferies, 1988). This criterion assures that 0 is an attractor for dx_i/dt, or in other words, that $x_I = 0$ some time after the disturbance.

Resilience is a measure of a feasible and stable system's ability to return to its original steady state following a disturbance. The return-time (RT) approximates the resilience of the system. Return-time is estimated as:

$$RT = -1/\lambda_{[max]} \qquad (12)$$

where [max] is the real part of the largest negative eigenvalue (negative value closest to zero).

Transitions in dynamic states

The dynamics of a food chain undergo transitions along a productivity gradient which are dependent on the energetic efficiencies of the species within the chain (May, 1976). We illustrate this point with the two-species primary producer food chains presented above.

We estimated the resilience of the food chains as the inverse of the return-time (RT) of the food chain, $RT = \pm 1/\lambda_{max}$, where λ_{max} is the real part of the least negative (dominant) eigenvalue.

The dominant eigenvalue for the two-species primary producer-based food chain is:

$$\lambda_{max} = \frac{\alpha_{11} + \sqrt{\alpha_{11}^2 + 4\alpha_{12}\alpha_{21}}}{2} \qquad (13)$$

where the α_{ij} are the elements of the Jacobian matrix. The real part of λ_{max} describes the direction and rate of decay of the perturbation; as $r_1 \rightarrow \infty$, $RT \rightarrow 2/\alpha_{11}$. The imaginary part of λ_{max} describes the oscillations that the system undergoes during the return to steady state. The term that is responsible for the introduction and increases in the oscillations with increased productivity is α_{21} within the discriminant, u, where $u = \alpha_{11}^2 + 4\alpha_{12}\alpha_{21}$. The α_{21} term describes the influence of the consumer on its prey, is negative, and is a function of the specific rate of increase (r_1) of the prey.

As $r_1 \rightarrow \infty$, u undergoes transitions from positive values, to zero, to negative values. These transitions in u across the productivity gradient generate three distinct types of recovery: (i) overdamped (monotonic damping); (ii) critically damped; and (iii) damped oscillations (Fig. 8.2). The overdamped recovery occurs when $u > 0$. In this situation the decay of the disturbance is governed by the real part of λ_{max} and the disturbance decays without oscillation to the original steady state. At some level of productivity, defined here as r_λ, $u = 0$ and the overdamped recovery gives way to the critically damped recovery where the disturbance decays in a monotonic manner but over-shoots the steady state once and then returns to the steady state without oscillations. At still greater levels of productivity when $u < 0$ the system develops damped oscillations with a quasi-period of $T_a = 2\neq/u$. For our two-species food chain, we define this point along the productivity gradient as $r_{\lambda 2}$.

Comparing Primary-producer with Detritus Food Chains

The stability and resilience of the simple detritus-based and primary producer-based food chains were analysed under the assumption that all the parameters in the model followed a uniform distribution, I (0,1). For each food chain, 10,000 iterations of the analysis described above were performed.

The feasibility of the detritus-based and primary producer-based food chains decreased with increased length (Fig. 8.3). The decline was most striking for the detritus-based food chains. The two-species detritus-based food chain model generated feasible equilibria for all parameter selections (100%), while only 19.12% of the two-species producer-based chains were feasible. At the three- and four-species lengths, both types of food chains exhibit precipitous declines in feasibility.

As expected, all of the feasible food chains were stable (May, 1973). From the distributions of the return times of the feasible stable food chains, we found longer food chains to be less resilient than shorter food chains (Fig. 8.3). For both types of models, few of the four-species food chains were feasible, and those that were possessed extremely long return-times.

We drew three conclusions from this exercise. First, our models have produced results that are consistent with the models that have shaped this field (May, 1973; Pimm and Lawton, 1977). Second, the decline in feasibility with increased food chain length is consistent with the thermodynamic reasons for short food chains (Hutchinson, 1959). Third, the distributions of return-times for detritus-based and producer-based models are similar, even though the detritus-based models are donor-controlled and, hence, differ from producer-based systems in a fundamental way which has been argued would affect dynamics (Pimm, 1982).

Modelling a Productivity Gradient

In the preceding exercise we restricted productivity within a single order of magnitude (productivity was measured by the variable R_d for the detritus model and by r_1 for the producer-based models) by assuming that all variables were distributed I (0,1). To assess the influence of productivity on the feasibility, stability and resilience of the detritus-based and primary producer-based food chains we repeated the analysis over different levels of productivity. The range of productivity included fixed levels that increased in order of magnitude beginning

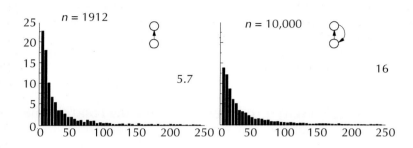

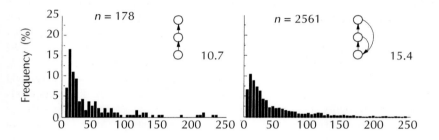

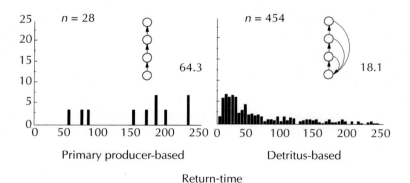

Primary producer-based   Detritus-based

Return-time

Fig. 8.3. Frequency distributions of return-times for the primary producer-based and detritus-based food chains of length 2, 3 and 4. All parameters in the models were assumed to possess a uniform distribution over the interval (0,1). The analysis included 10,000 iterations for each food chain. The number of runs that produced feasible food chains (*n*) is presented in the upper left corner of each graph. Feasibility is estimated by dividing *n* by 1000. The number to the right of the food chains represents the percentage of return-times that exceeded 250 time units (the selection of 250 as a stopping point was arbitrary).

with 0.01 units and ending with 10^5 units. The specific birth rates, specific death rates, consumption coefficients, assimilation efficiencies and production were assumed to be distributed I $(0,1)$. The analysis was repeated 1000 times at each level of productivity.

The level of productivity affected the feasibility and dynamic stability of the food chains (Fig. 8.4). The two-species detritus-based food chains were 100% feasible over the entire range of productivity. The feasibility of the remaining food chains increased with increased productivity (Fig. 8.4A). At 10^5 units of productivity all food chains were 100% feasible. The effect of productivity on feasibility is a direct result of satisfying the conditions for obtaining positive equilibrium densities, that is, inputs exceed outputs (Roberts, 1974).

The resilience of the food chains decreased with increased productivity (Fig. 8.4B). The reason for this relationship stems from the fact that the eigenvalues are a function of productivity. The critical eigenvalue of the two-species primary producer model is presented in Equation (13). Relatively high levels of productivity lead to negative determinants (u), and the real part of $\lambda_{(max)}$ approaches $\alpha_{11}/2 \to -1$, hence the return-time (RT) approaches $-2/\alpha_{11} \to 1$.

We drew three conclusions from this exercise. First, with respect to food chain length, productivity and dynamic stability are interrelated. At low levels of productivity few food chains are feasible, and those that are possess long return-times relative to food chains at higher levels of productivity. The one exception to this was the two-species detritus-based food chain. Second, the results support the productivity hypothesis as the feasibility of the food chains was a function of their length and productivity. This indicates that higher rates of energy inputs are required to support longer food chains. Third, the results agree with the general conclusion of Pimm and Lawton (1977) that dynamic stability might influence food chain length. However, the results question the specific conclusion that longer food chains possess longer return-times. Return-time is influenced by productivity and food chain length but in an interactive manner. At low levels of productivity the results corroborate those of Pimm and Lawton (1977), but at high levels of productivity, four-species food chains are more resilient than three-species food chains.

Energetic efficiencies

The efficiencies of the food chains were altered by changing the distributions from which we sampled the parameters. For this analysis the feasibility, stability and resilience of the detritus-based food chains (lengths 2–4) and primary producer-based food chains (lengths 2–4) were compared over the same range of productivity used above under the assumption that all variables were distributed Beta (2,9) for a low-efficiency food chain with a slow turnover rate (endotherms) and Beta (9,2) for a more efficient food chain with a rapid turnover rate (ectotherms).

For all but the two-species detritus-based food chains, the feasibility of the food chains was affected by the distribution of the variables over the range of

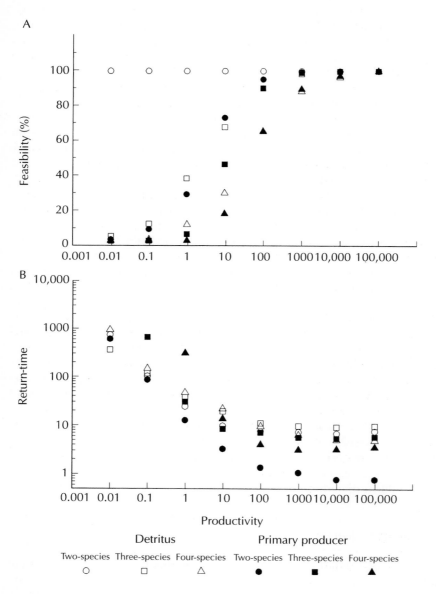

Fig. 8.4. Feasibility (A) and Return-times (B) as a function of productivity for detritus-based (open symbols) and primary producer-based (closed symbols) food chains. The analysis included 1000 iterations for each food chain. The food chains were evaluated at constant levels of productivity (r_1 for primary producer-based food chains and R_d for detritus-based food chains) beginning with 0.01 units and increasing by orders of magnitude to 100,000 units. All other parameters were assumed to possess a uniform distribution over the interval (0,1) (after Moore *et al.*, 1993).

productivity investigated (Fig. 8.5). The two-species detritus-based food chains were 100% feasible for all levels of productivity. For the remaining food chains, the more efficient chains (variables distributed as Beta (9,2)) were more likely to be feasible at the lower range of the productivity than the less efficient food chains (variables distributed as Beta (2,9)).

The maximum sill of a feasibility–productivity curve is 100%. With the exception of the low-efficiency four-species food chains, the maximum sill was reached at some point over the range of productivity investigated. Given the responses in the feasibility of the food chains to increased productivity, there was no reason to believe that the sill for the four-species chains would not be reached at some higher productivity. The efficiency of the food chain governed the level of productivity at which the sill was reached. By 10 units of productivity, all the high efficiency food chains were 100% feasible regardless of their length or type. The low efficiency food chains were an entirely different matter. For both types of food chains, the ranges differed by several orders of magnitude of productivity (Fig. 8.5).

The resilience of a feasible chain was a function of the type of food chain, its length, the productivity and the underlying distributions of the variables (Fig. 8.6). Direct comparisons of food chains of a given length and type reveal that below 10 units of productivity the more efficient food chains possessed shorter return-times than the less efficient food chains when both were feasible. Beyond 10 units of productivity, the results are less clear. For the detritus-based food chains, it appears that at some level of productivity the less efficient food chains become more resilient than the more efficient food chains. That is, the relationship observed below 10 units of productivity between the return-time and the efficiency of the food chains reverses. Furthermore, the level of productivity at which this reversal occurs increases with increased food chain length. For the primary producer-based food chains the reversal in the relationship between resilience and the efficiency of the food chains was observed for the two-species food chain. However, for the three- and four-species primary producer-based food chains the more efficient food chains were more resilient than the less efficient food chains over the entire range of productivity.

These results indicate that the underlying distribution(s) of the variables used to model the dynamics of the food chains affects the feasibility and resilience of the food chains. The results for feasibility support a corollary to the productivity hypothesis, that food chains dominated by energetically efficient organisms would be longer than those dominated by less efficient organisms (Hutchinson, 1959). Furthermore, the results for resilience confirm the general findings of Pimm and Lawton (1977) that food chain length may be influenced by dynamic stability, but not to the exclusion of productivity.

Coupling trophic structures and dynamic stability

There are clear limitations to the length of food chains within a community. The lower limits of food chain length can be explained to a large degree by arguments

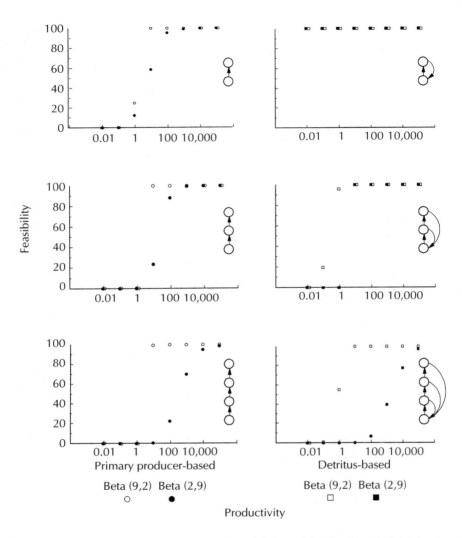

Fig. 8.5. Feasibility of primary producer-based (left) and detritus-based (right) food chains as a function of productivity. The analysis included 1000 iterations for each food chain. The food chains were evaluated at constant levels of productivity (r_1 for primary producer-based food chains and R_d for detritus-based food chains) beginning with 0.01 units and increasing by orders of magnitude to 100,000 units. The other parameters were assumed to be distributed as either Beta (9,2), open symbols or Beta (2,9), closed symbols.

advanced by thermodynamics. There exists a low threshold in productivity, below which there is not enough energy available to support another trophic level (Hutchinson, 1959). The upper limit has been set by dynamic constraints

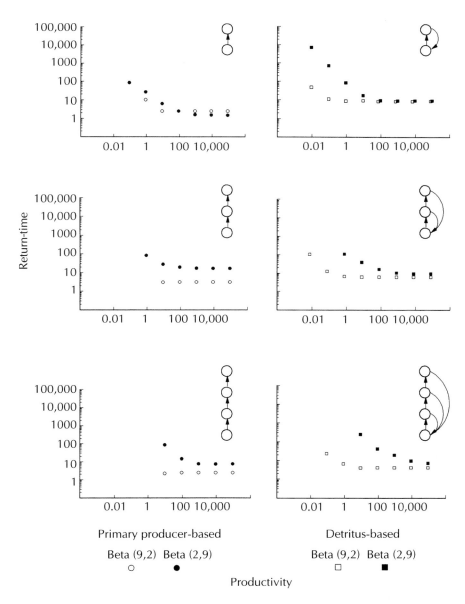

Fig. 8.6. Return-times of primary producer-based (left) and detritus-based (right) food chains as a function of productivity. The analysis included 1000 iterations for each food chain. The food chains were evaluated at constant levels of productivity (r_1 for primary producer-based food chains and R_d for detritus-based food chains) beginning with 0.01 units and increasing by orders of magnitude to 100,000 units. The other parameters were assumed to be distributed as either Beta (9,2), open symbols or Beta (2,9), closed symbols.

(Rosenzweig, 1971; Pimm, 1982), but we will argue that productivity can not be entirely ruled out.

To illustrate this point, we return to the two-species producer-based food chain. There is a third threshold along a gradient of productivity, $r_{\lambda 2}$, which marks the point at which oscillations in the dynamics of a community increase upon its return to steady state following a minor disturbance. Recall that the local stability of the community is governed by the signs of the eigenvalues of the community matrix (May, 1972), the resilience of a stable community (return-time) is governed by the size of the real parts of the eigenvalues (Pimm and Lawton, 1977), and the sinusoidal oscillations that follow a disturbance are controlled by the imaginary parts of the eigenvalues (Rosenzweig, 1971; May, 1973).

Over a wide range of productivity the return-times of detritus-based and producer-based food chains rapidly decayed with increased productivity, reached an inflection point and then gradually approached a limit (Fig. 8.4; Moore *et al.*, 1993). The inflection point ($r_{\lambda 2}$) represents the level of productivity where we see the onset of oscillatory dynamics. The placement of this inflection point ($r_{\lambda 2}$) relative to the points of feasibility (r_{F2}) and cascade (r_{C2}) is not readily apparent. We estimated $r_{\lambda 2}$ by determining the level of productivity that marks the point at which point $\lambda_{(max)}$ becomes complex. For the two-species primary producer-based food chain, $r_{\lambda 2}$ can be estimated as the level of productivity where the discriminant (u) of Equation (13) becomes negative. We obtained the following relationship:

$$r_{\lambda 2} = r_{C2} + \frac{c_{11}r_{F2}}{c_{12}a_2p_2} - \frac{d_2}{a_2p_2}. \qquad (14)$$

It is clear from Equation (13) that the onset of oscillations occurs before the cascade due to the death rate and ecological efficiency of the predator and the self-regulation of the prey. We find the following three possibilities:

1. if $c_{11}/c_{12}(a_2p_2)^{1/2} = 1$ then $r_{\lambda 2} < r_{C2}$ (15a)

2. if $c_{11}/c_{12}(a_2p_2)^{1/2} > 1$ then $r_{\lambda 2} < r_{C2}$ (15b)

3. if $c_{11}/c_{12}(a_2p_2)^{1/2} < 1$ then $r_{\lambda 2} > r_{C2}$.   (15c)

Since ecological efficiencies range from 0.03 for large mammals to a theoretical maximum of 0.80 for bacteria, and the self-limiting terms are often less than 0.01 (Phillipson, 1981), we can deduce that the following would occur with increased productivity: (i) the oscillatory behaviour in the dynamics which would follow a minor disturbance initiates within the window of productivity between feasibility (r_{F2}) and cascade (r_{C2}), i.e. $r_{F2} < r_{\lambda 2} < r_{C2}$; (ii) the frequency of the oscillatory behaviour increases and the period decreases; and (iii) the cascade of biomass for the two-species system would be more likely to experience wider fluctuations in dynamics in response to minor disturbances when compared with the pyramid of biomass.

Trophic Structures and Real Productivity Gradients

The modelling results suggest that communities would exhibit a continuum of trophic structures and dynamic states along a gradient of productivity. How these results apply to natural systems is still unclear because the units assigned to all the variables in this analysis were arbitrary. If the range of productivity observed in real systems covered the range of productivity presented in the models then it would simply be a matter of finding the appropriate mass, area and time units. If real world productivity were a subset of the range presented here then our interpretation of the results might be quite different. Data for soil communities from microcosm and field studies were used to scale the productivity units presented in Figs 8.4–8.6 to real world levels of productivity. The range of the specific death rates, d_i, and the predation coefficients, c_{ij}, for soil systems indicate that 1 unit of productivity scaled to between 4 and 36 g C m^{-2} year^{-1} (Moore *et al.*, 1993).

There is an important consequence of this positioning of real levels of productivity relative to those used in the model. If the assumptions presented here are correct (or even within an order of magnitude), then real world productivity overlaps regions of parameter space where the relationships between productivity, feasibility, return-time, food chain length and type of model are highly interactive and transitory. At low levels of productivity (< 10 units) the relationships are clear (Pimm and Lawton, 1977). At higher levels the relationships change. Hence, it should not be surprising that theory and observation are out of synchrony with one another. This suggests that the least ambiguous region of the real world with which to validate the models would be at the low end of the productivity gradient (Moore *et al.*, 1993).

The structure of the food chains within the caverns and sediments of Jewel Cave, South Dakota, follow the patterns suggested by our models. The cave ecosystem is supported by allochthonous input of carbon from different sources:

1. Dissolved organic material from surface water that drips throughout the caves.
2. The hair, skin and faeces of small rodents and bats near entrances, human hair, skin and clothing fibres along designated tour routes.
3. Airborne carbon debris that is dispersed via wind formed from differences in air pressure outside and inside the cave.

The correlation between input rates and food chain length is striking. Entrances and tour routes possess a simple assemblage of microbes, microbivores (protozoans, nematodes, arthropods) and predatory species (mites). Sediments within deeper reaches of the caverns that receive extremely low inputs of carbon (<1 g C m^{-2} year^{-1}) are limited to microbes and protozoans.

Food chains initiated by bacteria persist throughout most of the caves, while those initiated by fungi do not possess a fungivore link in regions of low input. This result is consistent with those of our models in that more efficient food chains are more likely to be feasible in low productivity habitats (Fig. 8.7).

For both the fungal and bacterial food chains, predators are found only at the entrances or in regions of higher energy inputs.

These results could be due in part to colonization since many of the regions of the cave that receive low inputs are also remote. We cannot entirely rule out colonization, but there are a couple of reasons to discount it. First, the caves are estimated to be at least 30 million years old, and the region above the caves was not glaciated during the Pleistocene. Given the age of the cave and the vagile

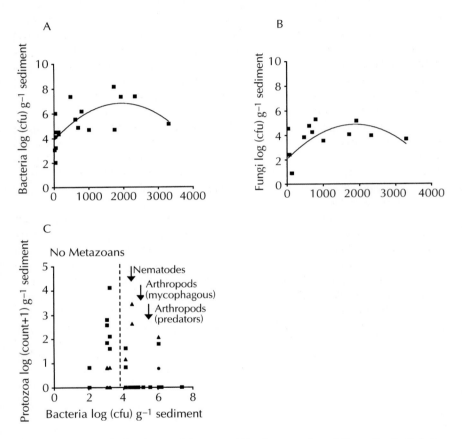

Fig. 8.7. Survey of microbes and fauna and lint inputs at Jewel Cave, South Dakota, USA. Bacterial (A) and fungal (B) densities expressed as colony/culture-forming units (cfu) g^{-1} sediment as a function of human lint input on to the cave floor (m lint m^{-2} $year^{-1}$). Both the bacteria and fungi exhibit a quadratic function with lint inputs ($r^2 = 0.47$ and 0.41, respectively). (C) Protozoan densities as a function of the bacterial densities: (●) ciliates, (■) flagellates and (▲) amoebae. The decline in bacterial and fungal densities at higher lint input rates is coincident with the appearance of higher trophic levels as indicated by the arrows. This is consistent with the predictions presented in Fig. 8.2, indicating a continuum of trophic structures along a productivity gradient.

nature of the biota that colonize the sediments, it is unlikely that these organisms have not had the opportunity to colonize the remote regions.

Further evidence against the colonization argument can be seen in the existing trophic structure. Nematodes are rare beyond the entrances of the cave and arthropods are a bit more prolific, yet predators from both taxa are all but absent beyond the entrances. If the trophic structures along the productivity gradient were the result of colonization then why have only the microbivorous forms of the nematodes and arthropods successfully established themselves along the gradient while predators have not? Microbivorous and predatory forms of these taxa do not differ significantly in their motility in ways that would preclude colonization. In the case of the arthropods, if motility were the issue then we would expect the more motile predatory mesostigmatid mites to be a dominant denizen of the cave sediments, yet they are absent.

Mammal activity is limited in the cave environment. The bats are utilizing the cave as a winter hibernaculum and not for sustenance. The rodents that occasion the cave forage for food outside the caverns as well as within, in similar fashion to shore birds foraging in the rocky intertidal zone. The models indicate that endothermy should be rare in low input systems, but more so in detritus-based systems than the producer-based systems (Figs 8.5 and 8.6). The caves would appear to confirm these predictions, but also serve as a metaphor for what has occurred on a global scale.

Current mammalian and avian diversity declines rapidly from the tropics to the polar regions. Productivity has been cited as one of the factors contributing to this trend. The productivity within the tropics ranges from an average of $4000 \, \text{g C m}^{-2} \, \text{year}^{-1}$ in coral reefs and $2500 \, \text{g C m}^{-2} \, \text{year}^{-1}$ in rainforests to a low of 1 and $10 \, \text{g C m}^{-2} \, \text{year}^{-1}$ in the arctic seas and tundra respectively. Oksanen *et al.* (1997) provide convincing evidence that the diversity and length of the endotherm-dominated grazer food chains of the Arctic mirror the patterns observed in Jewel Cave. The absence of a similar trend in the detritus-based soils of the Arctic (Doles and Moore, unpublished data) is not surprising given the quantitative differences in the feasibility and resilience of detritus-based food chains compared with those based on primary production.

Discussion

We raised several points in the introduction that undermined the theories on trophic structure and dynamics advanced by Hutchinson (1959), Hairston *et al.* (1960) and Oksanen *et al.* (1981). We address each below.

Non-linearity, trophic structure and dynamics

Our analyses suggest that trophic structures and dynamics change in a non-linear fashion along gradients of productivity. Furthermore, the analyses demonstrated how sensitive the models are to changes in initial conditions and assumptions. This conclusion can be seen at two levels. The first analysis where all variables were assumed distributed I $(0,1)$ was the closest analysis to those presented in the literature. Had the exercise stopped at that point the same conclusions would have been made about the factors that limit food chain length. The second analysis demonstrated that it was erroneous to assume that the results would extend over all ranges of productivity.

The models were extremely sensitive to changes in the distributions of the coefficients. Previous studies had assumed that the food chains were comprised of organisms that could differ in birth and death rates, consumption rates and efficiencies. The Monte Carlo trials used to study them sampled over the entire range of what was deemed feasible for these parameters. In these cases, the results represented an average of likely outcomes over the entire spectrum of possible physiologies and life histories. The final analysis presented here revealed that food chains dominated by organisms with different life strategies respond differently to changes in productivity. The differences in the response to changes in productivity between food chains modelled with the coefficients distributed Beta $(9,2)$ and Beta $(2,9)$, and between detritus-based and primary producer-based food chains raises some interesting questions and subjects for future studies.

These results may explain the observed difference in the responses of soil organisms to disturbances. Within soils, the nutrient dynamics associated with bacteria and fungi have been characterized as being 'fast' and 'slow' respectively (Coleman *et al.*, 1983). Bacteria and their consumers (the Beta $(9,2)$ food chains) on average are energetically more efficient, have shorter life spans and process matter more rapidly than fungi and their consumers (the Beta $(2,9)$ food chains; Coleman, 1994; Moore and de Ruiter, 1997). Our models predict and empirical evidence confirms that the bacterial assemblage would be more resilient than the fungal assemblage. The more rapid response and return to a prior steady state following disturbances of the bacterial assemblage has been shown following both freeze–thaw and wet–dry cycles (Allen-Morley and Coleman, 1989; Hunt *et al.*, 1989). Moreover, there is evidence to suggest that bacteria and their consumers colonize litter, recolonize disturbed soils, and recover from intensive agricultural practices more rapidly than fungi and their consumers (Hendrix *et al.*, 1986; Moore *et al.*, 1988; Moore and de Ruiter, 1991, 1997; Boyer, 1997).

Detritus versus primary producers

How different are systems based on detritus from systems based on primary producers? Our models suggest that in terms of trophic structure and dynamic states, both types of systems respond similarly, but differ in terms of where along the gradient of productivity transition from one trophic structure or dynamic state occurs. The results do suggest an interesting relationship between the energetic efficiencies of different organisms, and the likelihood of their being supported by detritus or primary production.

The food chains with coefficients distributed as Beta (9,2) were caricatures of chains dominated by ectotherms, while food chains with coefficients distributed as Beta (2,9) represented chains dominated by endotherms. Across a wide range of productivity (particularly at low productivity), detritus-based food chains would be more likely to be dominated by ectotherms than endotherms. Is there empirical evidence to support this? One may begin by asking how many endothermic species derive a large percentage of sustenance from detritus? These questions go far beyond the scope of this chapter, but our first responses would be 'not much' and 'not many', respectively. While feeding on carrion and standing dead plant material is common among endothermic vertebrates, these interactions represent a small fraction of all the trophic interactions that involve endotherms and a small fraction of total productivity. Feeding on detritus is largely the domain of bacteria, fungi, protozoa and invertebrates.

Trophic cascades

A trophic cascade of biomass is but one type of trophic structure that could occur along a gradient of productivity. Moving up a productivity gradient, trophic structures start as a pyramid of biomass and eventually give way to a cascade of biomass. The point at which this transition occurs depends on the physiologies of the organisms present and the availability of a new top predator entering the system. Systems dominated by ectotherms and not prone to colonization are more likely to exhibit a trophic cascade than systems dominated by vagile endotherms. How easy is it to invade lakes and streams compared with terrestrial systems? Put another way, what is the likelihood of a new predatory fish entering a lake or stream, or a predatory mite stumbling across an isolated community of microbes and invertebrates in a cave, compared with a migratory bird of prey or predatory mammal expanding its range?

Concluding Remarks

In closing, the answer to the question raised in the introduction as to what controls trophic structure and dynamics cannot be addressed by focusing on one factor. Rather, the determinants of community structure and dynamics are a complex of interactive components that most ecologists have an intuitive understanding of, yet have chosen to study in isolation and to the exclusion of one another.

We offer three guidelines. Firstly, studies designed to explore community structure and dynamics must establish the physiologies of the organisms and the productivity of the system under study. Both the transitions in trophic structures from pyramids to cascades, and dynamics from over-damped to oscillations are a function of energetic efficiencies and the level of productivity. Secondly, because of the non-linear relationship between energetic efficiency and productivity, the gradient of productivity should be sufficiently large and positioned properly relative to the organisms under study. Inducing a trophic cascade in a community dominated by endotherms may be a far more difficult task than it would be for one dominated by ectotherms. Lastly, systems that are based on detritus inputs are governed by the same principles as those based on primary producers. For a given level of input, the two types of systems will differ in the types of organisms that they support, the type of trophic structure and in their dynamics. However, detrital food webs dominated by microbes and invertebrates are not structured differently, nor behave differently from primary-producer food webs dominated by large mammals and birds. The observed differences between these systems is a reflection of how energetics governs both structure and dynamics, and differences in the transitions of both along gradients of productivity.

Acknowledgements

Special thanks to D.A. Crossley for many years of quiet mentoring and leader-ship to soil ecology, and D.C. Coleman and P.F. Hendrix for organizing the conference. This work was supported by grants from the National Science Foundation (DEB-9277510), US National Park Service and The University of Georgia.

References

Allen-Morley, C.R. and Coleman, D.C. (1989) Resilience of soil biota in various food webs to freezing perturbations. *Ecology* 70, 1127–1141.

Boyer, B.L. (1997) Nematode community structure and decomposition across adjacent mountain meadow ecosystems. Masters thesis, Department of Biological Sciences, University of Northern Colorado, Greeley, Colorado.

Briand, F. and Cohen, J.E. (1987) Environmental correlates to food chain length. *Science* 238, 956–960.

Carpenter, S.R., Kitchell, J.F. and Hodgeson, J.R. (1985) Cascading trophic interactions and lake productivity. *Bioscience* 35, 634–649.

Carpenter, S.R., Kitchell, J.F., Hodgeson, J.R., Cochran, P.A., Elser, J.J., Elser, M.M., Lodge, D.M., Kretchmer, D., He, X. and von Ende, C.N. (1987) Regulation of lake produc-tivity by food web structure. *Ecology* 56, 410–418.

Coleman, D.C. (1994) The microbial loop concept as used in terrestrial soil ecology studies. *Microbial Ecology* 28, 245–250.

Coleman, D.C., Reid, C.P.P. and Cole, C.V. (1983) Biological strategies of nutrient cycling in soil systems. In: Macfadyen, A. and Ford, E.D. (eds) *Advances in Ecological Research*, vol. 13. Academic Press, New York, pp. 1–55.

Hairston, N.G., Smith, F.E. and Slobodkin, L.B. (1960) Community structure, population control, and competition. *American Naturalist* 94, 421–425.

Hendrix, P.F., Parmelee, R.W., Crossley, D.A., Jr, Coleman, D.C., Odum, E.P. and Groffman, P.M. (1986) Detritus food webs in conventional and no-tillage agroecosystems. *BioScience* 36, 374–380.

Hunt, H.W., Elliott, E.T. and Walter, D.E. (1989) Inferring trophic interactions from pulse-dynamics in detrital food webs. *Plant and Soil* 115, 247–259.

Hutchinson, G.E. (1959) Homage to Santa Rosalia, or why are there so many kinds of animals? *American Naturalist* 93, 145–159.

Jefferies, C. (1988) *Mathematical Modeling in Ecology: a Workbook for Students*. Birkhäuser Press, Boston.

Lindemann, R.L. (1942) The trophic–dynamic aspect of ecology. *Ecology* 23, 399–418.

May, R.M. (1972) Will a large complex system be stable? *Nature* 238, 413–414.

May, R.M. (1973) *Stability and Complexity of Model Ecosystems*. Princeton University Press, Princeton, New Jersey.

May, R.M. (1976) Simple mathematical models with very complicated dynamics. *Nature* 261, 459–467.

Moore, J.C. and Hunt, H.W. (1988) Resource compartmentation and the stability of real ecosystems. *Nature* 333, 261–263.

Moore, J.C. and de Ruiter, P.C. (1991) Temporal and spatial heterogeneity of trophic interactions within belowground food webs: an analytical approach to understand multi-dimensional systems. *Agriculture, Ecosystems and Environment* 34, 371–397.

Moore, J.C. and de Ruiter, P.C. (1997) Compartmentalization of resource utilization in soil food webs. In: Gange, A.C. and Brown, V.K. (eds) *Multitrophic Interactions in Terrestrial Systems*. Blackwell Science, Oxford, pp. 375–393.

Moore, J.C., Walter, D.E. and Hunt, H.W. (1988) Arthropod regulation of micro- and mesobiota in belowground detrital food webs. *Annual Review of Entomology* 33, 419–439.

Moore, J.C., Walter, D.E. and Hunt, H.W. (1989) Habitat compartmentation and environmental correlates to food chain length. *Science* 243, 238–239.

Moore, J.C., de Ruiter, P.C. and Hunt, H.W. (1993) Influence of productivity on the stability of real and model ecosystems. *Science* 261, 906–908.

Naeem, S., Thompson, L.J., Lawler, S.P., Lawton, J.H. and Woodfin, R.M. (1994) Declining biodiversity can alter performance of ecosystems. *Nature* 368, 734–737.

Oksanen, L., Fretwell, S.D., Arruda, J. and Niemelä, P. (1981) Exploitative ecosystems in gradients of primary production. *American Naturalist* 118, 240–261.

Oksanen, L., Aunapuu, M., Oksanen, T., Schneider, M., Ekerholm, P., Lundberg, P., Armulik, T., Aruola, V. and Bondestad, L. (1997) Outlines of food webs in a low arctic tundra landscape in relation to three theories on trophic dynamics. In: Gange, A.C. and Brown, V.K. (eds) *Multitrophic Interactions in Terrestrial Systems*. Blackwell Science, Oxford, pp. 351–374.

Phillipson, J. (1981) Bioenergetic options and phylogeny. In: Townsend, C.R. and Calow, P. (eds) *Physiological Ecology: an Evolutionary Approach to Resource Use*. Blackwell Science, Oxford, pp. 20–45.

Pimm, S.L. (1982) *Food Webs*. Chapman and Hall, London.

Pimm, S.L. and Lawton, J.H. (1977) The number of trophic levels in ecological communities. *Nature* 268, 329–331.

Polis, G.A. (1994) Food webs, trophic cascades, and community structure. *Australian Journal of Ecology* 19, 121–136.

Polis, G.A. and Strong, D.R. (1996) Food web complexity and community dynamics. *American Naturalist* 147, 813–846.

Polis, G.A., Myers, C.A. and Holt, R.D. (1989) The ecology and evolution of intraguild predation: potential competitors that eat each other. *Annual Review of Ecology and Systematics* 20, 297–330.

Roberts, A. (1974) The stability of a feasible random system. *Nature* 251, 607–608.

Rosenzweig, M.L. (1971) The paradox of enrichment: destabilization of exploitation ecosystems in ecological time. *Science* 171, 385–387.

Rosenzweig, M.L. (1995) *Species Diversity in Space and Time.* Cambridge University Press, Cambridge.

Rosenzweig, M.L. and MacArthur, R.H. (1963) Graphical representation and stability conditions of predator–prey interactions. *American Naturalist* 97, 209–223.

de Ruiter, P.C., Neutel, A. and Moore, J.C. (1995) Energetics, patterns of interaction strengths, and stability in real ecosystems. *Science* 269, 1257–1260.

Strong, D.R. (1992) Are trophic cascades all wet? Differentiation and donor-control in speciose ecosystems. *Ecology* 73, 747–754.

Tilman, D., Wedin, D. and Knops, J. (1996) Productivity and sustainability influenced by biodiversity in grassland ecosystems. *Nature* 379, 718–720.

Webmasters and Ecosystem Diversity

Biodiversity of Oribatid Mites (Acari: Oribatida) in Tree Canopies and Litter

9

V. Behan-Pelletier[1] and D.E. Walter[2]

[1]*Biodiversity Program, Research Branch, Agriculture and Agri-Food Canada, Ottawa, Canada K1A 0C6;* [2]*Department of Zoology and Entomology, University of Queensland, St Lucia, Queensland 4072, Australia*

Introduction

Forest canopies are among the last biotic frontiers (Erwin, 1983). Habitats for much of the biosphere's undescribed diversity, many are in high-risk tropical biotopes, or occupy the last remaining tracts of old-growth forest in temperate zones (Winchester, 1997). Biodiversity research in forest canopies has focused on insects; until recently, mites (Acari), traditionally considered denizens of soil and litter, have been ignored (Walter and Behan-Pelletier, 1999). However, an abundant, free-living mite fauna inhabits temperate and tropical canopies and is sufficiently abundant to be considered as 'arboreal plankton' (Walter and Proctor, 1999). When adequate sampling methods are used, mites are consistently found to be more numerous than other arboreal arthropods, including insects, and often as species rich. For example, Nadkarni and Longino (1990) found mites to be the numerically dominant arthropods in cloud forest canopy, despite ignoring organisms less than 0.5 mm long, including probably most of the mites. Walter and O'Dowd (1995b) estimated that the leaves alone of a 10 m tall subtropical rainforest tree supported nearly 400,000 mites, mostly Oribatida. In a review of canopy fumigation studies, Watanabe (1997) noted that mites were recorded from 22 of 38 studies, and accounted for 2–82% of all arthropods collected.

Although more commonly associated with soil and litter, oribatid mites are often the dominant mite group collected from forest canopies, and it is this generally ignored group of canopy arthropods on which we focus in this review. We compare oribatid mite biodiversity in tree canopies and litter from three perspectives: components of the faunas, modifications for living in canopy habitats, and ecological roles of these faunas, including implications for ecological processes.

Components of the Faunas

Comparative diversity and abundance

Soil and litter are the habitats from which the greatest diversity and abundance of mites have been reported. In forests, mites are found in surface litter, rotting wood, moss, fungi, lichens and throughout the soil profile, though they are concentrated in the upper 0–20 cm (Petersen and Luxton, 1982). Mites have been quantitatively collected from litter and soil since the development of the Berlese funnel in 1905, but most studies use a high-gradient behavioural extraction method, or passive flotation (Edwards, 1991). Standard sampling methods are routinely used in studies of litter biodiversity (Edwards, 1991), and species richness and abundance are comparable across sites. In forest soil and litter, Oribatida are the most species rich and numerically dominant arthropod group. Even cursory examination of soil and litter will yield examples of 40–160 species, and their densities can reach several hundred thousand individuals per square metre (Norton, 1990). For example, 82 species of oribatid mites, with average population density of 61,500 adults m^{-2} have been collected from litter of beech moder (Wunderle, 1992b); and 96 oribatid species with average population density of 44,500 individuals m^{-2} were collected from oak forest litter and soil (Lamoncha and Crossley, 1998).

Wallwork (1983), reviewing the few data available on arboreal mites, considered arboreal habitats to be relatively simple systems compared with the forest floor. Today we recognize that canopy habitats present a significantly more complex picture, and the range of occupied microhabitats in the canopy include epicorticolous epiphytes, algae, bryophytes and lichen, suspended soils, under bark, in insect galleries, fungal sporocarps, phytotelmata and sap fluxes, on stems, leaves, flowers and fruit (Walter and Behan-Pelletier, 1999).

Diverse mite faunas have been collected from canopy habitats by fumigation (chemical-knockdown), beating canopy foliage, Tullgren funnel extraction, arboreal photoeclectors, bark brushing, lichen, twig and leaf-washing, branch traps and by hand. In many studies, oribatid mites are among the most abundant arthropods (Table 9.1). Oribatida were the numerically dominant and most species-rich arthropods in a canopy of 10-year-old black pine from Japan (Kurimoto, 1978), and were the dominant arthropods (after ants) in branch traps in oak canopy in Finland (Koponen *et al.*, 1997). Oribatid mites represented from 75 to 88% of arthropods extracted by Tullgren funnels from bark of conifers in Japan (Yamashita and Ohkubo, 1980), and 34% of arthropods extracted by brushing and Tullgren funnels from bark and epiphytes in a beech–hornbeam forest (André, 1984). Similarly, oribatid mites dominate the canopy acarofauna, representing 67% of acari on bark and epiphytes of conifers in Poland (Seniczak, 1998), and 61–99% of mites brushed from bark in floodplain forest in Brazil (Franklin *et al.*, 1998) and from leaf surfaces in subtropical Australia (Walter *et al.*, 1994).

Table 9.1. Relative abundance and species richness of oribatid mites in canopy habitats.

Country[a]	Canopy	Habitat	Methodology	Among Arthropoda: relative abundance of Acari or Oribatida	Among Acari: relative abundance of Oribatida (%)	Species richness
Canada	Sitka spruce	Canopy moss	Tullgren and Malaise trap	—	—	Oribatida: 43
USA: Minnesota	Oak forest	Bark	Hand	5–45% Oribatida	—	Oribatida: 16
Japan	Conifers	Bark	Tullgren	67.5–81.0% Acari	75.0–88.5	—
Finland	Oak forests	Branches	Branch trap	14% Oribatida	—	Oribatida: 33
Belgium	Beech, hornbeam	Bark and epiphytes	Brushing and Tullgren	33.6% Oribatida	—	Oribatida: 36
Poland	Scots pine forest	Epiphytes on bark	Tullgren	—	67	Oribatida: 29
Germany	Temperate forest	Bark	Hand, photoeclector	97% Oribatida	—	Oribatida: 34
Greece	*Olea, Pyrus* spp.	Bark and twigs	Tullgren	—	—	Acari: 20
Italy	Chestnut	Cortical cankers	Tullgren	—	—	Oribatida: 27
New Zealand	*Olearia*	Foliage	—	—	—	Oribatida: 18
Africa	Subtropical forest	Bark	Hand	3–87% Oribatida	—	Oribatida: 2
Brazil	Floodplain forest	Bark	Brushing	—	61–99	Oribatida: 21
Venezuela	Mixed forest	Epiphyte, bromeliad, suspended soils	Tullgren	—	—	Oribatida: 69
Peru	Mixed forest	Epiphyte, bromeliad, suspended soils	Tullgren	—	—	Oribatida:127
Australia	Subtropical	Canopy	Hand	—	—	Oribatida: 28

[a] References: Canada, Winchester *et al.* (1999); USA, Nicolai (1993); Japan, Yamashita and Ohkubo (1980); Finland, Koponen *et al.* (1997); Belgium, André (1984); Poland, Seniczak *et al.* (1997); Germany, Nicolai (1986); Greece, Emmanouel and Panou (1991); Italy, Nannelli and Turchetti (1993); New Zealand, Spain and Harrison (1968); Africa, Nicolai (1989); Brazil, Franklin *et al.* (1997); Venezuela, Behan-Pelletier *et al.* (1993); Peru, Wunderle (1992a); Australia, Walter *et al.* (1994).

As yet, canopy sampling for arthropods lacks standardization, and estimates of density by standard area or volume of canopy material are either not available or not comparable across sites. Each canopy microhabitat requires a different sampling method to estimate the diversity; for example, high-gradient extractors which effectively collect mites from canopy moss are ineffective for phylloplane acarofauna.

Most canopy studies yield undescribed species, and the recent discovery of a diverse (two genera; nine species) phthiracaroid fauna on the trunks and branches of mature trees in Tasmanian rain forest (Colloff and Niedbała, 1996) is not unusual. To date, species in at least 110 genera representing 51 oribatid families are known to live in arboreal habitats. Almost 91% of these genera found are members of the Brachypylina. Other than a species of Adelphacaridae recovered from beech bark (Wunderle, 1992b), members of the oribatid lineage Palaeosomata have not been collected from canopy habitats; however, we expect their presence, and that of members of the Parhyposomata, in drier, suspended soil. In contrast, 295 genera representing 106 families are known from soil and litter habitats in North America alone (Marshall *et al.*, 1987), with 74% of genera members of the Brachypylina, and the other oribatid lineages, Palaeosomata, Enarthronota, Parhyposomata, Mixonomata and Desmonomata, represented by 2%, 9%, 1%, 8% and 6% of genera, respectively. Species richness of canopy Oribatida, though not as high as that in soil and litter, ranges from 2 to 127 species (Table 9.1) and is dominated by members of the desmonomate families Crotoniidae, Camisiidae, and the brachypyline families Hammeriellidae, Eremaeidae, Peloppiidae, Cepheidae, Carabodidae, Cymbaeremaeidae, Micreremidae, Licneremaeidae, Adhaesozetidae, Neotrichozetidae, Haplozetidae, Scheloribatidae, Oribatulidae, Oripodidae, Mochlozetidae, Humerobatidae, Ceratozetidae, Mycobatidae and Galumnidae. Some genera in these families are restricted to arboreal habitats, and species have clearly diversified within these habitats; for example, *Crotonia, Darthvaderum, Labiogena, Dendrozetes, Scapheremaeus, Micreremus, Licneremaeus, Adhaesozetes, Dometorina, Eporibatula, Sellnickia, Neotrichozetes, Oripoda, Parapirnodus* and genera in the Mochlozetidae. Other genera are more eurytypic, and have few to many species in canopy habitats; for example, *Camisia, Eueremaeus* and *Carabodes*. Whether congeneric arboreal species represent single or multiple invasions of canopy habitats is unclear without phylogenetic analysis of their relationships.

Complementarity of the oribatid fauna of litter and canopy microhabitats

The litter oribatid mite fauna traditionally has been considered the source of canopy diversity, with specimens dispersing from the litter and reaching the canopy either by climbing the trunk highway or through wind dispersal (Dindal, 1990; Watanabe, 1997). This view overlooks the distinctness of the arboreal fauna, emphasized by Travé (1963), and ignores the disjunct nature of epiphytes on all but the oldest trees in rain forests. Even in old-growth Sitka spruce forest, where there is a continuous moss highway between forest floor and canopy, the moss-inhabiting oribatid faunas are distinct between canopy and forest floor (Winchester, 1997). In the few studies where canopy and litter faunas have been compared at the same site, complementarity or distinctness (*sensu* Colwell and Coddington, 1994) of these faunas is high (Table 9.2). Some soil mites do climb trees, and arboreal mites may fall into the litter but, by and large, canopy mite species are canopy specialists. Indeed, the earliest known oribatid fossils are of Devonian age, thus oribatid mites were already present and diverse when terrestrial vegetation began to flourish (Norton, 1986). However, there are no chronosequence studies comparing complementarity between forest floor and canopy faunas and, to date, most canopies studied have been in mature or old-growth forests.

Intuitively the tree trunk, the link between the canopy and soil, should support a varying assemblage of mites: dispersing soil and litter mites, wandering species (*sensu* Aoki, 1973), species associated with trunk epiphytes and dispersing canopy species. However, there are few studies of the continuum of habitats between and including the forest floor and the canopy. Sampling 11 height levels between 0.25 and 30 m along eight beech trees felled over a 2-year period, Wunderle (1992b) showed that species are usually dominant at particular trunk and canopy levels, although many species can be found at a range of levels. Thus, although *Carabodes labyrinthicus* (Carabodidae) was dominant throughout the trunk–canopy profile, *Cymbaeremaeus cymba* (Cymbaeremaeidae) was dominant at trunk heights between 8 and 28 m, and *Micreremus brevipes* (Micreremidae) was dominant in the high canopy at 28–30 m. In contrast, *Zygoribatula exilis* (Oribatulidae) was never found above 0.5 m.

Lawton (1986) proposed two hypotheses to account for the effects of plants on insect diversity: (i) size of plants: large, vertically standing plants will support greater diversity than smaller plants in ecological and evolutionary time, because their size presents a larger catchment area for dispersing organisms; and (ii) resource diversity: plants with greater varieties of resources support more species. There are no data to test the first hypothesis for canopy Oribatida, but the few studies that have compared the mite faunas of different resources (microhabitats) indicate that oribatid species richness is related to resource heterogeneity (Walter and Behan-Pelletier, 1999). In the same forest type, complementarity (dissimilarity) between species assemblages from different canopy microhabitats can be high (Table 9.3). Walter (1995) found that different habitats on the same tree species have higher complementarity than the same habitats on different tree species. Bark of different tree species can support distinct oribatid faunas, reflecting bark microstructure, microclimate and type of epiphytes (Nicolai, 1986). Similarly, André (1984, 1985) found that the distribution and species richness of oribatid mites on bark is related to type of epiphytic cover, i.e. crustose versus foliose versus fructose lichens. In floodplain forests in Brazil, Franklin *et al.* (1998) suggested that reduced lichens, mosses and epiphytes on bark of trees in 'várzea' forest explained their reduced species diversity and overall abundances compared with 'igapó' forest.

Table 9.2. Oribatid species richness and percentage complementarity in canopy and soil and litter at three tropical and three temperate forest sites (data from Behan-Pelletier et al., 1993; Ito, 1986; Walter et al., 1994; Behan-Pelletier and Winchester, 1998; Wunderle, 1992a, b).

Location	Canopy microhabitat	Canopy F:G:S[a]	Litter and soil F:G:S	% Complementarity[b] (no. species in common)
Tropical rain forest				
Venezuela: Parque Pittier	Epiphytes, bromeliads and suspended soils	30(4):50(18):69(35)	30:48:73	68.5 (34)
Peru: Panguana	Epiphytes, bromeliads and suspended soils	46(5):62(10):127(44)	49:79:160	59 (83)
Australia: Victoria	Leaves, stems, trunks	?:?:28(21)	?:?:40	88.5 (7)
Cool–temperate forest				
Germany	Canopy twigs and leaves; bark	34(3):47(8):78(10)	37:60:107	42 (68)
Canada: Carmanah	Canopy moss	25(5):35(11):43(19)	29:42:56	68 (24)
Japan	Canopy branches and lichens	23(8):29(18):36(27)	25:29:33	85 (9)

[a] Numbers of families (F), genera (G) and species (S). Numbers in parentheses are the numbers at each level restricted to the canopy.
[b] Complementarity (distinctness) as defined by Colwell and Coddington (1994); 100% represents complete dissimilarity.

Table 9.3. Species richness and percentage complementarity of oribatid mite faunas among canopy microhabitats in Australia and Germany. Matrix entries: percentage complementarity (number of species in common).

Australia: Lamington National Park (Walter, 1995)

Tree Habitat	Rose marara		Brown beech
	Leaves and stems	Suspended soils	Leaves and stems
Richness	7	12	20
Rose marara: suspended soils	100(0)	—	—
Brown beech: leaves and stems	87.5(3)	100(0)	—

Germany: Marburg, tree bark (Nicolai, 1986)

Tree Bark type	*Fagus* Smooth	*Quercus* Fissured	*Betula* White	*Acer* Scaly	*Salix* Fissured	*Ulmus* Fissured
Richness	10	22	8	15	14	9
Quercus	67(8)	—	—	—	—	—
Betula	61.5(5)	64(8)	—	—	—	—
Acer	61(7)	58(11)	65(6)	—	—	—
Salix	67(6)	71(8)	71(5)	55(9)	—	—
Ulmus	73(4)	71(7)	69(4)	66(6)	65(6)	—

Modifications for Arboreal Living

The canopy provides both architectural diversity (including type of leaf venation and shape and surface of leaf, twig and branch) and spatial diversity (including height in the canopy and horizontal distance from the trunk), and these in turn affect microclimate. We examine the evidence for behavioural, physiological and morphological modifications of Oribatida to living in the canopy.

Behavioural modifications

Oribatida, along with other arthropods in the canopy, must withstand desiccation, either by living in or moving to favourable microhabitats or by physiological tolerance, but a behavioural response to desiccation has proved to be difficult to demonstrate. For example, the canopy dwelling *Humerobates rostrolamellatus* (Humerobatidae) lives on the bark of trees, clustering in small groups among patches of the alga *Pleurococcus* (Madge, 1966). Specimens cluster and are inactive during daytime when the air is dry, and disperse and feed on the algae at night when the humidity rises (Madge, 1966). This may be a behavioural response to fluctuating moisture. But clusters are usually of adults only, and may represent mating aggregations, as has been suggested for a species of *Mochloribatula* (Mochlozetidae) clustering on exposed cement columns (Norton and Alberti, 1997).

Another intriguing but, as yet, unexplored difference in arboreal oribatids is that sexual dimorphism is better developed in canopy compared with litter mites. For example, arboreal males sometimes have processes on their posterior idiosoma (e.g. *Mochloribatula*) or horn-like setae (e.g. *Symbioribates* (Symbioribatidae)). Soil and litter oribatids are well known for exhibiting completely dissociated sperm transfer where males produce spermatophores without any pairing or contact with females (Proctor, 1998). However, anecdotal observations (D.E. Walter) suggest that male–female pairing is common in arboreal Oribatida, suggesting the hypothesis that more direct modes of sperm transfer may have evolved in arboreal taxa, perhaps in response to the drying effects of wind currents on the stalked spermatophores (Norton and Alberti, 1997).

Physiological modifications

There is no evidence of physiological modifications for living in canopy environments even for mites living on the phylloplane and twigs, which are highly subject to desiccation. Madge (1964, 1966) surveyed six oribatid species from the Enarthronota, Desmonomata and Brachypylina for survival under dry conditions. Though the brachypyline *Humerobates rostrolamellatus* withstood temperature and moisture extremes best, species representing all three lineages showed tolerance and survival at a range of relative humidities. Similarly, Jalil (1972) showed that desiccation resistance of *H. rostrolamellatus* was greater than that of either *Tectocepheus* spp. (Tectocepheidae) or *Platynothrus peltifer* (Camisiidae). Larvae, nymphs and adults of the brachypyline *Eupelops acromios* (Phenopelopoidea), often found on tree trunks, can survive for 3 days at low humidity (10–20%) (Atalla and Hobart, 1964). There is no evidence of specific adaptations, such as possible modifications of the cuticle to enhance resistance to desiccation (Atalla and Hobart, 1964; Madge, 1964). No members of Palaeosomata, Parhyposomata or Mixonomata have been tested for resistance to desiccation, so their limited representation in the canopy cannot be correlated with desiccation intolerance.

Morphological modifications

Most oribatid mites living in the canopy have a range of morphological modifications which apparently have evolved to address aspects of plant architecture and/or microclimate. These modifications seem to be independently derived in many taxa, but this needs verification through phylogenetic analyses of relationships. Adults and immatures of oribatid species living on leaves or bark generally have short, clubbed to globose bothridial setae, in contrast with the usually elongate setiform to plumose structures in soil mites, including those from suspended soils, such as species of Enarthronota (Aoki, 1973). Since bothridial setae sense air-currents, Norton and Palacios-Vargas (1982) proposed that the globular, reduced form in arboreal species is an adaptation to the stronger air movements in this habitat. Bothridial setal size reductions are also seen in species from other exposed, windswept habitats, for example *Maudheimia* spp. (Ceratozetidae) from nunataks in the Antarctic (Coetzee, 1997). As Travé (1963) and Wallwork (1976) emphasized, there are many similarities between arboreal and rock-dwelling oribatid mites, and bothridial setal size reduction found in genera common to both habitats, for example *Camisia*, *Scapheremaeus* and *Phauloppia* (Oribatulidae), is one similarity.

Arboreal members of the Brachypylina show a range of leg modifications, including distally expanded solenidia, leg sacculi, tarsal pulvilli and modified claws. Expanded tibial solenidia found in the arboreal genera *Nasozetes* (Scheloribatidae), *Micreremus* and *Siculobata* (Oribatulidae) may enhance chemosensory functions in habitats exposed to wind (Grandjean, 1959). Leg porose organs, whose function is respiratory, are often invaginated as sacculi or brachytracheae in oribatid species from canopy or other exposed environments, such as rocks. These invaginations of respiratory surfaces are hypothesized to reduce water loss across these surfaces in desiccating environments (Norton and Alberti, 1997). Pad-like pulvilli are found in several genera that live on smooth leaf surfaces, for example *Adhaesozetes* (Adhaesozetidae) and *Ametroproctus* (Cymbaeremaeidae), and these apparently represent an adaptation to maintain grip (Walter and Behan-Pelletier, 1993). Another adaptation to maintain grip is the bidentate lateral claws found only in the arboreal oribatulid genera *Phauloppia*, *Lucoppia*, *Dometorina* and *Sellnickia*; the mochlozetid genera *Unguizetes*, *Terrazetes*, *Drymobatoides*, *Drymobates* and *Mochlozetes*; and the ameronothrid genus *Podacarus* known from exposed ground habitats on the windswept Kerguelen Island, Southern Ocean (Grandjean, 1960).

Modifications of notogastral porose organs in oribatid mites include striking sexual dimorphism (as noted above), as well as enlargement and multiplication of porose areas, and porose organs occupying protuberances. These modifications are mainly restricted to species subject to periodic desiccation, a common feature of canopy habitats (Norton and Alberti, 1997). The modifications are possibly for cuticular maintenance in microhabitats with fluctuating moisture. Alternatively they may produce and facilitate dispersal of semiochemicals, such as aggregation pheromones. Certainly, the latter explanation is most probable in arboreal genera such as *Adhaesozetes*, *Sellnickia*, *Grandjeania* (Oribatulidae), *Jugatala* (Ceratozetidae) and *Notogalumna* (Galumnidae), which have notogastral porose organs on protuberances (Norton and Alberti, 1997).

Role of Canopy Fauna

The earliest oribatid mites were probably particulate-mycophages, the same feeding habit as most extant oribatid species (Norton, 1986; Labandeira *et al.*,

1997). In forest soil and litter, Oribatida are actively involved in decomposition of organic material, in nutrient cycling and in soil formation. Active instars of these mites can feed on a wide variety of material, including living and dead vascular plant and fungal material, algae, moss, lichens and soft-bodied invertebrates, such as nematodes and rotifers. Oribatid mites are the most important group of arachnids from the standpoint of direct and indirect effects on the formation and maintenance of soil structure (Moore *et al.*, 1988). They disperse bacteria and fungi, both externally on their body surface and by feeding on spores that survive passage through their alimentary tracts. An important catalytic function of oribatid mites in decomposition and nutrient cycling is their role as stimulants of dormant microflora through their grazing. They can adjust to forced changes in diet, and the gut contents of a given species can vary with site and season (Anderson, 1975) or at different stages in the life cycle (Siepel, 1990). This probably reflects changes in their active gut microflora which produce many digestive enzymes, including cellulase and chitinase (Norton, 1986).

Gut content analyses of a few species indicate that the canopy oribatid fauna utilizes resources that are broadly similar to those exploited by species in forest floor litter (André and Voegtlin, 1981; Walter and Behan-Pelletier, 1993). However, though resources may be similar, ecological roles of some canopy oribatid mites may differ from their litter relatives. An example of this is the interaction of arboreal mites with phytopathogenic fungi on leaves and bark. Grandjean (1963) discussed the possibly beneficial role of adults and immatures of *Podoribates gratus* (Mochlozetidae), which feed on the mycelia and spores of parasitic fungi on the leaves of plants. However, there are no unambiguous studies showing clear benefits to the plants from oribatid grazing on phyto-pathogenic fungi on leaves, such as has been shown for the mycophagous prostigmatic mite, *Orthotydeus lambi*, which reduces the incidence of powdery mildew on the riverbank grape, *Vitis riparia* (Norton *et al.*, 1998).

In fact, a simple functional classification of mycophagy can be misleading as it does not distinguish beneficial from detrimental feeding. For example, if spores of these phytopathogenic fungi survive passage through the mite gut, *P. gratus* may be a dispersal agent of these plant parasites. Jacot (1934) suggested such a possible role for Oribatida in the transmission of Dutch elm disease. Wendt (1983) found oribatid mites associated with *Endothia parasitica*, the causative agent of canker in American chestnut, and considered that the mites may be carriers and disseminators of the fungus. Nannelli and Turchetti (1993) found 74 species of mites associated with cortical cankers of the European chestnut, of which 17, primarily Oribatida, are mycophagous. Some of these, for example *Scheloribates latipes, Liebstadia humerata* (Scheloribatidae) and *Zygoribatula* sp. (Oribatulidae), have been successfully reared for many genera-tions on virulent and hypovirulent strains of *Cryphonectria parasitica*, the European chestnut canker (Nannelli and Turchetti, 1993; Nannelli *et al.*, 1998). Furthermore, the virulence of the fungal parasite remains unchanged as it passes through the gut of the mites, and these authors consider these species to be dispersal agents of the parasite. Similarly, adults and immatures of *Dentizetes ledensis* (Ceratozetidae), a mycophagous oribatid living on the phylloplane of *Ledum groenlandicum*, feed on the rust *Chrysomyxa ledi* (V. Behan-Pelletier, personal observation). Whether fungal spores maintain viability in passage through the gut of individuals, however, is unknown. Undoubtedly, air currents or rain drops are much more effective dispersal agents of fungal spores in the canopy than oribatid mites; however, the latter may play a role in spore dispersal on leaves with a thick tomentum on the lower surface which would slow air movement.

Possibly, as in forest floor litter, oribatid mites may function to stimulate the growth of dormant phytopathogenic fungi. Certainly, a mutually beneficial interaction occurs between the lichen *Evernia prunastri* and grazing microarthropods, such as the oribatids *Carabodes labyrinthicus* and *Phauloppia lucorum* (Prinzing and Wirtz, 1997). Grazing mites induce variability in the growth pattern of lichen thalli, which can benefit the lichen especially when it is growing on wind-exposed trees.

Phytophagy, or feeding on fresh plant material, is rare in oribatid mites. Among 18 oribatid species feeding on the aquatic weed, *Eichhornia crassipes*, Sumangala and Haq (1995) considered 12 to be surface feeders on epidermal cells, resulting in exposure of mesophyll tissue. This, in turn, would enhance access by phytopathogens, which may be the preferred food of the mites. One species, *Orthogalumna terebrantis* (Galumnidae), tunnels in the leaves, contributing to leaf breakdown and decay. Females cut oviposition holes, deposit eggs in the mesophyll, and the immatures eat the leaf tissue, forming linear galleries in the leaves. However, faecal pellets are brown to black, and the primary food may be phytopathogenic fungi. But, whether oribatids graze on or burrow in fresh leaf tissue in aboveground foliage or canopies is unknown.

Density estimates for oribatid mites in the canopy are lower than those recorded from soil and litter. Oribatid densities of $2200–6100$ m^{-2} were recorded from the trunk of Japanese cedar and cypress (Yamashita and Ohkubo, 1980). Based on extractions of soil and litter microarthropods and chemical knockdown of canopy microarthropods in Japanese cedar plantations, Hijii (1987, 1989) estimated that canopy oribatid populations of up to 1158 m^{-2} are approximately a hundred times smaller than those in soil and litter. In a comparison of guild structure (including herbivores, detritivores, predators and parasitoids) in these conifer canopies, Hijii (1989) found that the detritivore guild (all Oribatida and Collembola) dominated abundance estimates on all sampling dates, and biomass estimates on two of 12 sampling dates. Clearly, these Oribatida, along with Collembola, affect nutrient dynamics in the canopy, providing nutrients for canopy roots (Nadkarni and Primack, 1989) and the roots of epiphytes. As early as 1980, Carroll postulated that mites grazing microepiphytes on needles and small twigs of Douglas fir indirectly affect patterns of nutrient exchange within the canopy. Based on the numbers of oribatid mites on leaves (up to 94 on a leaf, average of 4 per leaf), Walter and O'Dowd (1995a) postulated that these leaf microbivores can directly or indirectly affect plant fitness. Old epiphyll-covered leaves may be microbial nutrient gardens for tropical trees, with the mites, through grazing and excretion, freeing up nutrients sequestered by these epiphylls. However, as yet studies designed to tease apart possible effects of canopy oribatid mites on decomposition and nutrient cycling, using methods such as litterbags or radioisotopes (Seastedt, 1984) are rare (Fagan and Winchester, 1999).

Regardless of feeding habits in the canopy, oribatid faecal pellets would provide organic nutrients essential for the growth of phylloplane microflora (Jensen, 1974). Tree leaves are invaded by microorganisms rapidly (within 1–2 days) after unfolding and most of these, being non-pathogenic, require a source of nutrients (Jensen, 1974). These microorganisms are, in turn, grazed by oribatid mites. Clearly, a succession of microorganisms with associated grazers occurs on leaves (Pugh, 1974) and decomposition of leaves starts in the canopy.

Summary

We compare oribatid mite biodiversity in tree canopies and litter from three perspectives: components of the faunas, modifications for living in canopy microhabitats, and ecological roles of tree-canopy and litter faunas, including implications for ecological processes. In tree canopies, as in litter, Acari dominate the microarthropod fauna, usually exceeding all other arthropod groups in abundance, and competing with insects in terms of species richness. The mite fauna of tree canopies is not just a subset of the litter fauna; it comprises many lineages specific to arboreal microhabitats. Many species appear to have adaptations for living in canopies, perhaps reflecting the different architecture, air movements and general microclimate of canopies in comparison with litter. The range of canopy microhabitats occupied by mites rivals that of the forest floor. The role of litter Oribatida in ecosystem processes includes decomposition, mineralization and development of soil microstructure. The role of components of the canopy oribatid fauna in ecosystem processes may not differ significantly from that in litter, and gut content analyses indicate that resources utilized are broadly similar to those exploited by species in litter. Canopy Oribatida probably help release nutrients sequestered in epiphytes and epiphylls for uptake by the plant. They may also have more complex ecological roles, such as controlling or dispersing leaf fungal pathogens.

Acknowledgements

Dac Crossley, Jr knows and loves mites and has introduced a generation of soil ecologists to these charismatic arthropods; this chapter is dedicated to him. We thank Roy Norton, SUNY, Syracuse, and Evert Lindquist, Agriculture and Agri-Food Canada, for their useful suggestions and comments on earlier drafts of this manuscript.

References

Anderson, J.M. (1975) Succession, diversity and trophic relationships of some soil animals in decomposing leaf litter. *Journal of Animal Ecology* 44, 475–495.

André, H.M. (1984) Notes on the ecology of corticolous epiphyte dwellers. 3. Oribatida. *Acarologia* 25, 385–396.

André, H.M. (1985) Associations between corticolous microarthropod communities and epiphytic cover on bark. *Holarctic Ecology* 8, 113–119.

André, H.M. and Voegtlin, D.J. (1981) Some observations on the biology of *Camisia carrolli* (Acari: Oribatida). *Acarologia* 23, 81–89.

Aoki, J.-I. (1973) Soil mites (oribatids) climbing trees. *Proceedings of the 3rd International Congress of Acarology*. Academia, Prague, pp. 59–65.

Atalla, E.A.R. and Hobart, J. (1964) The survival of some soil mites at different humidities and their reaction to humidity gradients. *Entomologia Experimentalis Applicata* 7, 215–228.

Behan-Pelletier, V.M. and Winchester, N.N. (1998) Arboreal oribatid mites diversity: colonizing the canopy. *Applied Soil Ecology* 9, 45–51.

Behan-Pelletier, V.M., Paoletti, M.G., Bissett, B. and Stinner, B.R. (1993) Oribatid mites of forest habitats in northern Venezuela. *Tropical Zoology, Special Issue* 1, 39–54.

Carroll, G.C. (1980) Forest canopies: complex and independent subsystems. In: Waring, R.H. (ed.) *Forests: Fresh Perspectives from Ecosystem Analysis*. Oregon State University, Corvallis, Oregon, pp. 87–108.

Coetzee, L. (1997) The Antarctic mite genus *Maudheimia* (Acari, Oribatida). *Navorsinge van die Nasionale Museum Bloemfontein* 13, 101–135.

Colloff, M. and Niedbała, W. (1996) Arboreal and terrestrial habitats of phthiracaroid mites (Oribatida) in Tasmanian rainforests. In: Mitchell, R., Horn, D.J., Needham, G.R. and Welbourn, W.C. (eds) *Acarology IX – Proceedings*, Vol. 1. Ohio Biological Survey, Columbus, Ohio, pp. 607–611.

Colwell, R.K. and Coddington, J.A. (1994) Estimating terrestrial biodiversity through extrapolation. *Philosophical Transactions of the Royal Society of London Series B* 345, 101–118.

Dindal, D.L. (1990) *Soil Biology Guide*. John Wiley & Sons, New York.

Edwards, C.A. (1991) The assessment of populations of soil-inhabiting invertebrates. *Agriculture, Ecosystems and Environment* 34, 145–176.

Emmanouel, N.G. and Panou, H. (1991) A study on mites associated with bark and twigs of various trees in Attica (Greece). In: Dusbábek, F. and Bukva, V. (eds) *Modern Acarology*, Vol. 1. SPB Academic Publications, The Hague, pp. 523–532.

Erwin, T.L. (1983) Tropical forest canopies: the last biotic frontier. *Bulletin of the Entomological Society of America* 29, 14–19.

Fagan, L.L. and Winchester, N.N. (1999) Arboreal arthropods: diversity and rates of colonization in a temperate montane forest. *Selbyana* (in press).

Franklin, E.N., Woas, S., Schubart, H.O.R. and Adis, J.U. (1998) Ácaros oribatídeos (Acari: Oribatida) arboríolas de duas florestas inundáveis da Amazônia central. *Revista Brasilia Biologia* 58, 317–335.

Grandjean, F. (1959) Observations sur les Oribates (40e série). *Bulletin Muséum National d'Histoire Naturelle Paris*, 2e série 31, 359–366.

Grandjean, F. (1960) Les Mochhlozetidae n. fam. (Oribates). *Acarologia* 2, 101–148.

Grandjean, F. (1963) Concernant *Sphaerobates gratus*, les Mochlozetidae et les Ceratozetidae (Oribates). *Acarologia* 5, 284–305.

Hijii, N. (1987) Seasonal changes in abundance and spatial distribution of the soil arthropods in a Japanese cedar (*Cryptomeria japonica* D. Don) plantation, with special reference to Collembola and Acarina. *Ecological Research* 2, 159–173.

Hijii, N. (1989) Arthropod communities in a Japanese cedar (*Cryptomeria japonica* D. Don) plantation: abundance, biomass and some properties. *Ecological Research* 4, 243–260.

Ito, M. (1986) An ecological survey on arboreal oribatid mites (Acari: Oribatida) in a subalpine coniferous forest of Shiga-Kogen, Central Japan. *Edaphologia* 35, 19–26.

Jacot, A.P. (1934) Acarina as possible vectors of the Dutch elm disease. *Journal of Economic Entomology* 27, 858–859.

Jalil, M. (1972) The effect of desiccation on some oribatid mites. *Proceedings of the Entomological Society of Washington* 74, 406–410.

Jensen, V. (1974) Decomposition of angiosperm tree leaf litter. In: Dickinson, D.H. and Pugh, G.J.F. (eds) *Biology of Plant Litter Decomposition*, vols 1 and 2. Academic Press, London, pp. 69–104.

Koponen, S., Rinne, V. and Clayhills, T. (1997) Arthropods on oak branches in SW Finland, collected by a new trap type. *Entomologica Fennica* 8, 177–183.

Kurimoto, Y. (1978) Arthropods in black pine crowns. *Transactions 26th Management Chubu Branch Japan Forestry Society* 26, 175–179.

Labandeira, C.C., Philips, T.L. and Norton, R.A. (1997) Oribatid mites and the decomposition of plant tissues in paleozoic coal swamp forests. *Palaios* 12, 319–353.

Lamoncha, K.L. and Crossley, D.A., Jr (1998) Oribatid mite diversity along an elevation gradient in a southeastern appalachian forest. *Pedobiologia* 42, 43–55.

Lawton, J.H. (1986) Surface availability and insect community structure: the effects of architecture and fractal dimension of plants. In: Juniper, B. and Southwood, R. (eds) *Insects and the Plant Surface*. Edward Arnold, London, pp. 317–331.

Madge, D.S. (1964) The humidity reactions of oribatid mites. *Acarologia* 6, 566–591.

Madge, D.S. (1966) The significance of the sensory physiology of oribatid mites in their natural environment. *Acarologia* 8, 155–160.

Marshall, V.G., Reeves, R.M. and Norton, R.A. (1987) Catalogue of the Oribatida (Acari) of Continental United States and Canada. *Memoirs Entomological Society Canada* 139.

Moore, J.C., Hunt, H.W. and Walter, D.E. (1988) Arthropod regulation of micro- and mesobiota in below ground detrital food webs. *Annual Review of Entomology* 33, 419–439.

Nadkarni, N.M. and Longino, J.T. (1990) Invertebrates in canopy and ground organic matter in a neotropical montane forest, Costa Rica. *Biotropica* 22, 286–289.

Nadkarni, N.M. and Primack, R.B. (1989) A comparison of mineral uptake and translocation by above-ground and below-ground systems of *Salix syringiana*. *Plant and Soil* 113, 39–45.

Nannelli, R. and Turchetti, T. (1993) Interazioni acari corticoli – *Cryptonectria parasitica*: probabile ruolo delgi acari nella diffusione degli isolati del parassita. In: Covassi, M. (ed.) *Piante Forestali*, Firenze 1992, Istituto Sperimentale Patologia Vegetale, Roma, pp. 157–163.

Nannelli, R., Turchetti, T. and Maresi, G. (1998) Corticolous mites (Acari) as potential vectors of *Cryptonectria parasitica* (Murr.) Barr hypovirulent strains. *International Journal of Acarology* 24, 237–244.

Nicolai, V. (1986) The bark of trees: thermal properties, microclimate and fauna. *Oecologia* 69, 148–160.

Nicolai, V. (1989) Thermal properties and fauna on the bark of trees in two different African ecosystems. *Oecologia* 80, 421–430.

Nicolai, V. (1993) The arthropod fauna on the bark of deciduous and coniferous trees in a mixed forest of the Itasca State Park, MN, USA. *Spixiana* 16, 61–69.

Norton, A.P., English-Loeb, G., Gadoury, D.M. and Seem, R.C. (1998) Mycophagous mites and foliar pathogens: leaf domatia mediate a plant–arthropod mutualism. In: Abstracts, *Ecological Society of America, Annual Meeting, Baltimore, MD*, p. 101.

Norton, R.A. (1986) Aspects of the biology and systematics of soil arachnids particularly saprophagous and mycophagous mites. *Quaestiones Entomologicae* 21, 523–541.

Norton, R.A. (1990) Acarina: Oribatida. In: Dindal, D.L. (ed.) *Soil Biology Guide.* John Wiley & Sons, New York, pp. 779–803.

Norton, R.A. and Alberti, G. (1997) Porose integumental organs of oribatid mites (Acari, Oribatida). 3. Evolutionary and ecological aspects. *Zoologica* 146, 115–143.

Norton, R.A. and Palacios-Vargas, J.G. (1982) Nueva *Belba* (Oribatei : Damaeidae) de musgos epifitos de Mexico. *Folia Entomolica Mexico* 52, 61–73.

Petersen, H. and Luxton, M. (1982) A comparative analysis of soil fauna populations and their role in decomposition processes. *Oikos* 39, 288–388.

Prinzing, A. and Wirtz, H.-P. (1997) The epiphytic lichen, *Evernia prunastri* L., as a habitat for arthropods: shelter from desiccation, food limitation and indirect mutualism. In: Stork, N.E., Adis, J. and Didham, R.K. (eds) *Canopy Arthropods.* Chapman & Hall, London, pp. 477–494.

Proctor, H.C. (1998) Indirect sperm transfer in arthropods: behavioral and evolutionary trends. *Annual Review of Entomology* 43, 153–174.

Pugh, G.J.F. (1974) Terrestrial fungi. In: Dickinson, D.H. and Pugh, G.J.F. (eds) *Biology of Plant Litter Decomposition,* vols 1 and 2. Academic Press, London, pp. 303–336.

Seastedt, T.R. (1984) The role of microarthropods in decomposition and mineralization processes. *Annual Review of Entomology* 29, 25–46.

Seniczak, S., Dabrowski, J., Klimek, A. and Kaczmarek, S. (1998) The mites associated with young Scots pine forests polluted by a copper smelting works in Glogow, Poland. In: Mitchell, R., Horn, D., Needham, G.R. and Welbourn W.C. (eds) *Acarology IX,* Vol. 1, *Proceedings.* Ohio Biological Survey, Columbus, Ohio, pp. 573–574.

Siepel, H. (1990) Niche relationships between two panphytophagous soil mites, *Nothrus silvestris* Nicolet (Acari, Oribatida, Nothridae) and *Platynothrus peltifer* (Koch) (Acari, Oribatida, Camisiidae). *Biology and Fertility of Soils* 9, 139–144.

Spain, A.V. and Harrison, R.A. (1968) Some aspects of the ecology of arboreal Cryptostigmata (Acari) in New Zealand with special reference to the species associated with *Olearia colensoi* Hoof. f. *New Zealand Journal of Science* 11, 452–458.

Sumangala, K. and Haq, M.A. (1995) Nutritional diversity of Acari infesting *Eichhornia crassipes. Journal of Ecobiology* 7, 289–297.

Travé, J. (1963) Écologie et biologie des Oribates (Acariens) saxicoles et arbicoles. *Vie et Milieu, Supplement* 14, 1–267.

Wallwork, J.A. (1976) *The Distribution and Diversity of Soil Fauna.* Academic Press, London.

Wallwork, J.A. (1983) Oribatids in forest ecosystems. *Annual Review of Entomology* 28, 109–130.

Walter, D.E. (1995) Dancing on the head of a pin: mites in the rainforest canopy. *Records of the Western Australian Museum, Supplement* 52, 49–53.

Walter, D.E. and Behan-Pelletier, V. (1993) Systematics and ecology of *Adhaesozetes polyphyllos*, sp. nov. (Acari: Oribatida: Licneremaeoidea) leaf-inhabiting mites from Australian rainforests. *Canadian Journal of Zoology* 71, 1024–1040.

Walter, D.E. and Behan-Pelletier, V. (1999) Mites in forest canopies: filling the size distribution shortfall? *Annual Review of Entomology* 44, 1–19.

Walter, D.E. and O'Dowd, D.J. (1995a) Life on the forest phylloplane: hairs, little houses, and myriad mites. In: Lowman, M.D. and Nadkarni, N. (eds) *Forest Canopies*. Academic Press, New York, pp. 325–351.

Walter, D.E. and O'Dowd, D.J. (1995b) Beneath biodiversity: factors influencing the diversity and abundance of canopy mites. *Selbyana* 16, 12–20.

Walter, D.E. and Proctor, H.C. (1999) *Mites: Ecology, Evolution and Behavior*. New South Wales Press, Sydney, 352 pp.

Walter, D.E., O'Dowd, D.J. and Barnes, V. (1994) The forgotten arthropods: foliar mites in the forest canopy. *Memoirs of the Queensland Museum* 36, 221–226.

Watanabe, H. (1997) Estimation of arboreal and terrestrial arthropod densities in the forest canopy as measured by insecticide smoking. In: Stork, N.E., Adis, J. and Didham, R.K. (eds) *Canopy Arthropods*. Chapman & Hall, London, pp. 401–414.

Wendt, R. (1983) Association of *Endothia parasitica* with mites isolated from cankers on American chestnut trees. *Plant Disease* 67, 757–758.

Winchester, N.N. (1997) Canopy arthropods of coastal Sitka spruce trees on Vancouver Island, British Columbia, Canada. In: Stork, N.E., Adis, J. and Didham, R.K. (eds) *Canopy Arthropods*. Chapman & Hall, London, pp. 151–168.

Winchester, N.N., Behan-Pelletier, V. and Ring, R.A. (1999) Arboreal specificity, diversity and abundance of canopy-dwelling oribatid mites (Acari: Oribatida). *Pedobiologia* (in press).

Wunderle, I. (1992a) Die baum- und bodenbewohnenden Oribatiden (Acari) im Tieflandregenwald von Panguana, Peru. *Amazoniana* 12, 119–142.

Wunderle, I. (1992b) Die Oribatiden-Gemeinschaften (Acari) der verschiedenen Habitate eines Buchenwaldes. *Carolinea* 50, 79–144.

Yamashita, Z. and Ohkubo, N. (1980) Arthropod communities on the barks of Sugi, *Cryptomeria japonica* D. Don and Hinoki, *Chamaecyparis obtusa* Endl. transplanted in the compounds of the Grand Shrine of Ise. *Insect Survey Report for the Compounds of the Grand Shrine of Ise*. Natural Science Institute, Mie Prefecture.

Diversity in the Decomposing Landscape

10

R.A. Hansen

Department of Biological Sciences, University of South Carolina, Columbia, SC 29208, USA

Introduction

At practically any terrestrial site, the vast majority of the animal species are invertebrate members of the decomposer community. In temperate forests, these animals, primarily arthropods and nematodes, are concentrated in the decomposing organic layers that make up the top few centimetres of the soil. As thin and fragile as this layer is, it is the zone through which most of the productivity of the system, as leaves, roots, wood and animal debris, passes and is transformed. Through their activity as microbial grazers and saprophages, the decomposer fauna is the gate-keeper to the flow of material through the system.

In ecosystem-level models of nutrient cycling, this zone and its inhabitants are most commonly represented by several compartments through which nutrients flow, entering as a large pulse input at litterfall and leaving through plant uptake, leaching, denitrification and respiration. But the mechanisms that regulate decomposer interaction webs, their structure and biotic diversity, are nested on a finer scale of resolution, one that discerns structure within the litter profile and the dynamics of its annual fluctuations. A mechanistic understanding of the organization of soil assemblages and their mediation of processes will require study on this scale.

In this chapter, I will focus on the most diverse of the litter-dwelling mesofaunal groups, the non-astigmatic oribatid mites, and how the morphology and dynamics of the litter profile determine their diversity, species composition and their functional impact. As a case study, I will discuss an experiment that compared the oribatid assemblage in profiles developed from monotypic litter, the typical medium for studying litter dynamics and a common consequence of human activities, with those in profiles developed from natural, mixed litter.

The Profile as a Successional Landscape

We are accustomed to viewing the organic profile as a chronosequence of decomposition stages (Fig. 10.1). Since most decomposition studies are of litterbags of one litter type, our data reinforce this simple, linear schematic. The dynamics in the profile as a whole, the stacked strata represented in litterbags, is complicated by interactions between strata, the dynamics of which have a strong seasonal component. At litterfall, the new stratum begins to leach nutrients to those below and there is translocation of nutrients to the upper layers from below (Wagener and Schimel, 1998). The new litter changes conditions in the layers below by buffering them from moisture and tempera-ture fluctuations. The condition of the litter below can be, in turn, a strong determinant of the decomposition rate of the new litter cohort (Knutson, 1997). Depending on climate and litter quality, litter profiles vary widely in their overall stability and complexity. Rapidly decomposing litters will dwindle to provide little habitat before the next litterfall. Slowly decomposing leaves will form complicated profiles with multiple cohorts represented.

The mixed litter that characterizes most natural forests adds yet another dimension of complexity to the model (Fig. 10.1). Each stratum is a mosaic of patches which differ in chemistry, architecture and microbial species composi-tion. As each layer moves down the profile, each patch undergoes a series of successional stages, both microbial and structural, from fresh litter to granular humus. Litters of different initial quality will vary in the timing and duration and in the particular character of each stage. The heterogeneity of substrates, in terms of their quality, architecture and microclimate, increases as fresh litter fragments and litter types that differ in quality diverge in their decomposition trajectories. The habitat becomes more uniform again at the base of the profile as fragmented litter gives way to granular humus and microbial diversity declines (Swift, 1976).

Variety in initial substrate quality contributes to the profile as a habitat in several ways. It increases microhabitat and resource heterogeneity. The co-occurrence of labile and recalcitrant litters should stagger the timing of decomposition stages and thus the availability of resources associated with them. There is also the potential for interactions among different substrate types. A growing number of studies are demonstrating strong, non-linear effects of mixing litters on response variables including decomposition and respiration rates and N dynamics (Taylor *et al.*, 1989; Blair *et al.*, 1990; Fyles and Fyles, 1993; McArthur *et al.*, 1994; Briones and Ineson, 1996; McTiernan *et al.*, 1997; Wardle *et al.*, 1997; Salamanca *et al.*, 1998). These interactions may be chemical: the translocation of nutrients (Staaf, 1980; Chapman *et al.*, 1988; McTiernan *et al.*, 1997) or inhibitory compounds (McArthur *et al.*, 1994) from one substrate to another. They may also be physical: litters of different water potentials (Dix, 1984) changing each other's moisture regime, for example. Finally, such substrate interactions may affect the decomposer fauna (Blair *et al.*, 1990), and these effects may, in turn, alter decomposition dynamics.

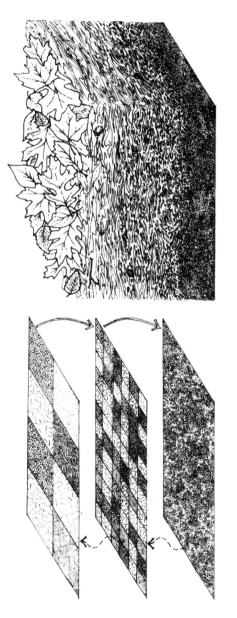

Fig. 10.1. The decomposing landscape. The organic profile formed by a mixed litter can be seen as a series of strata, each of which is a mosaic of patches. As the strata descend in the profile, each patch undergoes structural and chemical changes and microbial succession. Potential interactions between patches and strata include translocation of nutrients by fungi and particulate matter by animals and gravity, physical sheltering of lower strata by upper strata and support of upper substrates by lower ones.

The Maintenance of Faunal Diversity in the Litter-to-soil Profile

How the prodigious diversity within decomposer assemblages composed of apparently similar species is maintained and the degree to which assemblages are organized by attributes of the local habitat has been a long-standing problem (Anderson, 1973; Giller, 1996). It is useful to phrase the question in terms of where assemblages lie on the gradient from an equilibrium state, in which niche space is partitioned and species composition is predictable within a habitat type, to a non-equilibrium state, in which spatial and temporal heterogeneity, in the form of patchiness and disturbance for example, keep competition from driving niche partitioning and reducing diversity (Giller, 1996). In mechanistic models of species coexistence on either side of this spectrum, how the organisms perceive habitat heterogeneity and the relative scales of temporal and spatial variation in habitat dynamics and of the organisms' behaviour and life history are crucial parameters. In the litter habitat, where units of substrate undergo rapid transformations relative to the life cycles of many inhabitants and can have strong interactions over time, understanding habitat structure and dynamics is particularly crucial.

Most studies of litter dynamics that include animals do so because of interest in how the fauna act upon substrates, and so our vision of the decomposer fauna is based largely on animals recruited to aliquots of a single litter type at a particular decomposition stage. These data inform us about habitat selection and activity, but represent only snapshots of activity in a subset of the habitat. Progress towards answering questions about assemblage diversity and organization will require data on assemblage responses in their full habitat and over a sufficiently long time-frame to encompass temporal and spatial dynamics of the full profile.

A Case Study

Experimental design and hypotheses

What follows is a discussion of a field experiment designed to explore how aspects of the profile habitat maintain their assemblages. At a single forested site at the Coweeta Hydrological Laboratory in the Nantahala Mountain Range of western North Carolina, USA, latitude 35° 03′N, longitude 83° 25′W, natural litterfall was excluded from a series of 1 m^2 plots and replaced with treatment litters that varied in their composition and complexity. Plots of litter from each of the dominant tree species on the site, red oak, yellow birch and sugar maple, composed the simple litter treatments. Two complex litter treatments were a mixture of these three litter species and a mixture containing seven litter species and a standard weight of dry twigs. Plots were enclosed to 15 cm depth to seal them from immigration and emigration. The pretreatment litter layer and its assemblage were left undisturbed and a standard weight of air-dried treatment litter, based on the average litterfall for the site, was applied to the plots. Treatments were applied for 3 consecutive years so that, in each, the full litter profile characteristic of that litter type was allowed to form and go through multiple annual cycles. Differences in profile morphology and in the diversity and composition of the oribatid mite assemblage were tracked over 3 years. The general questions were these:

1. For a diverse group of similar species, what are the consequences of litter simplification for diversity? Since habitat heterogeneity, patchiness and variety of resources and microhabitats, is an established determinant of diversity, with some direct evidence of this for oribatid mites (Anderson, 1978; Hansen and Coleman, 1998), diversity is expected to decline in simple litter profiles.

2. Do species that differ in their habitat use respond differently to the treatments? Smaller species, which dominate the lower profile where litters become less distinct from one another, may respond less strongly to differences in litter type. Species that use both the litter and the woody microhabitats may respond to both litter characteristics and the presence of woody material.

3. Does each litter type develop a characteristic species composition? To the extent that the assemblage is organized by oribatid affiliations to particular litter types, a characteristic species composition should emerge within each litter type.

4. Can we relate the responses of the oribatid assemblage to profile morphology and dynamics? How does profile morphology vary among the treatment litters? Initial differences in litter complexity are expected to endure and perhaps amplify in the full profile, with potential interaction effects in the mixed litters.

Some background on oribatid mites

The oribatid mites were chosen as the group to follow in this study because they are the most abundant and diverse of the mesofaunal litter-dwellers. At the study site, there are 170 identified species of adult oribatid mites from the litter and soil with as many as 40 species represented in a 20 cm^2 core through the profile (Hansen, 1997). Adult densities range from 22×10^3 to 92×10^3 m^{-2} depending on the season.

The primary food resource for oribatid mites is microbial, gleaned by direct consumption of fungal hyphae and, in a subset of species, through the consumption of decaying plant matter permeated by microbial growth. Other diet items can include algae, moss, lichen, pollen and nematodes. Endophagy is a distinctive feeding mode in a subset of oribatid mites. Endophagous species as juveniles are obligate burrowers in woody microhabitats, including twigs and branches, acorns, polyporous fungi, conifer needles and the petioles of leaves. Many of these species spend their adulthood grazing on a varied diet in the litter. Because of their comminution of the most recalcitrant substrates, these may represent a particularly important functional group within the oribatid mites.

Oribatid feeding habits have been found to be general and opportunistic, varying within a species with locality and season (Norton, 1985). There is some evidence that species have some fidelity to strata within the profile (Anderson, 1971; Luxton, 1981a). In general, the lower profile is dominated by smaller species while the fauna of the upper litter layers is characterized by larger species (Lebrun, 1971; Pande and Berthet, 1975; Luxton, 1981a; Wallwork, 1983).

While there is certainly variation in life history within the oribatid mites, in general, they possess the traits traditionally characterized as 'K-selected': slow development, low reproductive potential, long adult life and iteroparity. Most species initiate one to three generations per year and adult longevities of 1–2 years may not be uncommon in temperate soils (Norton, 1994). Thus, oribatid individuals can experience one to several annual cycles in the profile, integrating much of the temporal variation in the profile. This is in contrast to many litter taxa that operate in a faster time-frame, collembola and nematodes for example, and which may mirror temporal habitat fluctuations more directly in their abundance.

Oribatid response to habitat simplification

A look at the response of the oribatid assemblage after 3 years in treatment profiles reveals several surprising results. First, as predicted, oribatid species richness declined in each of the simple litters by about 28% while the richness in mixed litters remained constant. There was also an unanticipated decline in total oribatid abundance to a similar degree in each of the simple litters, to about 72% of that in mixed litters (Fig. 10.2). This erosion of diversity and abundance was an aggregate of many species responses and was significant in both large and small mites and in both endophagous and non-endophagous species.

It is surprising that the decline in diversity and abundance of the assemblage was similar in all the simple litters, since oak, birch and maple differ widely in their quality and architecture. The general negative response to monotypic litter leads one to look for commonalities in the experience of the animals to simplification *per se*. Examination of the profile development in simple and mixed litters does yield some generalizations about simplification of the litter habitat.

Consequences of simplification for the litter habitat

Litter complexity and habitat structure
As expected, a census of microhabitat types at several depths in the treatment profiles showed that each stratum of mixed litter contained more different

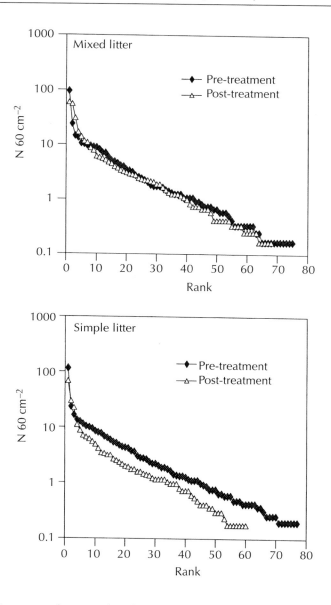

Fig. 10.2. Average rank versus abundance curves for the oribatid assemblages pretreatment and 3 years post-treatment in simple and mixed litter plots. Species were ranked according to their mean abundance in each of the five litter types. Data here are the abundance at each rank averaged among the three simple litters and between the two mixed litters. Assemblages in simple litters underwent a broad-based decline in both average abundance of many species and in total richness, while in the mixed litter the relative abundance structure of the assemblages remained intact.

microhabitats, defined by substrate types and fungal growth forms, than did simple litters (Hansen and Coleman, 1998). There were also some unanticipated differences between mixed and simple litters that might help explain the overall loss of oribatid abundance in simplified habitats. Over a full annual cycle, the litter habitat, defined as the strata that contain recognizable leaf fragments, included less material, measured as dry weight, in simple litter treatments than in mixed litters. This was particularly true in the late summer when the profile was at its thinnest. The pure maple profile dwindled particularly quickly. The microhabitat census also revealed that two non-litter substrates, roots and arthropod faeces and humic material were more prevalent in mixed than in simple litters (Hansen, 2000).

There are several possible mechanisms at work here, and underlying each of them is a complementarity among litters of different quality. First, there is complementarity in physical structure. In litters where substrates with different quality and decomposition stage are layered, the recalcitrant constituents might create stable shelves for more labile ones. In this case, red oak leaves maintain unfragmented leaf planes late in decomposition and birch leaves, rather than fragmenting, become skeletonized nets. Both might lend support to friable litter types like maple and herb litter. In contrast, in pure maple litter, faecal pellets and comminuted litter particles filter unimpeded into the humus layer. In mixed litter, pockets of readily exploitable substrate suspended above the humic layer proper may encourage higher root invasion and roots, in turn, may further increase the structure, stability and retention of material. Retention of comminuted particles and excreta may stimulate microbial growth in the litter (Lussenhop, 1992). For species that are concentrated in or limited to the litter profile, slower transport of material and nutrition out of their envelope of habitat means an increase in their resource base.

Variety in litter quality may also create a more favourable physical habitat for litter-dwellers. In this case, each of the single litter types, on its own, has structural attributes that created deficits in physical habitat. Birch litter forms a compact mat which is waterlogged much of the time (Hansen and Coleman, 1998) and contains relatively little habitable pore space. Fresh oak leaves are thick and hydrophobic and, early in decomposition, create a dry habitat with relatively little surface area per unit mass of litter. Because the friable and high quality maple litter disappears so quickly, the profile becomes very thin and poorly buffered against heat and drying in the late summer. When these litters are mixed, complementary architecture should buffer the habitat from extremes in moisture and create structure, with thick, rigid leaves preventing compaction and thinner ones providing surface area.

The loss of structure and microhabitat variety in monotypic litters may put pressure on litter-dwellers by reducing refugia from predation. Estimates of juvenile mortality of oribatid species are substantial (Norton, 1994). The relatively unsclerotized immatures are vulnerable to a wide variety of arthropod predators (Norton, 1985; Walter, 1989). Structural microhabitats for juvenile development which offer some protection from predators may be an important component of the habitat.

Litter complexity and the microbial resource base

There is also potential complementarity among litters in the pattern of microbial growth. A number of studies have found higher respiration rates in a preponderance of mixtures of tree leaf litters (Briones and Ineson, 1996; McTiernan *et al.*, 1997; Wardle *et al.*, 1997; Salamanca *et al.*, 1998). The general hypothesis emerging from these studies is that nutrients from labile substrates facilitate microbial exploitation of recalcitrant ones.

Also, because it contains a greater number of successional stages at any given time, mixed litter should provide a more stable supply of microbial resources. In contrast, in simple litters, synchronized decomposition of uniform substrates should produce flushes and troughs of particular resources, the timing of which depends upon the decomposition trajectory of the litters. For example, maple litter loses mass rapidly in the first months after leaf fall (Hansen and Coleman, 1998) and is likely to sponsor a flush of early successional fungi followed by a trough as labile substrates are exhausted. Oak litter, in contrast, is slow to begin decomposition and decomposes more rapidly later in the season (Hansen and Coleman, 1998). This lag time should manifest as a trough in microbial growth followed by an increase later in the year.

Such differences in timing of resource availability could shape the community structure of litter-dwellers by favouring some life-history strategies over others. Siepel (1996) has demonstrated the filtering of microarthropod species according to life-history traits over a gradient of increasing disturbance. I suggest that the same kind of filtering may go on at a smaller scale in litter layers with increasing instability in structural and microbial resources. In addition to slow life histories, oribatid mites have slow digestive processes and low metabolisms relative to other microarthropod groups (Luxton, 1975; Mitchell, 1979; Thomas, 1979). While long adult lives and iteroparity should buffer populations from catastrophic losses during resource scarcity, periods of low food quality should lower adult fecundity and juvenile development rate and survival. Given their slow metabolisms, long development and generation times, oribatid mites are ill-suited to take advantage of flushes of microbial growth (Mitchell, 1977).

Within the guild of microbial feeders in the litter, oribatid mites, in general, are on the least opportunistic end of the spectrum. Other animals supported by the microbial resource base include collembola, protura, fungivorous prostigmatid mites, nematodes and enchytraeids. Many members of these groups have faster development times than oribatid mites and would be better able to exploit relatively short-lived flushes of resource. In particular, collembola, fungal grazers that vie for numerical dominance with oribatids, have higher ingestion and metabolic rates and shorter generation times (Crossley, 1977). Collembolan species demonstrate phenotypic plasticity such that their fecundity and rate of development can increase a great deal depending on the quality of available food (Von Amelsvoort and Usher, 1989). These groups may have fared better than the oribatid mites in simplified habitats with higher amplitude of resource availability.

In this context it is interesting to note that the dominant oribatid species in the assemblage, *Oppiella nova*, was unaffected by litter simplification. *O. nova* is the most cosmopolitan of all oribatid species, being abundant in many habitats on every continent (Marshall *et al.*, 1987). It is often cited as a pioneer and dominant species in highly disturbed systems (Beckmann, 1988; Skubala, 1995; Siepel, 1996) and is highly tolerant of drought (Wauthy and Vannier, 1988). It is parthenogenic (Norton *et al.*, 1993) and has the most rapid development time of all species for which there are data (Luxton, 1981b; Kaneko, 1988). *O. nova*'s success in simplified habitats relative to the remainder of the assemblage may be due to greater plasticity and a faster life cycle which allows it to capitalize more effectively on periods of abundance.

It is not difficult to assemble a list of potential mixing effects that may have contributed to the oribatid responses to litter complexity. I hasten to add that studies of mixing effects on tree leaf litter decomposition dynamics have shown a great deal of variation among mixtures of different leaf species. While faster mass loss, N loss and higher respiration are common mixing effects, there are plenty of counter-examples. The mechanisms of complementarity outlined above rely on a range of variation in quality and structure among the constituent litters. The range and proportion of different litter qualities in mixtures may be a sturdy guide for systematizing mixing effects. Finzi and Canham (1998), for example, have developed models from field measurements in different forest stands in which N mineralization rates in mixed litter are depressed by low-quality litter until the mixture is less than 30% low-quality litter.

Oribatid species composition in litter types

The oribatid responses in terms of overall abundance and diversity were similar among the three simple litters and between the two mixtures. At the species level, however, the responses to each litter type were distinctive. I analysed the species composition of three divisions of the assemblages that differ in their habitat use: large litter-dwellers which dominate the upper litter layers, small litter-dwellers which dominate the lower profile, and endophagous and other wood-associated species whose habitat use is complex, including litter and woody microhabitats. For the large litter-dwellers and the endophages, replicate plots within each litter type did converge in their species composition, although small litter-dwellers showed no such response (Table 10.1). This emergence of a consistent species composition of large litter-dwellers and endophages in each litter type indicates that these sectors of the assemblage are organized by species affiliations with particular substrates. The lack of such a pattern in the small litter-dwellers may be a reflection of decreasing structural and microbial distinctiveness of different litter types at depth where these species dominate.

Table 10.1. Percentage similarity (PS) among plots of different treatments and among replicates within treatments, pretreatment and 3 years post-treatment. PS is calculated as $PS = 200 \sum_{j=1}^{s} \min\left(x_{j1}, x_{j2}\right) \Big/ \sum_{j=1}^{s}\left(x_{j1}, x_{j2}\right)$ where x_{j1} is the abundance of species j in plot 1 ($j = 1, 2, \ldots s$) and x_{j2} is its abundance in plot 2 for all species found in either plot. The mean PS for a treatment was obtained by averaging all pairwise comparisons of replicate plots. Between-treatment similarities are an average of all possible pairwise comparisons between plots of the two treatments. Data here are averages across treatments and their standard deviations.

Oribatid group	Initial PS among treatments (%)	Final PS among treatments (%)	Initial PS within treatments (%)	Final PS within treatments (%)
Large litter-dwellers	47 ± 1	40 ± 2	46 ± 4	60 ± 4
Small litter-dwellers	51 ± 2	53 ± 4	51 ± 5	54 ± 8
Endophages	24 ± 1	28 ± 3	25 ± 3	53 ± 8

In the large litter-dwellers and endophages, there are a number of optional modes for associations with particular litter types. Species may be specialized on microbial species or structural microhabitats that are distinctive to a litter type. The microbial species composition is somewhat distinctive on different litter types (Swift, 1976; Lodge, 1997). There is little in the literature, however, to suggest specialization of oribatid species to particular microbial resources. There is reason to expect, in fact, that specialization in microbial food resources might be limited given the pattern of occurrence of microbial species in litter. Decomposing substrates as habitat patches change quickly relative to the pace of oribatid life cycles. Thus, any given patch, characterized by chemistry and microbial species, may last for only a brief portion of an individual's lifetime. In the range within which an individual can forage, resources become increasingly rare as they become more specific. This, coupled with the ephemerality of resources and conditions should represent a limit to specialization, particularly in taxa with limited mobility.

There is also the possibility that changes in habitat structure alter species composition by shifting predator–prey relationships. If oribatid species are specialized upon litter-specific structural microhabitats in the sites for juvenile development, the deletion of a subset of these refugia in each of the monotypic litters could have increased predation on species associated with them and thus shaped species composition. Little is known about predation on oribatid mites and such a mechanism cannot be ruled out.

Alternatively, oribatid species may be more generally associated with the class of microhabitats and resources associated with a stage in decomposition. Such associations could be the mechanism behind the distinctive species composition within each litter type since those decomposition stages are represented in different proportions and with different timing in each litter type. There is plenty of evidence for affiliation of oribatid species with particular decomposition stages both from studies of mite succession in leaf litter (Crossley and Hoglund, 1962; Lebrun and Mignolet, 1974; Anderson, 1975) and from associations of species with strata of the profile (Anderson, 1971; Usher *et al.*, 1979; Luxton, 1981a; Ponge, 1991a). Berg *et al.* (1998), in an elegant study using stratified litterbags, showed that microarthropod species composition in each stratum, as it reaches a particular decomposition stage, approaches the species composition that the stratum below it had at that stage.

Functional consequences of oribatid responses

In the study described in this chapter, simple litter profiles were unfavourable habitat for oribatid mites. The same conditions that did not favour oribatid mites may have shunted resources towards other taxa, perhaps those with more rapid and plastic life-histories or humus and soil dwellers. Shifts in the balance between faunal groups, even those that share the same food, can have functional consequences through the differences among fauna in digestive processes and excreta. Practically all litter material passes through an animal gut at least once and faunal excreta can make up almost all of the humus layer (Bal, 1970; Rusek, 1975). The physical and chemical attributes of excreta can thus be crucial to the structure of the soil that develops and its properties with respect to nutrient retention and release (Martin and Marinissen, 1993). For example, collembola, enchytraeids, fly larvae and oribatid mites feeding on the same litter material produce very different excreta with those of oribatid mites being much longer lived than others (Bal, 1970; Rusek, 1975; Ponge, 1991b). They may contribute to the formation of water-stable aggregates (Lussenhop, 1992).

Also apparent from this case study is that particular litter types select a particular species composition, at least within the oribatid mites. Are there functional consequences for variation in fauna on this finer taxonomic scale? Several recent studies indicate that variation at the level of species composition of microarthropods can affect litter decomposition and nutrient cycling. Faber and Verhoef (1991) demonstrated differences in collembolan species' effects on nitrogen mobilization. Schulz and Scheu (1994) have shown species-specific effects on cellulose breakdown among oribatid mites involved in the decomposition of wood. Microarthropod assemblages of different species composition have been shown to yield different rates of leaching of mineral nutrients (Heneghan and Bolger, 1996). Siepel and Maaskamp (1994) have shown that different feeding guilds of oribatid mites, defined by carbohydrase activity, vary in their effects on microbial respiration.

Another example emerges from the case study described here. Endophagous species demonstrated a strong affiliation with the woody portions of oak leaves, and that affiliation appears to have had a significant impact on the litter's decomposition rate (Hansen, 1999). Decomposition rates of treatment litters were measured from litterbags in the plots for two consecutive litter cohorts, while the shifts in the assemblage were taking place. In all treatment litters, except oak, litter from both years decomposed at similar rates. The second cohort of oak litter disappeared significantly faster than the first. Inspection of the litter revealed endophages burrowing in the oak's petioles and leaf veins. Numbers and species richness of endophagous adults, in fact, nearly doubled in the second year's litter (Hansen, 1997). Oak petioles represent a particularly stable microhabitat type which can provide refuge and food for juvenile endophages during their long development. Because it entrained a particularly functionally significant group, the oak litter's selection of a fauna has functional consequences.

These results have interesting implications for the effects of disturbance on faunal contributions to processes. To the extent that local habitat characteristics determine the faunal composition, undisturbed organic profiles will develop characteristic fauna. If there are, as is suggested by the effect of increasing endophage populations in oak litter, measurable differences in the process rates between assemblages that have adjusted to the profile type and those that have not, then disturbances that alter the fauna, even fairly subtly, can measurably alter process rates. If, as is the case for endophagous species, the fauna are slow to recover from population reductions (Scheu and Schulz, 1996; Crossley *et al.*, 1998), these changes could be persistent.

Conclusion

Characterization of the temporal and spatial dynamics of the three-dimensional litter profile will be necessary both to build an understanding of maintenance of diversity of soil assemblages and to understand the regulation of decomposer food webs and link their structure to system functions. A litterbasket (Blair *et al.*, 1991) that encloses a full profile, delineates between strata and allows the measurement of changes in their nutrient contents over time, should replace the litterbag as the experimental unit of choice for these kinds of studies. Particularly in temperate and boreal forests, in which litter is typically mixed and profiles of multiple cohorts of litter form, understanding the interactions among litter types in profile development is a high priority. Most studies of litter mixing effects have used litterbags as the unit of study and have focused on chemical interactions among substrates. Many have been done in the absence of natural assemblages or even without animals entirely. The study described in this chapter illustrates that diversity of initial litter input can have effects which are manifest only in the fully developed profile in natural field conditions. It suggests that interactions among substrates in the three-dimensional matrix can have strong effects on structural habitat variables. These, in turn, can be strong determinants of faunal populations and their mediation of ecosystem processes.

References

Anderson, J.M. (1971) Observations on the vertical distribution of Oribatei (Acarina) in two woodland soils. In: *Fourth Colloquium of the International Zoological Society, INRA*. INRA Publishing, Paris, pp. 71–77.

Anderson, J.M. (1973) The enigma of soil animal species diversity. In: *Progress in Soil Zoology: Proceedings of the 5th International Colloquium of Soil Zoology*. Academic, Czechoslovak Academy of Sciences, Prague, pp. 51–58.

Anderson, J.M. (1975) Succession, diversity and trophic relationships of some soil animals in decomposing leaf litter. *Journal of Animal Ecology* 44, 475–495.

Anderson, J.M. (1978) Inter- and intra-habitat relationships between woodland cryptostigmata species diversity and the diversity of soil and litter microhabitats. *Oecologia* 32, 341–348.

Bal, L. (1970) Morphological investigation of two modor-humus profiles and the role of the soil fauna in their genesis. *Geoderma* 4, 5–35.

Beckmann, M. (1988) The development of soil mesofauna in a ruderal ecosystem as influenced by reclamation measures I: Oribatei Acari. *Pedobiologia* 31, 391–408.

Berg, M.P., Kniese, J.P., Bedaoux, J.J.M. and Verhoef, H.A. (1998) Dynamics and stratification of functional groups of micro- and mesoarthropods in the organic layer of a Scots pine forest. *Biology and Fertility of Soils* 26, 268–284.

Blair, J.M., Parmelee, R.W. and Beare, M.H. (1990) Decay rates, nitrogen fluxes, and decomposer communities of single- and mixed-species foliar litter. *Ecology* 71, 1976–1985.

Blair, J.M., Crossley, D.A., Jr and Callaham, L.C. (1991) A litterbasket technique for measurement of nutrient dynamics in forest floors. *Agriculture, Ecosystems and Environment* 34, 465–471.

Briones, M.J.I. and Ineson, P. (1996) Decomposition of eucalyptus leaves in litter mixtures. *Soil Biology and Biochemistry* 28, 1381–1388.

Chapman, K., Whittaker, J.B. and Heal, O.W. (1988) Metabolic and faunal activity in litters of tree mixtures compared with pure stands. *Agriculture, Ecosystems and Environment* 24, 33–40.

Crossley, D.A., Jr (1977) Oribatid mites and nutrient cycling. In: Dindal, D.L. (ed.) *Biology of Oribatid Mites*. SUNY-CESF, Syracuse, New York, pp. 71–85.

Crossley, D.A., Jr and Hoglund, M.P. (1962) A litterbag method for the study of microarthropods inhabiting leaf litter. *Ecology* 43, 571–573.

Crossley, D.A., Jr, Hansen, R.A. and Lamoncha, K.L. (1999) Response of forest floor microarthropods to a forest regeneration burn of a southern Appalachian water-shed. In: Oswald, B. (ed.) *Proceedings First Biennial North American Forest Ecology Workshop*, Raleigh, North Carolina, 24–26 June, 1997, 419 pp.

Dix, N.J. (1984) Moisture content and water potential of abscised leaves in relation to decay. *Soil Biology and Biochemistry* 16, 367–370.

Faber, J.H. and Verhoef, H.A. (1991) Functional differences between closely-related soil arthropods with respect to decomposition processes in the presence or absence of pine tree roots. *Soil Biology and Biochemistry* 23, 15–23.

Finzi, A.C. and Canham, C.D. (1998) Non-additive effects of litter mixtures on net N mineralization in a southern New England forest. *Forest Ecology and Management* 105, 129–136.

Fyles, J.W. and Fyles, I.H. (1993) Interaction of douglas-fir with red alder and salal foliage litter during decomposition. *Canadian Journal of Forest Research* 23, 358–361.

Giller, P.S. (1996) The diversity of soil communities, the 'poor man's tropical rainforest'. *Biodiversity and Conservation* 5, 135–168.

Hansen, R.A. (1999) Red oak litter promotes a microarthropod functional group that accelerates its decomposition. *Plant and Soil* 209, 37–45.

Hansen, R.A. (2000) Effects of habitat complexity and composition on a diverse litter microarthropod assemblage. *Ecology* (in press).

Hansen, R.A. and Coleman, D.C. (1998) Litter complexity and composition are determinants of the diversity and species composition of oribatid mites (Acari: Oribatida) in litterbags. *Applied Soil Ecology* 9, 17–23.

Heneghan, L. and Bolger, T. (1996) Effect of components of 'acid rain' on the contribution of soil microarthropods to ecosystem function. *Journal of Applied Ecology* 33, 1329–1344.

Kaneko, N. (1988) Life history of *Oppiella nova* (Oudemans) (Oribatei) in cool temperate forest soils in Japan. *Acarologia* 29, 215–220.

Knutson, R.M. (1997) An 18-year study of litterfall and litter decomposition in a northeast Iowa deciduous forest. *American Midland Naturalist* 138, 77–83.

Lebrun, P. (1971) Contribution à l'étude écologique des oribates de la litière dans une fôret de Moyenne-Belgique. *Mém. Inst. Roy. Sci. Nat. Belg.* 153, 1–96.

Lebrun, P. and Mignolet, R. (1974) Phenologie des population d'Oribates en relation avec la vitesse de décomposition des litières. *Proceedings of the 4th International Congress of Acarology*, Saalfelden, pp. 93–100.

Lodge, D.J. (1997) Factors relating to diversity of decomposer fungi in tropical forests. *Biodiversity and Conservation* 6, 681–688.

Lussenhop, J. (1992) Mechanisms of microarthropod–microbial interactions in soil. *Advances in Ecological Research* 23, 4–33.

Luxton, M. (1975) Studies on the oribatid mites of a Danish beech wood soil: II. Biomass, calorimetry, and respirometry. *Pedobiologia* 15, 161–200.

Luxton, M. (1981a) Studies on the oribatid mites of a Danish beech wood soil: V. Vertical distribution. *Pedobiologia* 21, 365–386.

Luxton, M. (1981b) Studies on the oribatid mites of a Danish beech wood soil: IV. Developmental biology. *Pedobiologia* 21, 312–340.

Marshall, V.G., Reeves, R.M. and Norton, R.A. (1987) Catalogue of the Oribatida (Acari) of Continental United States and Canada. *Memoirs of the Entomological Society of Canada* 139.

Martin, A. and Marinissen, J.C.Y. (1993) Biological and physio-chemical processes in excrements of soil animals. *Geoderma* 56, 331–347.

McArthur, J.V., Aho, J.M., Rader, R.B. and Mills, G.L. (1994) Interspecific leaf interactions during decomposition in aquatic and floodplain ecosystems. *Journal of the North American Benthological Society* 13, 57–67.

McTiernan, K.B., Ineson, P. and Coward, P.A. (1997) Respiration and nutrient release from tree leaf litter mixtures. *Oikos* 78, 527–538.

Mitchell, M.J. (1977) Life history strategies of oribatid mites. In: Dindal, D.L. (ed.) *Biology of Oribatid Mites*. SUNY-CESF, Syracuse, New York, pp. 65–69.

Mitchell, M.J. (1979) Energetics of oribatid mites in an aspen woodland soil. *Pedobiologia* 17, 305–319.

Norton, R.A. (1985) Aspects of the biology and systematics of soil arachnids, particularly saprophagous and mycophagous mites. *Quaestiones Entomologicae* 21, 523–541.

Norton, R. (1994) Evolutionary aspects of oribatid mite life histories and consequences for the origin of the Astigmata. In: Houck, M. (ed.) *Mites: Ecological and Evolutionary Analyses of Life History Patterns*. Chapman & Hall, New York, p. 357.

Norton, R.A., Kethley, J.B., Johnston, D.E. and O'Conner, B.M. (1993) Phylogenic perspectives on genetic systems and reproductive modes of mites. In: Wrench, D.L.

and Ebbert, M.A. (eds) *Evolution and Diversity of Sex Ratio in Insects and Mites*. Chapman & Hall, New York.

Pande, Y.D. and Berthet, P. (1975) Observations on the vertical distribution of soil Oribatei in a woodland soil. *Transactions of the Royal Entomological Society of London* 127, 259–275.

Ponge, J.F. (1991a) Succession of fungi and fauna during decomposition of needles in a small area of Scots pine litter. *Plant and Soil* 138, 99–113.

Ponge, J.F. (1991b) Food resources and diets of soil animals in a small area of Scots pine litter. *Geoderma* 49, 33–62.

Rusek, J. (1975) Die bodenbildende funktion von Collembolen und Acarina. *Pedobiologia* 15, 299–308.

Salamanca, E.F., Kaneko, N. and Katagiri, S. (1998) Effect of litter mixtures on the decomposition of *Quercus serrata* and *Pinus densiflora* using field and lab microcosm methods. *Ecological Engineering* 10, 53–73.

Scheu, S. and Schulz, E. (1996) Secondary succession, soil formation and development of a diverse community of oribatids and saprophagous soil macro-invertebrates. *Biodiversity and Conservation* 5, 235–250.

Schulz, E. and Scheu, S. (1994) Oribatid mite mediated changes in litter decomposition: model experiments with ^{14}C-labelled holocellulose. *Pedobiologia* 38, 344–352.

Siepel, H. (1996) Biodiversity of soil microarthropods: the filtering of species. *Biodiversity and Conservation* 5, 251–260.

Siepel, H. and Maaskamp, F. (1994) Mites of different feeding guilds affect decomposition of organic matter. *Soil Biology and Biochemistry* 26, 1389–1394.

Skubala, P. (1995) Moss mites (Acarina: Oribatida) on industrial dumps of different ages. *Pedobiologia* 39, 170–184.

Staaf, H. (1980) Influence of chemical composition, addition of raspberry leaves, and nitrogen supply on decomposition rate and dynamics of nitrogen and phosphorus in beech leaf litter. *Oikos* 35, 55–62.

Swift, M.J. (1976) Species diversity and the structure of microbial communities in terrestrial habitats. In: Anderson, J.M. and Macfadyen, A. (eds) *The Role of Terrestrial and Aquatic Organisms in Decomposition Processes*. Blackwell Scientific, Oxford.

Taylor, B.R., Parsons, W.M.F. and Parkinson, D. (1989) Decomposition of *Populus tremuloides* leaf litter accelerated by addition of *Alnus crispa* litter. *Canadian Journal of Forest Research* 19, 674–679.

Thomas, J.O.M. (1979) An energy budget for a woodland population of oribatid mites. *Pedobiologia* 19, 346–378.

Usher, M.B., Davis, P.R., Harris, J.R.W. and Longstaff, B.C. (1979) A Profusion of Species? Approaches towards understanding the dynamics of the populations of the microarthropods in decomposer communities. In: Anderson, R.M., Turner, B.D. and Taylor, L.R. (eds) *Population Dynamics. The 20th Symposium of the British Ecological Society, London, April, 1978*. Blackwell Scientific, London, pp. 359–384.

Von Amelsvoort, P.A.M. and Usher, M.B. (1989) Egg production related to food quality in *Folsomia candida* (Collembola: Isotomidae): effects on life history strategies. *Pedobiologia* 33, 61–66.

Wagener, S.M. and Schimel, J.P. (1998) Stratification of soil ecological processes: a study of the birch forest floor in the Alaskan taiga. *Oikos* 81, 63–74.

Wallwork, J.A. (1983) Oribatids in forest ecosystems. *Annual Review of Entomology* 28, 109–130.

Walter, D.E. (1989) Trophic behaviour of mycophagous microarthropods. *Ecology* 68, 226–229.

Wardle, D.A., Bonner, K.I. and Nicholson, K.S. (1997) Biodiversity and plant litter: experimental evidence which does not support the view that enhanced species richness improves ecosystem function. *Oikos* 79, 247–258.

Wauthy, G. and Vannier, G. (1988) Application of an analytical model measuring tolerance of an oribatid community (Acari) to continuously increasing drought. *Experimental and Applied Acarology* 5, 137–150.

The Pervasive Ecological Effects of Invasive Species: Exotic and Native Fire Ants

11

C.R. Carroll and C.A. Hoffman

Institute of Ecology, University of Georgia, Athens, GA 30602–2202, USA

Introduction

A common consequence of the establishment of invasive species is the displacement of local species, especially, but not always, of those that use similar resources. The mechanisms invoked to explain the displacement of native species include high population densities of invasive species, and absence of predators, effective competitors, or diseases that might reduce the abundance and, hence, the impact of the invasive species. Commonly, these interactions play out in disturbed environments, with the invasive species having the dispersal and growth characteristics of 'weedy' species that would be expected to be good colonizers of disturbed environments (Meffe and Carroll, 1997). However, as we discuss below, *Solenopsis invicta* (red imported fire ant) is able to colonize successfully undisturbed native grassland in Texas and longleaf pine/wiregrass savanna in the Georgia coastal plain.

Not every non-native species becomes invasive and displaces local species. Undoubtedly, the vast majority of colonization attempts by non-native species fail. Of the two South American fire ant species that have successfully invaded the southern United States, *S. invicta* has become widespread and has displaced native species. The other non-native fire ant, *Solenopsis richteri* (the black imported fire ant), remains localized in northern Alabama.

In this chapter we review the ecological effects of native and exotic fire ants in the southern United States and Mexico with respect to their effects on biota and the soil. We review the significance of potential control agents, especially the largely subterranean foraging (hypogeic) complex, *Solenopsis* (*Diplorhoptrum*) spp., and we draw attention to important ecological differences between monogynous and polygynous forms of *S. invicta*.

The Fire Ant Complex

Among the many species in the genus *Solenopsis*, there are 20 morphologically and behaviourally similar species known collectively as 'fire ants' (Trager, 1991). They are native to the New World, although some species, notably *Solenopsis invicta* and *Solenopsis geminata* (the tropical fire ant) have established populations well outside their native range. In the southern United States and Mexico there are three common native fire ants. *Solenopsis xyloni* (the southern fire ant) is found throughout the southern tier of states, mostly inland from the coastal plain, and also in moister regions of southern California. *Solenopsis quinquecuspis* (the desert fire ant) is found in deserts of the south-west. The third native fire ant in the southern states is *Solenopsis geminata*, with a distribution that ranges from the coastal plain of the southern states down into Mexico and Central America (Vinson, 1997).

The two exotic (non-native) fire ants in the United States are *Solenopsis richteri* and *S. invicta*. Both species arrived more than 70 years ago, probably as small colonies in soil used as ballast in ships arriving in Mobile, Alabama, from South America (Vinson, 1997). *Solenopsis richteri* is restricted to north-eastern Alabama. *S. invicta* is widespread throughout the south but has not extended its range significantly into Mexico or Central America. In recent years, *S. invicta* has become established as local populations as far west as southern California (California Department of Food and Agriculture, 1999).

Monogyny and Polygyny

Monogyny, one egg-laying queen per colony, is the most common reproductive pattern in ants (Hölldobler and Wilson, 1990). Fire ants are unusual in that the polygynous condition appears to be widespread, occurring in *S. invicta*, *S. richteri*, *S. xyloni*, *S. geminata*, and probably in other fire ant species as well (MacKay *et al.*, 1990). Monogynous and polygynous fire ants are genetically very similar (Ross *et al.*, 1996), but there are important behavioural differences between these two forms. Workers in monogynous colonies attack and kill workers and queens from any other colony. Even when monogynous colony queens initiate colonies as groups of young queens (claustral colony foundation), they eventually fight until all but one queen is killed. Workers in polygynous colonies accept workers from other polygynous colonies but not from monogynous colonies. Polygynous colony workers discriminate between polygynous queens and monogynous queens and kill the latter. (See Tschinkel (1998) for an extensive review of monogyny and polygyny in *S. invicta*.)

There are important ecological consequences of polygynous colony structure. Mound densities of polygynous colonies can be much higher than densities of monogynous colonies and this has obvious consequences for soil structure and turnover from mound construction and for populations of terrestrial invertebrates, especially native ants (Porter and Savagnano, 1990).

An important message is that the small genetic differences between monogyne and polygyne forms are associated with large ecological effects.

Effects of *S. invicta* on Soil Structure and Fertility

In the heavy clay soils of central Texas, the eroded piedmont of Georgia and the sandy silt soils of the southern coastal plain, fire ant mounds are recognizable at least 3 years after they have been abandoned (Wolf, 1993, personal observation). In marked mounds, the macropore structure of the ant tunnels is still discernable when a few centimetres of superficial soil, usually less than 3 cm, are removed. The average occupancy time of a monogynous *S. invicta* mound is 2 years in the piedmont of Georgia before the ants move to a new mound site. However, the variance around this occupancy time is large, with some mounds abandoned within a few months after initiation and a few being continuously occupied for at least 4 years (personal observation). Because densities of active mounds in monogynous populations may reach 100 ha^{-1} or more and mounds in polygynous populations several hundred ha^{-1}, the many active and abandoned mounds can significantly affect soil structure and fertility. The large macropores in colonies obviously increase infiltration rates and, depending on the nutrient content of the subsoil, the construction of mounds can influence soil fertility.

In the heavy clay soils of central Texas, spring season grasses and forbs growing on abandoned *S. invicta* mounds have much larger infructescences than conspecifics growing adjacent to the mounds (Table 11.1), but these differences disappear by summer. In this case, the increase in spring plant productivity is probably a consequence of the mounds providing better drainage and warmer early spring temperatures. In the coastal plain of southern Georgia, upland soils have low native fertility and the subsoils have very low levels of nitrogen and carbon (Table 11.2). In this location, mounds have less nitrogen and carbon than adjacent soil (Table 11.3) and the effect of mound construction is to lower soil fertility by transporting infertile subsoil to the surface. In the lower coastal plain longleaf pine/wiregrass savanna, Wolfe (1993) found little difference between the composition of plant species colonizing abandoned *S. invicta* mounds and plants occurring on nearby control soil. However, plant abundance on abandoned mounds is strikingly lower than on control soil and three large mounds that are known to have been abandoned for at least 6 years (1992–1998) have collapsed, forming shallow depressions with less than 2% plant cover, contrasting with 20–30% plant cover adjacent to these abandoned mound sites (personal observation).

We can get a sense of the magnitude of the effects of mound construction on soil structure and fertility in the Georgia coastal plain with the following example. A 'typical' monogyne *S. invicta* colony is active at one mound site for 2 years, although there is considerable variation. The macropore structure is intact for at least 3 years after abandonment. Given an average mound surface

Table 11.1. Mean length of infructescences (mm) on abandoned *Solenopsis invicta* mounds and adjacent control sites in central Texas. One plant was selected from the central-most location from each of six or eight mounds. Control plants were 2 m from the mound in a random direction.

	Mound	Control
Lolium perenne (ryegrass)	19.18 ± 3.20	7.42 ± 0.30 ***
	(*n* = 6)	(*n* = 6)
Hordeum jubatum	8.3 ± 0.88	2.55 ± 0.21***
(foxtail barley)	(*n* = 6)	(*n* = 6)
Plantago lanceolata	3.81 ± 0.41	2.19 ± 0.23 ***
(English plantain)	(*n* = 8)	(*n* = 8)

*** *P* < 0.001.

Table 11.2. Carbon and nitrogen content of upland soil at several depths at the Joseph Jones Ecological Research Center, Newton, Baker County, Georgia, southern coastal plain.

Depth (cm)	C (%)	N (%)
0–10	0.88	0.04
10–20	0.44	0.03
20–30	0.35	0.02

Table 11.3. Mean percentages of nitrogen and carbon found in the top 5 cm soil of active *Solenopsis invicta* mounds, abandoned *S. invicta* mounds and adjacent control sites without mounds.

	Nitrogen	Carbon
Active mound	0.043 ± 0.006 a	1.122 ± 0.174 a
(*n* = 5)		
Abandoned mound	0.042 ± 0.010 a	0.981 ± 0.118 a
(*n* = 5)		
Control	0.078 ± 0.011 b *	2.563 ± 0.441 b **
(*n* = 5)		

Column values marked with different letters are significantly different at *P* < 0.05 or ** *P* < 0.01.

area of 0.5 m², 100 active mounds ha^{-1} result in 1.5% of the area affected by the pore structure of mounds. Over a 20-year period, approximately 16% of the superficial soil colonized at this density will be turned over by *S. invicta* workers and the level of nitrogen and carbon in this 16% reduced by half. This example is only meant to be illustrative of a reasonable scenario. The actual effects of *S. invicta* mounds on soil fertility and structure will depend on many site-specific factors. Some of these site-specific factors include the following: nutrient content of subsoils versus superficial soils, the extent to which mounds once abandoned are recolonized, the average and variance of mound occupancy time, number of years the area has been colonized, and the spatial distribution of mounds.

Biodiversity Impact of Native and Non-native Fire Ants

A high density of fire ants, their ability to recruit large numbers of workers rapidly, and their aggressive behaviour combine to reduce the diversity and abundance of other invertebrates (Vinson, 1994), especially ants (Porter and Savignano, 1990). Seed harvesting fire ants, mainly *S. geminata*, can affect the composition of forbs and grasses by selective removal of seeds from the soil (Risch and Carroll, 1986). Other fire ants, notably *S. invicta*, may influence plant species composition indirectly by displacing seed dispersing ant species and by modifying soil structure and fertility as we discuss below.

Impact on invertebrate populations

In south-eastern Mexico and Central America, *S. geminata* nests and forages in coffee plantations, pastures, agricultural fields and other disturbed areas. Workers do not forage in closed forest or in heavily shaded second growth. Before the forested zone of south-eastern Mexico was transformed by large-scale forest clearing, *S. geminata* was probably restricted to small natural clearings and clearings created by traditional *milpa* (shifting) agriculture. In contemporary *milpa* agriculture, *S. geminata* significantly reduces insect abundance and diversity on maize (*Zea mays*) and squash (*Cucurbita moschata*) plants (Carroll, 1990). In shaded coffee plantations, *S. geminata* foraging behaviour is restricted by the ant *Pheidole radoszkowskii* (Perfecto, 1994), probably reflecting the marginality of this shaded environment for *S. geminata*. The foraging activity of *S. geminata* is also reduced in the presence of parasitic phorid flies (Feener and Brown, 1992). Of the other fire ants native to the United States, Mexico and Central America, we have not found any reports on their effects on invertebrate populations in agricultural or natural systems. An early observation (Barber, 1933) that *S. xyloni* was a major predator of corn stalk borers suggests that this native fire ant species may have played an important predatory role in southern agroecosystems.

Of the two non-native fire ants, most investigations have emphasized *S. invicta*. As mentioned above, *S. invicta* greatly reduces the biodiversity of the native ant fauna and probably of terrestrial invertebrates in general. Some ant species, however, appear to coexist with *S. invicta*. Stein and Thorvilson (1989) found 14 ant species in sites infested with *S. invicta*, but these 14 species represented less than 2% of the total ants collected, the remaining 98% being *S. invicta*. In longleaf pine/wiregrass savanna (*Pinus palustris, Aristida stricta*), we found *S. invicta* to be the most common ant in unbaited pitfall traps (55% of all ants). In addition to *S. invicta*, 23 other species of ants were collected in these traps (Table 11.4). In order to investigate spatial patterns of foraging, one pitfall trap (following the design of Majer, 1978) was placed in the centre of 30 cm × 70 cm artificial disturbances, created by superficial rototilling, and in non-disturbed control sites within 5 m of the disturbance patches. In total, 32 disturbed and 32 non-disturbed sites were sampled over a 24-h period on eight dates in 1991 and 1992. The species abundance curves for undisturbed and disturbed sites are similar (Figs 11.1 and 11.2), which suggests that at the community level, foraging ants are not discriminating between these two patch types. This is not to say that discrimination would not occur at some larger soil disturbance scale or type of disturbance. In contrast to the general literature which describes *S. invicta* (and other invasive non-native species) as invaders of disturbed environments, *S. invicta* is the most common species on disturbed and undisturbed sites. The abundance of *S. invicta* in this intact savanna illustrates that the ecological impact of this invasive species is not restricted to agricultural or other disturbed land uses.

Table 11.4. Ant species collected in unbaited pitfall traps at the Joseph Jones Center, Newton, Baker County, Georgia. On each sampling date (7/91, 9/91, 11/91, 3/92, 5/92, 7/92, 10/92, 12/92) 64 pitfall traps were left open for 24-h periods.

Brachymyrmex depilis
Camponotus abdominalis
Camponotus sp.
Conomyrma bossuta
Conomyrma bureni
Conomyrma flavopectus
Conomyrma pyramicus
Crematogaster lineolata
Cyphomyrmex rimosus
Forelius pruinosus
Formica sp.
Iridomyrmex pruinosum
Paratrechina concinna
Paratrechina parvula
Paratrechina vividula
Pheidole sp.
Pogonomyrmex badius
Proceratium sp.
Smithistruma sp.
Solenopsis diplorhoptrum (spp.)
Solenopsis invicta
Strumigenys louisianae
Tapinoma sp.
Trachymyrmex septentrionalis

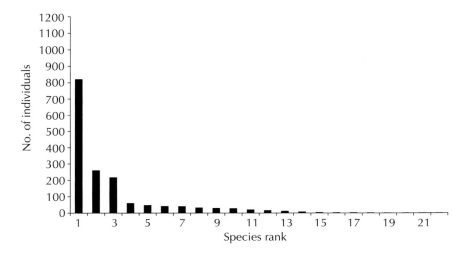

Fig. 11.1. Ant species abundance curve for disturbance plots at Joseph Jones Center for Ecological Research, Newton, Georgia, southern coastal plain.

In central Texas, at two sites on the Blacklands Research Station (Soil Conservation Service) near Riesel, Falls County, Texas, *S. invicta* was rare or absent in 1984, though fire ants were common in the vicinity of these sites. One

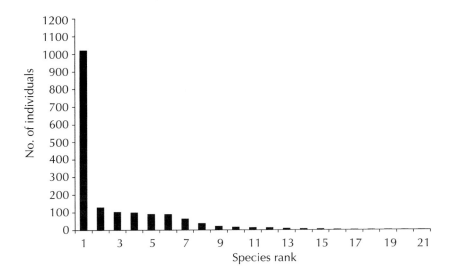

Fig. 11.2. Ant species abundance curve for undisturbed plots at Joseph Jones Center for Ecological Research, Newton, Georgia, southern coastal plain.

site was a typical heavily grazed Bermuda grass (*Cynodon dactylon*) pasture, common in the area. The other was a native grass 'prairie' that had not been grazed, tilled, fertilized or treated with pesticides for at least 50 years. Ant abundance was investigated by placing two filter paper discs (one saturated with sugar water, the other with tuna fish oil) in the centre of 1-m quadrants on 100 m × 100 m grids. Sampling took place between 9 a.m. and noon on 5 days during October and November in 1984 and again in 1987. As shown in Figs 11.3 and 11.4, both sites had become heavily infested by 1987. The fact that some other ant species increased during the post-invasion period in one site but not the other (e.g. *Pheidole dentata* increased in the Bermuda grass site but decreased in native prairie site) suggests that invasion by *S. invicta* may not have been the only influence on the native ant community. The two main points we wish to emphasize, however, are that, as with the natural longleaf pine/wiregrass savanna case, *S. invicta* can invade natural communities and that this invasion can take place quickly.

In the longleaf pine/wiregrass savanna, several ant species apparently coexist with *S. invicta*. However, we have no information about their former relative or absolute abundances prior to invasion by *S. invicta*. *Conomyrma bureni* is common, and the genus is reported to coexist with *S. invicta* (Bhatkar, 1988).

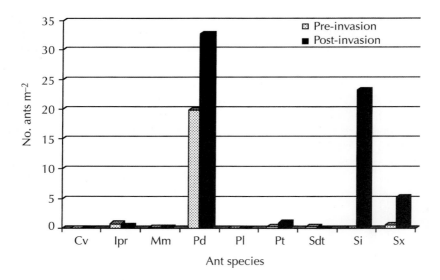

Fig. 11.3. Ant community changes in Bermuda grass pastures. Number of ants m^{-2} is the average number of ants collected over the 5-day sampling period. Pre-invasion sampling period was October and November, 1984. Post-invasion was October and November, 1987. Key to ant species: Cv, *Conomyrma*; Ipr, *Iridomyrmex pruinosus*; Mm, *Monomorium minimum*; Pd, *Pheidole dentatum*; Pl, *Pheidole lamia*; Pt, *Paratrechina terricola*; Sdt, *Solenopsis diplorhoptrum texana*; Si, *Solenopsis invicta*; Sx, *Solenopsis xyloni*.

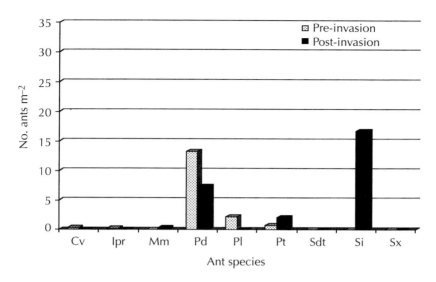

Fig. 11.4. Ant community changes in native prairie pastures. Sampling regimes as in Fig. 11.3. Key to ant species: Cv, *Conomyrma*; Ipr, *Iridomyrmex pruinosus*; Mm, *Monomorium minimum*; Pd, *Pheidole dentatum*; Pl, *Pheidole lamia*; Pt, *Paratrechina terricola*; Sdt, *Solenopsis diplorhoptrum texana*; Si, *Solenopsis invicta*; Sx, *Solenopsis xyloni*.

One species, *Brachymyrmex depilis*, commonly nests in association with large colonies of *Pogonomyrmex badius* and may gain some protection against fire ants from this harvester ant. For example, of 12 *Pogonomyrmex* nests examined, all contained at least one *Brachymyrmex* colony on the bare disc surrounding the colony entrance of the harvester ants. Because ant queens commonly live for several years (see Hölldobler and Wilson (1990) for examples), the effects of fire ants on native ant abundance through predation on young queens may be expressed slowly over years. The probability of young queens being killed by fire ants is key information needed in order to understand the effects of fire ants on native ants.

Loss of an Important Predator of Fire Ants

Many animals eat fire ants. Armadillos, insectivorous birds, lizards, spiders and other predators eat fire ants, especially brood and winged reproductives. However, their population level effect on fire ant demography does not appear to be significant. Recently, phorid flies, parasitic mites and microsporidians have been employed as possible control agents against fire ants with some success (see Vinson (1997) for examples and Jouvenaz (1990) for a review). One group of native predators, which are poorly known but may have considerable effect on fire ant demography, is the subgenus *Solenopsis* (*Diplorhoptrum*). There are several species of *Diplorhoptrum* in the southern United States and they are characterized by the very small size of the workers, subterranean foraging and small colony size. Although little ecological information is available, they are thought to be predators on young ant queens during the process of colony foundation. Thus, they impose mortality during a vulnerable stage of colony development.

In central Texas, *S. diplorhoptrum* (probably *S. diplorhoptrum texana*) is an important predator on young ant queens of *S. invicta*, *Pogonomyrmex badius* and probably of other ant species. Lammers (1987) found that young, post-reproductive flight queens of *S. invicta* were rapidly found and destroyed by *S. diplorhoptrum* when placed in artificial burrows. *S. diplorhoptrum* would be expected to have a much stronger direct effect on the demography of mono-gynous than on polygynous forms of *S. invicta*, because the majority of mono-gynous colonies are initiated by single queens. Because polygynous colonies persist by accepting new queens into established colonies, these colonies are rarely in the vulnerable single founding queen stage and may effectively escape predation by *S. diplorhoptrum*.

S. diplorhoptrum is highly sensitive to at least some insecticides that are commonly applied to control fire ants, and efforts to control *S. invicta* through pesticide application may have contributed to its spread. In central Texas (Tradinghouse Creek Farm of Lammers, 1987), two pesticides, Prodrone® (a synthetic insect development hormone) and Amdro® (a widely used insecticide on fire ants) were applied in a field that had low populations of fire ants (see Lammers (1987) for details of the field trial). Prior to the application, *S. diplorhoptrum* was found in 38 of 60 sampling stations. Two weeks after the Amdro® applications, fire ants and many native ant species largely disappeared. Prodrone® was less effective, eliminating ants at half of the previously occupied locations (6/20 stations had ants). As late as 2 years following the application of the pesticides, fire ants were common in the field but *S. diplorhoptrum* was completely absent, despite intensive attempts to locate foragers or colonies. This observation suggests that the use of pesticides to control fire ants may inadver-tently eliminate *S. diplorhoptrum* predators of fire ants. Furthermore, it is likely that pesticides are more effective on monogyne than polygyne colonies of fire ants for two reasons. First, polygyne colonies commonly accept new colonizing queens. Second, if pesticides kill even a large fraction of the queens in a polygyne colony, the remaining queens may continue the colony. Thus, an interesting consequence of the chemical control programme for *S. invicta* may be to simultaneously remove *S. diplorhoptrum* predators while favouring the spread of polygyne over monogyne colonies. This issue needs further investigation.

Summary

The fire ant complex in the United States contains both introduced and native species. Although ants in this complex are primarily associated with disturbed habitat types, the red imported fire ant (*S. invicta*) has successfully invaded two stable habitats, native prairie in Texas and longleaf pine/wiregrass savanna in Georgia. Studies have demonstrated some effects on soil characteristics and on plant and invertebrate communities, although integrated, detailed studies on all these components in a single location still remain to be done.

References

Barber, G.W. (1933) On the probable reason for the scarcity of the southern corn stalk borer (*Diatraea crambidoides* Grote) in southeastern Georgia. *Journal of Economic Entomology* 26, 1174.

Bhatkar, A.P. (1988) Confrontation behavior between *Solenopsis invicta* and *S. geminata*, and competitiveness of certain Florida ant species against *S. invicta*. In: Trager, J.C. (ed.) *Advances in Myrmecology*. E.J. Brill, New York.

California Department of Food and Agriculture (1999) *California Red Imported Fire Ant Action Plan*, March 19, 1999. California Department of Food and Agriculture, 1220 N Street, Suite 409, Sacramento, CA 95814.

Carroll, C.R. (1990) The ecological role of ants in annual tropical agroecosystems. In: Gliessman, S.R. (ed.) *Approaches in the Ecological Study of Agriculture*. Ecological Studies Series, Springer-Verlag, Dordrecht.

Feener, D.H. and Brown, B.V. (1992) Reduced foraging of *Solenopsis geminata* (Hymenoptera: Formicidae) in the presence of parasitic *Pseudacteon* sp. (Phoridae: Diptera). *Annals of the Entomological Society of America* 85, 80–84.

Hölldobler, B. and Wilson, E.O. (1990) *The Ants*, 1st edn. Belknap/Harvard, Cambridge.

Jouvenaz, D.P. (1990) Approaches to biological control of fire ants in the United States. In: Vander Meer, R.K., Jaffe, K. and Cedeno, A. (eds) *Applied Myrmecology*. Westview Press, Boulder, Colorado, pp. 621–627.

Lammers, J.M. (1987) Mortality factors associated with the founding queens of *Solenopsis invicta* Buren, the red imported fire ant: a study of the native ant community in central Texas. MS thesis, Texas A & M University, College Station, Texas.

MacKay, W.P., Porter, S.D., Gonzales, D., Rodriguez, A., Amendedo, H., Rebeles, A. and Vinson, S.B. (1990) A comparison of monogyne and polygyne populations of the tropical fire ant, *Solenopsis geminata* (Hymenoptera: Formicidae), in Mexico. *Journal of the Kansas Entomological Society* 63, 611–615.

Majer, J.D. (1978) An improved pitfall trap for sampling ants and other epigaeic invertebrates. *Journal of the Australian Entomological Society* 17, 261–262.

Meffe, G.K. and Carroll, C.R. (eds) (1997) *Principles of Conservation Biology*, 2nd edn. Sinauer, Sunderland, Massachusetts.

Perfecto, I. (1994) Foraging behavior as a determinant of asymmetric competitive interaction between two ant species in a tropical agroecosystem. *Oecologia* 98, 184–192.

Porter, S.D. and Savagnano, D.A. (1990) Invasion of polygyne fire ants decimate native ants and disrupts arthropod communities. *Ecology* 71, 2095–2106.

Risch, S.J. and Carroll, C.R. (1986) Effects of seed predation by a tropical ant on competition among weeds. *Ecology* 67, 1319–1327.

Ross, K.G., Vargo, E.L. and Keller, L. (1996) Simple genetic basis for important social traits in the fire ant *Solenopsis invicta*. *Evolution* 50, 2387–2399.

Stein, M.B. and Thorvilson, H.G. (1989) Ant species sympatric with the red imported fire ant in southeastern Texas. *Southwestern Entomologist* 14, 225–231.

Trager, J.C. (1991) A revision of the fire ants, *Solenopsis geminata*, group (Hymenoptera: Formicidae). *Journal of the New York Entomological Society* 99, 141–198.

Tschinkel, W.R. (1998) The reproductive biology of fire ant societies. *BioScience* 48, 593–605.

Vinson, S.B. (1994) Impact of the invasion of *Solenopsis invicta* on native food webs. In: Williams, D.F. (ed.) *Exotic Ants: Biology, Impact, and Control of Introduced Species.* Westview Press, Boulder, Colorado, pp. 240–258.

Vinson, S.B. (1997) Invasion of the red imported fire ant. *American Entomologist* 43, 23–39.

Wolfe, D.W. (1993) The effects of soil disturbance on wiregrass (*Aristida stricta* Michx.) and the associated herbaceous community. MS thesis, University of Georgia, Athens, Georgia.

12

Soil Invertebrate Species Diversity in Natural and Disturbed Environments

J. Rusek

Institute of Soil Biology, Academy of Sciences of the Czech Republic, Nasádkách 7, 370 05 České Budějovice, Czech Republic

Introduction

Long-term changes of Central European soil invertebrate communities in alpine ecosystems and arable soils connected with global change, pollution and land use are described and analysed in this chapter.

Soil and soil biota are integral parts of terrestrial ecosystems. They communicate, to a greater or lesser extent, with the surface system. The above-ground portion of the system contributes many atmospheric pollutants, dead organic matter, plus a great portion of inactive soil organisms, e.g. spores, cysts, anhydrobiotic stages and eggs, to the subterranean system portion. Many aspects of the relationships between above and below soil surface parts of the ecosystems are not yet well known. Feedback mechanisms relating to above-ground and subsurface biota are an important research priority (Freckman *et al.*, 1997). In the past few decades ecosystems have been damaged on a global scale by various human activities such as mining, forest cutting, agricultural practices and land transformation, all of which can directly affect ecosystems and their components, including soil biota. The most important factors affecting soil biota are chemical compounds, which enter ecosystems in different ways. Airborne pollutants affect vegetation both on a continental and on a global scale, and they enter the soil subsystem where they influence soil biota and decomposition processes (Kopeszki, 1992).

Acid rain and other airborne pollutants are complex, containing a multitude of substances besides H^+, NO_x, SO_3^- and SO_4^{2-} (Miller, 1984), for example large amounts of heavy metals, organic compounds and fly ash. Aquatic soil animals come into closer contact with airborne pollutants dissolved in soil water and react to them directly and earlier than the terrestrial soil fauna.

Earthworms, enchytraeids and other invertebrates with permeable skin may suffer more from these pollutants than other terrestrial soil fauna. They may take up pollutants, e.g. copper, through their epidermis as well as in their food (Streit, 1984). Enchytraeids and earthworms come into direct contact with soil water occasionally and this may explain some contradictory field observations where fluctuating densities of these groups could be a reaction to acidification. But there are also data in the literature documenting a direct influence of pollutants on terrestrial representatives of soil fauna. Jaeger and Eisenbeis (1984) established a direct effect of acidic solution on Collembola, decreasing the water uptake from the experimental environment. Fecundity and longevity of many species of Collembola are also negatively affected by low pH (Hutson, 1978).

Springtails are not the only group of soil invertebrates actively taking free water from soil through water transporting tissues (ventral tube) to saturate water loss. Such morpho-physiological systems are known to occur in terrestrial isopods (water conducting system) (Hoesse, 1981), Diplopoda (rectal water uptake) (Meyer and Eisenbeis, 1987), Thysanura, Diplura, Protura, Pauropoda, Symphyla and others (Eisenbeis and Wichard, 1985). Pollutants could also indirectly affect soil invertebrates through damage to food resources such as the microflora for microphytophagous arthropods. The disappearance of Protura from spruce forest soils in North Bohemia in the 1960s was connected with forest and mycorrhizal fungi die-off (Stumpp, 1991).

Methods

On each sampling date, ten random soil samples (10 cm^2 surface area and 5 cm deep, and in arable soils up to 30 cm depth) were taken from permanent plots for soil mesofauna studies and extracted in Tullgren funnels (Rusek *et al.*, 1975). Following extraction, specimens were kept in 96% ethanol before being sorted into taxonomic groups. Collembola and Protura were determined to species level and counted. The coenotical relations among Collembola communities from different sites were calculated by different types of agglomerate classification using three types of data analysis: (i) presence and absence of species in the samples for measuring the faunistic similarity of communities (Sørensen's index); (ii) log ($x + 1$) transformed number of individuals of each species in samples for measuring the coenotic dissimilarity of communities (Orlóci, 1978); and (iii) number of individuals (not transformed) of each species in samples for measuring functional similarity of communities. The average linkage method was used for cluster analysis. The TWINSPAN program (Hill, 1979) was used for hierarchical division classification of samples and for establishing indicator and preference species. Detrended correspondence analysis of log ($x + 1$) transformed data, program CANOCO (Ter Braak, 1987), was used for sample ordination analysis.

Site Description

Tomanova Dolina Valley in the Tatra National Park

The long-term studies on acid deposition impact of, and global change effects on, soil fauna and other components and parameters of alpine, subalpine and montane ecosystems were made in the Tomanova Dolina Valley, western part of Tatra National Park (49°13′N, 19°55′E), Slovak Republic. The study area comprises the Tomanova Dolina Valley with surrounding mountain ridges and the adjacent smaller valley Hvizdalka and Žlab spod Diery. The main valley is about 1.5 km long, with a total area of about 5 km². The lowest point corresponds to 1225 m above sea level (asl) while the highest one (the Kresanica summit) is 2122 m asl. The valley is drained by the Tomanovy Potok Stream and by several subterranean karst creeks. A small glacial lake and smaller bogs are situated in the upper valley basin (1595 m asl).

This area was chosen for the studies because of its contrasting geological composition and strong ecological differentiation into altitudinal vegetation belts and many well-developed plant communities. The geology of the area is characterized by granite and pegmatites south of the Tomanovy Potok Stream and by limestone with smaller outcrops of werfenic shales in the vicinity of the stream forming the northern part. The underlying rocks determine the soil types. Ranker soils, brown soils, podzols and boggy soils are formed on the acidic bedrocks, and different types of montainous rendzinas on the limestone. Kubiena's system of alpine soil classification is used here (Kubiena, 1950).

The climate of the entire area is the cold alpine type, with an average annual temperature of 0°C and an average July temperature of 10°C, and with yearly precipitation of 1750 mm. Typically, strong winds blow over the mountain ridges.

The flora of the Tomanova Dolina Valley is very rich, and more than 500 species of vascular plants have been recorded there. Phytocoenological studies resulted in the identification and mapping of 45 communities (Unar *et al.*, 1984, 1985). The upper spruce forest boundary is about 1400 m asl, close to the valley mouth. The subalpine belt consists of dwarf pine (*Pinus mugo* Turra ssp. *mugo*) up to an elevation of 1700 m. Only islands of dwarf pine are present at higher elevations in the alpine zone, which is extensively covered with alpine grassland plant communities.

Data from the following 11 permanent plots in the Tatra National Park were used in this study (the site numbers are used from the unified permanent plots numbering):

Granite bedrock sites
Site no.:

1. *Eriophoretum angustifolii.* East of Tomanove Sedlo Pass, 1591 m asl, swampy remnants of a former alpine lake.

5. *Vaccinietum myrtilli subalpinum.* Northern slope of Polska Tomanova, 1660 m asl.
6. *Festucetum pictae.* Mild eastern slope below the Tomanove Sedlo Pass, 1610 m asl.
8. *Juncetum trifidi.* Ridge east of Polska Tomanova Summit, 1950 m asl.
9. *Sphagno-Empetretum hermaphroditi.* Northern slope of Polska Tomanova, 1780 m asl, 60 cm thick mat of *Sphagnum* moss.
10. *Calamagrostietum villosae tatricum.* Water runoff gully on the northern slope of Polska Tomanova, 1850 m asl.

Limestone bedrock sites
Site no.:

12. *Festucetum versicoloris.* Eastern slope of Hvizdalka Valley, 1750 m asl.
13. *Festucetum carpaticae.* North-eastern slope of Hvizdalka Valley, 1700 m asl.
14. *Caricetum firmae carpaticum.* Northern slope of Stoly in Hvizdalka Valley, on rocky terraces, 1750 m asl.
15. *Geranio-Alchemilletum crinitae.* South-eastern slope of Hvizdalka Valley, in a shallow gully, 1750 m asl.
16. *Saxifragetum perdurantis.* Open climax plant community in deep karst sinkhole at the bottom of Hvizdalka Valley, 1650 m asl.

Arable plots in the Czech and Slovak Republics

Site no.:

1. South Moravia, Nejdek and Bulhary near Lednice na Moravě, maize fields on chernozem soil, 1 September, 1964; 23 September, 1967; 7 May, 1990.
2. South Slovakia, Palárikovo, maize field on chernozem soil, 2 September, 1964.
3. Bohemian Karst, Mořina, wheat field on brown rendzina, 7 June, 1977.
4. Central Bohemia, Hrnčíře, barley field on cambisol, 29 April and 11 May, 1974.
5. South Moravia, Troubsko near Brno, lucerne field (first year of cultivation) on chernozem soil, 30 May and 2 July, 1974.
6. Northwest Bohemia, Lahošť near Teplice, barley field on cambisol, 15 June, 1992.
7. South-west Bohemia, Nezamyslice near Sušice, barley field on cambisol, 15 June, 1992.
8. Central Slovakia, Pohranice near Nitra, wheat field (after harvest) on cambisol, 4 September, 1985.
9. East Bohemia, Pelhřimov, potato field on cambisol, 22 May, 1986.

Lindane-type insecticides (3.4 kg of effective substance ha^{-1}) were used in 1962–1966 on sites 1 and 2, herbicides were used in the 1960s to 1980s every year on all sites (dinosebacetate on sites 1 and 5, simazine and other herbicide types on remaining sites), NPK fertilizers were applied in doses of 125 kg ha^{-1} in the 1960s and 250 or more kg ha^{-1} in the 1970–80s every year on all sites.

Global Changes and their Impacts on Soil Invertebrates

Soil fauna and their relation to vegetation and bedrock

In 1968 studies were made on a complex of nine alpine Collembola and Protura communities representing samples from acid granite substrate soils (sites 1–10) plus samples from limestone (sites 12–16) in Tatra National Park. The 1968 samples were selected for the community analysis of undisturbed sites occurring here up to 1977. The soil samples from all alpine plant communities under study contained 46 species of Collembola and three species of Protura (Table 12.1). The number of species in the ecosystems on granite was 5–13 (at densities of 1700–81,700 individuals m^{-2}), and on limestone 8–17 (at densities of 4400–18,800 individuals m^{-2}). The Shannon–Wiener index of diversity H′ reached 1.392–1.697 on granite and 1.471–1.992 on limestone. The equitability E was 0.562–0.902 on granite and 0.681–0.766 on limestone.

The faunistic similarity of collembolan communities is given in Fig. 12.1. Sites 1, 16, 5 and 14 are quite different from the others. Site 1 is from an initial stage of a hydric succession sere, 16 is a climax relic arcto-alpine community on limestone, 5 is a climax on granite and 14 a climax on limestone with a solifluction rendzina soil. The highest similarity was between sites 15 (limestone) and 8 (granite). The coenotic similarity of collembolan communities is given in Fig. 12.2. All samples had distance values higher than 0.5 indicating that the communities of Collembola were different for each plant community and ecosystem for both dominant and rare species. The functional similarity of Collembola communities showed (Fig. 12.3) that only two pairs of samples from the limestone sites were similar, i.e. sites 16 and 13, plus 14 and 15. The remaining samples from the granite bedrock sites could not be considered as functionally similar because of a lower amount of shared dominant species. The high dissimilarity of Collembola communities among all sites documented was also evident in the ordination analysis (Fig. 12.4). Sample 1 lies to the far right of axis 2, and sample 16 far left of axis 2. The samples from the sites on granite lie below axis 1 and to the left of axis 2 except sample 5 which is slightly above axis 1. All samples from limestone lie above axis 1.

The results of the division classification of samples are shown in Fig. 12.5. The first level of classification separated sample 16 (+1), with one indicator species (*Ceratophysella succinea*), from the remaining eight samples (negative sub-cluster −0). The first negative sub-cluster did not contain indicator species, but it had 15 species with high preference. Sample 16 was preferred by *Paratullbergia callipygos*, *Tullbergia simplex*, *Anurida* sp.3 and *Tomocerus* sp. The cluster −0 is divided at the second level of classification into sample 1 (−00)

Table 12.1. Communities of Collembola and Protura in alpine ecosystems in the Tatra National Park. A, abundance individuals 100 cm^{-2}; D, dominance (location numbers see text).

Species	1 A	1 D	5 A	5 D	6 A	6 D	8 A	8 D	10 A	10 D	13 A	13 D	14 A	14 D	15 A	15 D	16 A	16 D
Sminthurides aquaticus	6	35,3	—	—	—	—	—	—	—	—	—	—	—	—	—	—	—	—
Sphaeridia pumilis	5	29,4	—	—	—	—	—	—	—	—	—	—	—	—	—	—	—	—
Protaphorura armata	3	17,6	—	—	10	2,5	11	3,3	108	13,3	10	10,2	20	10,6	23	13,9	—	—
Isotoma violacea	2	11,8	1	1,9	32	8,1	37	11,1	—	—	—	—	54	28,7	42	25,3	—	—
Tetrodontophora bielan	1	5,9	—	—	—	—	—	—	—	—	—	—	—	—	—	—	—	—
Folsomia manolachei	—	—	6	11,3	165	41,7	27	8,1	391	47,9	48	49,0	54	28,7	42	25,3	24	54,5
Tetracanthella fjellbergi	—	—	4	7,5	139	35,1	169	50,6	11	1,3	—	—	—	—	—	—	—	—
Oreogastrura parva	—	—	—	—	22	5,6	—	—	—	—	1	1,0	—	—	—	—	—	—
Isotomiella minor	—	—	—	—	10	2,5	16	4,8	8	1,0	1	1,0	2	1,1	2	1,2	5	11,4
Parisotoma notabilis	—	—	—	—	8	2,0	3	0,9	17	2,1	2	2,0	—	—	2	1,2	—	—
Lepidocyrtus lignorum	—	—	—	—	4	1,0	32	9,6	15	1,8	3	3,1	2	1,1	3	1,8	—	—
Friesea truncata	—	—	16	30,2	2	0,5	13	3,9	2	0,4	—	—	5	2,7	5	3,0	3	6,8
Pognognathellus flavescens	—	—	—	—	1	0,3	—	—	—	—	—	—	—	—	—	—	—	—
Pseudanurophorus binoculatus	—	—	—	—	1	0,3	—	—	—	—	2	2,0	14	7,4	—	—	—	—
Proisotoma recta	—	—	17	32,1	—	—	—	—	—	—	—	—	—	—	—	—	—	—
Isotoma neglecta	—	—	5	9,4	—	—	—	—	—	—	—	—	—	—	—	—	—	—
Mesaphorua tenuisensillata	—	—	1	1,9	—	—	—	—	—	—	—	—	—	—	5	3,0	—	—
Ceratophysella sigillata	—	—	1	1,9	—	—	—	—	—	—	1	1,0	—	—	—	—	—	—
Tomocerus minor	—	—	1	1,9	—	—	2	0,6	—	—	2	2,0	14	7,4	—	—	—	—
Folsomia penicula	—	—	—	—	—	—	—	—	239	29,3	—	—	—	—	—	—	—	—
Pseudisotoma monochaeta	—	—	—	—	—	—	—	—	13	1,6	—	—	—	—	—	—	—	—
Folsomia sensibilis	—	—	—	—	—	—	—	—	8	1,0	—	—	—	—	—	—	—	—
Ceratophysella granulata	—	—	—	—	—	—	—	—	4	0,5	—	—	—	—	—	—	—	—
Metaphorura affinis	—	—	—	—	—	—	—	—	1	0,1	—	—	—	—	—	—	—	—
Acerentomon sp.	—	—	—	—	—	—	2	0,6	—	—	—	—	—	—	1	0,6	—	—

Species	17	53	396	334	817	98	188	166	44
Entomobrya juv	—	—	—	21 / 6,3	—	—	—	—	—
Isotoma fennica	—	—	—	1 / 0,3	—	—	—	—	—
Ceratophysella succinea	—	—	—	—	—	—	—	—	6 / 13,6
Paratullbergia callipygos	—	—	—	—	—	—	—	—	3 / 6,8
Tullbergia simplex	—	—	—	—	—	—	—	—	1 / 2,3
Anurida juv	—	—	—	—	—	—	—	—	1 / 2,3
Tomocerus juv	—	—	—	—	—	—	—	—	1 / 2,3
Pseudosinella zygophora	—	—	—	—	—	9 / 9,2	—	—	—
Mesaphorura sylvatica	—	—	—	—	—	8 / 8,2	2 / 2,1	—	—
Eosentomon bohemicum	—	—	1 / 0,3	—	—	3 / 3,1	—	5 / 3,0	—
Eosentomon sp. juv	—	—	—	—	—	—	—	—	—
Pseudachorutes subcrassus	—	—	—	—	—	2 / 2,0	—	—	—
Pseudosinella alba	—	—	—	—	—	2 / 2,0	—	—	—
Willemia intermedia	—	—	—	—	—	2 / 2,0	—	—	—
Megalothorax minimus	—	—	—	—	—	1 / 1,0	—	—	—
Isotoma sp. juv	—	—	—	—	—	—	8 / 4,3	—	—
Folsomia alpina	—	—	—	—	—	—	7 / 3,7	—	—
Ceratophysella armata	—	—	—	—	—	—	3 / 1,6	—	—
Lepidocyrtus paradoxus	—	—	—	—	—	—	1 / 0,5	—	—
Micranurida forsslundi	—	—	—	—	—	—	1 / 0,5	—	—
Pseudachorutes asigillatus	—	—	—	—	—	—	1 / 0,5	—	—
Neanura sp.	—	—	1 / 0,3	—	—	—	—	—	—
Ceratophysella engadinensis	—	1 / 1,9	—	—	—	—	—	32 / 19,3	—
Willemia anophthalma	—	—	—	—	—	—	—	4 / 2,4	—
Σ	17	53	396	334	817	98	188	166	44

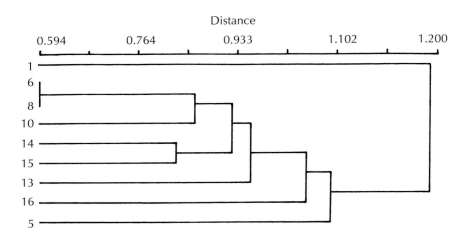

Fig. 12.1. Faunistic similarity of Collembola communities in the alpine zone of the Tomanova Dolina Valley, Tatra National Park. See text for site numbers.

without an indicator species and *Sminthurides aquaticus*, *Sphaeridia pumilis* and *Tetrodontophora bielanensis* as preference species, whereas the remaining seven samples (sub-cluster +01) have one indicator species, *Folsomia quadrioculata*, and 15 preference species. The third level of division separates two samples from limestone 13 and 14 (−010) with one indicator, *Mesaphorura sylvatica*, and 18 preference species. The positive sub-cluster (+010) with six samples from granite (except no. 15 from limestone) has 18 preference species and many indicators.

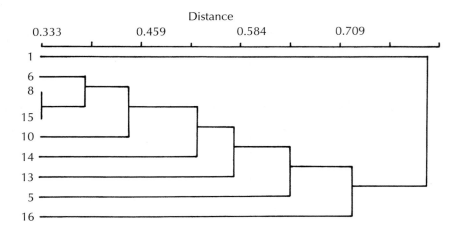

Fig. 12.2. Coenotic similarity of Collembola communities in the alpine zone of the Tomanova Dolina Valley, Tatra National Park. See text for site numbers.

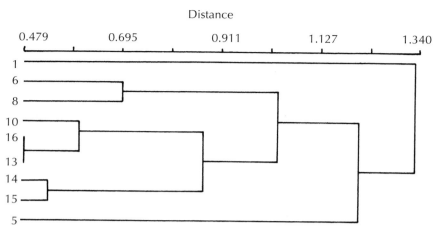

Fig. 12.3. Functional similarity of Collembola communities in the alpine zone of the Tomanova Dolina Valley, Tatra National Park. See text for site numbers.

The above results from different types of community analyses indicate high specificity of soil fauna (Collembola) communities and individual species in relation to acid (granite) versus alkaline (limestone) bedrock soils, plant communities and ecosystems in undisturbed landscapes such as that which existed in the Tomanova Dolina Valley, at least until 1977.

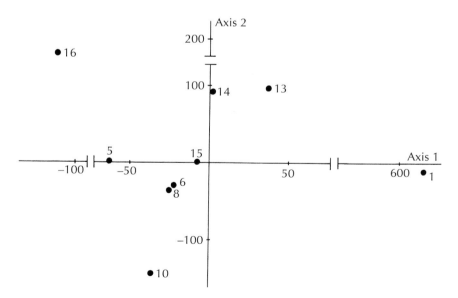

Fig. 12.4. Ordination analysis of Collembola communities in the alpine zone of the Tomanova Dolina Valley, Tatra National Park. See text for site numbers.

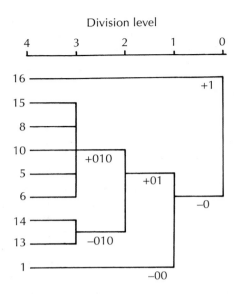

Fig. 12.5. Division classification of Collembola communities in the alpine zone of the Tomanova Dolina Valley, Tatra National Park. See text for site numbers.

The impact of acidification, eutrophication and global warming

No substantial changes in Collembola community parameters, soil pH, microstructure of soils, or other parameters were found at the permanent plots in the Tomanova Dolina Valley up until 1977 when the studies ceased. However, when they were resumed in 1990–1998 substantial changes were recorded in many ecosystem parameters after 1990 and beyond. Changes in Collembola communities (Table 12.2) related to changes in plant communities and soil parameters, plus the micro- and meso-climatic conditions in the Tomanova Dolina Valley during the 1977 to 1990 period.

The density of Collembola communities reached 3600–81,700 ind. m^{-2} in the past while in 1990 (1992) there were 22,000–301,600 ind. m^{-2} (Table 12.2). In the past the lowest density was in the endemic plant community *Saxifragetum perdurantis* (16) on limestone, the highest in the endemic *Calamagrostietum villosae tatricum* (10) on granite. Recently the lowest density was recorded in *Festucetum carpaticae* (13) on limestone and the highest again in the *Calamagrostietum villosae tatricum*. On all sites there was a 1.6- to 47.9-fold increase in Collembola density. The highest increase was recorded at site 9 with the second lowest density in the past and recently the second highest.

In the past the number of species in the collembolan communities was 5–17, but in the recent studies this had increased to 9–24. The lowest number of species was established in *Saxifragetum perdurantis* for both past and recent samples (5 and 9, respectively). In the past the largest species number was in

Festucetum carpaticae (17 spp.) and recently the highest species number was in the *Callamagrostietum villosae tatricum* (24 spp.). There was no decrease in the number of species in the recent samples. The species increase at each site was not a simple process of additional species immigration, but it involved a decrease or disappearance of some other species. Some species disappeared in recent years from the entire Tomanova Dolina Valley ecosystems under study. *Folsomia alpina* was typical for the plant communities *Festucetum versicoloris* (site 12) and *Caricetum firmae* (14), but in recent samples it was not recorded even outside the permanent plots. The same occurred for *Proisotoma recta*, which in the past was typical for *Sphagno-Empetretum*. However, it was still present in the adjacent Rozpadliny valley *Adenostyletum alliariae* plant community.

Other species which have disappeared from one or more ecosystems have invaded other ecosystems where they did not occur in the past, e.g. *Folsomia penicula* disappeared from *Festucetum pictae* (site 6), and decreased in density in *Calamagrostietum villosae* (site 10) and established new high dominant populations in the *Festucetum carpaticae* (site 13) and *Geranio-Alchemilletum crinitae* (site 15) ecosystems. To the same category of species belong *Friesea truncata*, *Tetracanthella fjellbergi*, *Oreogastrura parva*, *Folsomia sensibilis*, *Megalothorax minimus*, *Pseudisotoma monochaeta* and *Ceratophysella engadinensis*.

Another category of species which was not originally found in the samples, but which occurred in recent samples, although often in low densities and only with a single specimen at a site, included *Mesaphorura italica*, *Jevania fageticola*, *Karlstejnia rusekiana*, *Friesea albida*, *Friesea mirabilis*, *Odontella* sp., *Mesaphorura macrochaeta*, *Sminthurinus signatus*, *Mesaphorura hylophila*, *Protaphorura illaborata*, *Protaphorura cancellata*, *Onychiurus granulosus*, *Xenylla boerneri*, *Entomobrya lanuginosa*, *Arhopalites principalis*, *Sminthurinus ochropus*, *Isotoma divergens* and *Willemia scandinavica*. These species are typical of lowland and mountain forests and did not occur in the alpine zone in the past. The upward movement of lowland species is connected with global warming. Of special interest are species which previously occurred only on granite sites, but which were recently found on limestone sites, e.g. *Tetracanthella fjellbergi*.

Is it possible to deduce what caused the changes in the alpine ecosystems? It was shown that some species of Collembola were restricted to limestone or granite bedrock sites in the past. No species restricted to limestone in the past has been recorded recently on the granite sites, whereas some species from granite were recorded recently on limestone sites. This is the case for *Tetracanthella fjellbergi*, a dominant species in many ecosystems on granite (Rusek, 1993), which established high density populations recently in some ecosystems on limestone. Examples are sites 12 and 16. These sites were heavily acidified, soil pH dropped from 7.5 to 5.7 and from 7.7 to 6.45, respectively (Rusek, 1993). Acidification could be the most important factor causing migration of acidophilic species bound to acid bedrock (granite) soils in the past, but free microhabitats could also play a role in colonization of new sites. Without doubt, *Folsomia penicula* took over the free microhabitat left by the *Folsomia alpina* in *Festucetum versicoloris* (site 12).

Table 12.2. Density of Collembola and Protura (individuals 100 cm^{-2}) in selected ecosystems in the Tatra National Park (location numbers see text).

Species	6 1968	6 1990	9 1968	9 1990	10 1968	10 1992	12 1968	12 1990	13 1968	13 1990	15 1968	15 1990	16 1968	16 1990
Folsomia manolachei Bagnall, 1939	165	609	6	458	391	1753	6	165	48	53	42	244	24	178
Protaphorura armata (Tullberg, 1869)	10	65	—	—	108	—	6	11	10	22	23	31	—	11
Isotomiella minor (Schäffer, 1896)	10	22	1	152	8	390	1	11	1	20	42	37	5	—
Isotoma violacea Tullberg, 1876	32	1												
Friesea truncata Cassagnau, 1958	2	67	16	111	2	—		15		2	5	14		28
Tetracanthella fjellbergi Deharveng, 1987	139	—	4	1726	11	6		170		7			3	189
Oreogastrura parva (Gisin, 1949)	22	—			17			1	1	19		1		
Parisotoma notabilis (Schäffer, 1896)	8	7			17	58			2	15	2	6		1
Lepidocyrtus lignorum (Fabricius, 1781)	4	4						2	3		3	26		
Pseudanurophorus binoculatus (Ksen.,1934)	1	—						38	2	4		6		
Folsomia penicula Bagnall, 1939	111	—			239	50				27				
Folsomia sensibilis Kseneman, 1936	115	1			8	57	61	—	21	1				
Megalothorax minimus Willem, 1900		1		2		28			1	1				
Orchesella frontimaculata Gisin, 1946	8					28						8		
Folsomia alpina Kseneman, 1936														
Entomobrya juv							17							1
Ceratophysella sigillata (Uzel, 1891)			1				11							
Mesaphorura tenuisensillata Rusek, 1974			1			1	2	2	1	2	5	3		
Isotoma juv				1			2	2	1					
Willemia aspinata Stach, 1949							1							
Tomocerus minor (Lubbock, 1862)	1		1				1		1				1	
Tomocerus flavescens (Tullberg, 1871)														
Mesaphorura sylvatica (Rusek, 1971)								4	2					
Micranurida sp.				2				1						
Sphaeridia pumilis (Krausbauer, 1898)														
Pseudosinella zygophora (Schille, 1908)									9	1				
Pseudachorutes subcrassus Tullberg, 1871									2					
Pseudosinella alba (Packard, 1873)				1					2	29	4			
Willemia anophthalma Börner, 1901									2	1		3		
Mesaphorura italica (Rusek, 1971)									1					
Jevania fageticola Rusek, 1987										1				
Karlstejnia rusekiana Weiner, 1983										9	32	37		
Ceratophysella engadinensis (Gisin, 1949)										5				24
Friesea albida Stach, 1949										1				
Friesea mirabilis (Tullberg, 1871)						1								

Odontella sp.	—	—	—	—	—	—	—	—	—	—	—	—	—	—
Mesaphorura macrochaeta Rusek, 1976	—	—	—	—	—	—	—	—	—	—	—	—	—	—
Tullbergia simplex Gisin, 1958	—	—	—	—	—	—	—	—	—	—	—	1	—	—
Ceratophysella succinea (Gisin, 1949)	—	—	—	—	—	—	—	—	—	—	6	2	3	—
Stenaphorura quadrispina Börner, 1901	—	—	—	—	—	—	—	—	—	—	1	—	—	1
Anurida sp.	—	—	—	—	—	—	—	—	—	—	—	—	—	—
Pseudisotoma monochaeta (Kos, 1942)	•	—	—	28	13	—	—	—	—	—	—	1	—	—
Sminthurinus signatus (Krausbauer, 1898)	—	—	—	2	—	50	—	—	—	—	—	1	—	—
Mesaphorura hylophila Rusek, 1982	•	—	—	—	—	—	—	—	—	—	—	—	—	—
Proisotoma recta Stach, 1930	—	17	—	—	—	—	—	—	—	—	—	—	—	—
Isotoma neglecta Schäffer, 1900	—	5	—	—	—	—	—	—	—	—	—	—	—	—
Isotoma nivalis Carl, 1910	31	—	—	—	—	—	—	—	—	—	—	—	—	—
Neanura alba Törne, 1956	—	—	1	—	—	—	—	—	—	—	—	—	—	—
Protaphorura illaborata (Gisin, 1952)	—	—	—	5	—	1	—	—	—	1	—	—	—	—
Protaphorura cancellata (Gisin, 1956)	—	—	—	21	—	1	—	—	—	—	—	—	—	—
Onychiurus granulosus Stach, 1930	—	—	—	14	—	—	—	—	—	—	—	—	—	—
Lepidocyrtus peisonis Traser and Christian, 1992	—	—	—	1	15	50	—	—	—	—	—	—	—	—
Xenylla boerneri Axelson, 1905	—	—	—	10	—	—	—	—	—	—	—	—	—	—
Pseudachorutes parvulus Börner, 1901	—	—	—	1	—	—	—	—	—	—	—	—	—	—
Entomobrya lanuginosa (Nicolet, 1841)	—	—	—	1	—	—	—	—	—	—	—	—	—	—
Ceratophysella granulata (Stach, 1949)	23	—	—	—	4	14	—	—	—	—	—	—	—	—
Oligaphorura absoloni (Börner, 1901)	2	—	—	—	—	—	—	—	—	—	—	—	—	1
Metaphorura affinis (Börner, 1902)	—	—	—	—	1	1	—	—	—	—	—	—	—	—
Arrhopalites principalis Stach, 1945	—	—	—	—	—	12	—	—	—	—	—	—	—	—
Sminthurinus ochropus (Reuter, 1891)	—	—	—	—	—	20	—	—	—	—	—	—	—	—
Isotoma divergens Axelson, 1900	—	—	—	—	—	2	—	—	—	—	—	—	—	—
Protaphorura unari Rusek, 1995	—	—	—	—	—	224	—	—	—	—	—	—	—	—
Protaphorura procampata (Gisin, 1956)	—	—	—	—	—	291	—	—	—	—	—	—	—	—
Willemia scandinavica Stach, 1949	—	—	—	—	—	1	—	—	—	—	—	—	—	—
Fosentomon bloszyki Szeptycki, 1985	1	—	—	—	—	—	—	—	—	—	—	—	—	—
Σ	628	834	53	2537	817	3016	109	417	109	220	168	422	36	434
Increase in density	1.3			47.9		3.7		3.8		2.0		2.5		12
Species no	14	13	10	18	12	24	11	12	17	19	13	17	5	9

Two sites on granite bedrock soils recently reached very high density values of Collembola which are unusual for alpine ecosystems. These were in *Sphagno-Empetretum hermaphroditi* with a 47.9-fold increase to 253,700 ind. m^{-2} and in *Calamagrostietum villosae tatricum* with a 3.7-fold increase only, but to the highest established density of 301,600 ind. m^{-2}. The density increase in these two ecosystems was also accompanied by a high species number increase, from 10 to 18 and 12 to 24 species, respectively. In the case of *Sphagno-Empetretum hermaphroditi* there was a change at ecosystem level. The former plant community died off and its area was occupied by *Oreochloetum distichae*. This was accompanied by a soil type change from tundra-moss to a ranker. The Collembola community also changed completely there (Table 12.2). It is well known that *Sphagnum*-moss dominated plant communities are oligotrophic and dependent on a low nitrogen content. Ruck and Adams (1988) have shown the mutual influence of forest stands and orography on each other, and explained the deposition increase on hillside regions caused by increased turbulence levels. Also on mountains, acid deposition is higher in places with increased snow accumulation or in water runoff gullies (Rusek, 1993). This may explain the higher nitrogen input into the *Sphagno-Empetretum* ecosystem leading to assimilation of stored carbon in *Sphagnum* by soil microflora and decomposition processes.

The *Calamagrostietum villosae tatricum* ecosystem in the water runoff gullies was able to use the higher nitrogen supply for increasing primary and litter production. The alpine mull-like ranker soil type had high biological activity in the past and was able to increase it recently under the influence of higher primary production. These two ecosystems are examples of the eutrophication effect in alpine soils. This is probably a combined effect of acidification and eutrophication leading to collembolan (soil biota) density increases in some limestone sites where indigenous or invaded plants had higher primary production.

Many collembolan species typical of lowland and montane forests are now occurring in the alpine zone ecosystems. Some of them are already dominant, most were recorded by one or a few individuals. Numbers of such species are relatively high, and are responsible for the greatest increase in species richness in many communities (Table 12.2). The upward movement of lowland and montane forest soil fauna to the alpine zone is comparable with that of the alpine-nival flora observed by Grabherr *et al.* (1994). The rate in the case of some collembolan species is even higher than in plants. Global warming, without doubt, causes this process of upward movement which did not occur in the past in studies up to 1977.

Soil Invertebrates and Land Use

Changes in soil fauna diversity and density in arable soils

Samples of meso- and macrofauna in arable soils in different parts of the Czech and Slovak Republics studied since 1963 have revealed a significant decrease in animal density and diversity. This was most obvious at Bulhary in South Moravia and Palárikovo in South Slovakia. Changes at these localities have been so dramatic as to be classed as an ecological catastrophe in the most fertile soils in Central Europe (Rusek, 1998). These soils had very high density and biodiversity in the past, but changed between 1964 and 1967 (see Table 12.3). The collembolan communities at Bulhary had a density of up to 62,200 ind. m^{-2} and comprised 34 species in 1964. By 1967 this had dropped to 800–4200 ind. m^{-2} and four or five species. The sites at Palárikovo were first studied in 1964 and the studies continued there in 1997. The values for Palárikovo were 139,600 ind. m^{-2} and 28 species in 1964 (the samples from 1997 are sorted, but not yet determined and counted). From the vertical distribution of different microarthropod groups it is evident that only a small proportion of soil mesofauna lives below 20 cm depth in arable chernozem soils (Table 12.4).

At Palárikovo some species were found, often in high densities, which were previously unreported, or rare, for the former Czechoslovakia. These include *Pseudosinella imparipunctata, Heteromurus major, Entomobrya handschini* and *Willemia buddenbrocki*, while at Bulhary *Brachystomella curvula, Xenyllodes bayeri, Pseudosinella ksenemani* and *Pseudanurophorus boerneri* were recorded. These data are presented here (Tables 12.3 and 12.4) to demonstrate how rich the soil fauna in arable soils was in the past. At Bulhary, in 1967, there were only four or five species remaining from this formerly very rich community (*Parisotoma notabilis, Pseudosinella alba, Sminthurinus elegans, Ceratophysella succinea, Folsomides parvulus*). The low populations and number of species had not recovered to their former community parameters in 1990 (Table 12.3). Common representatives of soil macrofauna such as *Catajapyx confusus, Plusiocampa* sp., *Campodea chionea,* Symphyla, geophilid myriapods, anecic earthworms and even the injurious elaterid larvae of *Agriotes brevis* and mesofauna (Protura, *Acerentulus traghardi;* Pauropoda, *Stylopauropus* spp.) which were present in the past are now missing.

Table 12.3. Density and number of Collembola species in arable soils in the Czech and Slovak Republics. For site number see text.

Locality no.	Year	Individuals m^{-2}	Species number
1	1964	40,600–62,200	34
	1967	800–4,200	4–5
	1990	800–4,100	4–5
2	1964	139,600	28
3	1977	36,400	21
4	1974	2,500–25,600	6–17
5	1974	8,100–26,900	12–13
6	1992	16,400	17
7	1992	9,000	12
8	1985	7,000	6
9	1986	2,200	3

Table 12.4. Densities (individuals m^{-2}) of Collembola and other microarthropods in a maize field at Palárikovo (1964) in different soil horizons.

	Soil horizon			
	0–10 cm	10–20 cm	20–30 cm	Σ
Collembola	118,600	21,000	0	139,600
Oribatida	366,100	41,200	100	407,400
Prostigmata	257,000	22,200	100	279,300
Mesostigmata	82,400	29,500	100	112,000
Pauropoda	5,800	5,300	0	11,100
Protura	0	300	0	300
Σ	829,900	119,500	300	949,700

Causes and consequences of biodiversity and density collapse in arable soils

Kováč (1994a,b) and Kováč and Miklisová (1995) published data on collembolan density and biodiversity from different arable soils in a large area of East Slovakia. Their population levels were extremely low compared with other studies dealing with Collembola in arable soils (cf. Alejnikova and Martynova, 1966; Akkerhuis *et al.*, 1988), reaching, on average, 660–2540 ind. m^{-2}. Kováč's low density numbers evoked a discussion at the XII International Colloquium on Soil Zoology at Dublin, 1996. He thought that these low values were attributed to very low extractor efficiency. Data in this chapter exclude low extraction efficiency and, in contrast, long-term investigations have recorded high biodiversity and density of Collembola and other soil faunal communities as well as their collapse in the 1960s. Rusek (1998) presented density and species number values of Collembola from soil samples taken at nine localities (Table 12.3, see also p. 247) in Czechoslovakia between 1964 and 1990. The data exhibited great variability between sites over time. High density and species diversity values from soil samples taken in 1964 in South Slovakia (Palárikovo) and South Moravia (Bulhary, Nejdek), as well as in the Bohemian Karst (Mořina) in 1977 could be considered as values for undisturbed soils. They are comparable with collembolan community parameters of 'natural' grassland soils (cf. Rusek, 1984). Edwards and Lofty (1975) summarized the general changes in microarthropod communities in arable soils affected by agriculture practices. Cultivation, rotation, monoculture and application of pesticides eliminate species susceptible to damage, desiccation and destruction of their microhabitats. Intensive farming using high doses of fertilizers, pesticides and heavy machines was introduced in the 1960s in the former Czechoslovakia.

Herbicides were applied on 60% of arable fields every year. Herbicides and other pesticides and pollutants have direct and indirect negative effects on organisms and processes in soil (Rusek, 1986). All these factors, together with maize and cereal monocultures, high doses of mineral fertilizers and no manure were responsible for a gradual destruction of soil meso- and macrofauna communities in arable soils, as documented by the sites with moderate and low densities of Collembola. Data documenting a parallel decrease of other groups of soil micro-, meso-, macro- and megafauna and even soil microflora are available (Balík, Testacea, unpublished; Hattori and Rusek, bacteria, unpublished; Rusek, macroarthropods, unpublished; Houšková, 1991).

Intensive farming is without doubt the most important factor causing a gradual decrease in density and species diversity of Collembola and other groups of soil organisms. But it does not explain the catastrophic decrease in density and species richness of different groups of soil organisms during a period of 2 years in chernozem soils in South Moravia. I started to study these soils in 1963 in connection with heavy infestation by wireworms (Rusek, 1972). Very high doses of DDT (in the 1950s) and HCH were repeatedly applied against elaterid larvae on the soil surface and as seed dressings on the same fields (Rusek, 1974). The combined effects of the harmful practices together with the repeated application of high doses of insecticides probably destroyed the ecosystem stability at South Moravia (site 1). Although not so well documented, these same effects occurred at other sites (established in the following years) with low soil fauna density. Soils with high clay mineral content (loess in chernozem soils) are probably affected sooner than are sandy soils and soils with a higher proportion of larger mineral particles. This, together with the extreme soil water regime, may explain the very low densities in luvisols and gleic soils established by Kováč (1994a) in East Slovakia.

The dramatic destruction of soil faunal communities led to the demise of such important keystone species as the anecic earthworms. It also resulted in soil microstructure collapse, because of interruption of their tunnelling activity and below and above soil surface cast production. The air-filled pore space, an important microhabitat for terrestrial and aquatic soil organisms, was strongly reduced. This has negative consequences for soil aeration, water infiltration, root penetration into deeper soil layers, soil fertility and other important soil functions. Soil thin sections from arable chernozems in adjacent Austria, 15–20 km away from the Bulhary sites, have well-developed soil microstructure including free air space (Rusek, unpublished). The soils in Austria were managed by private landowners better than the large cooperatives in the former Czechoslovakia.

Conclusions

- Long-term studies of alpine ecosystems in Tatra National Park (Slovakia) indicate high specificity of soil fauna (Collembola) communities and individual species in relation to acid (granite) versus alkaline (limestone) bedrock soils, plant communities and ecosystems in undisturbed landscapes.
- Substantial changes in alpine collembolan communities were related to changes in plant communities, soil and climatic parameters during the 1977–1990 period.

- On all sites in the alpine zone (Tatra National Park) there was a 1.6–47.9-fold increase in Collembola density and an increase from 5–17 to 9–24 in species number in recent samples. There was no decrease in the species number on the studied sites. The changes in alpine ecosystems are related to soil acidification, eutrophication and global warming. There are groups of Collembola species that react differently to these factors.
- Considerable differences were established in density and species number in Collembola communities in arable soils over time.
- High density and species number were recorded in arable soils of the former Czechoslovakia in the past (1964–1977).
- Severe decrease in density and number of soil fauna species (Collembola) of one or two orders between 1964 and 1967 at Bulhary, and the parallel destruction of other soil biota, is explained by the combined effect of intensive farming and pesticide application.
- Density and biodiversity decreases in arable soil fauna had negative consequences for soil microstructure collapse and soil compaction, soil aeration, water infiltration, soil fertility and other basic functions in soil.
- The dramatic changes in arable chernozem soils could be classified as an ecological catastrophe of the most fertile soils in Central Europe.

Acknowledgements

Part of the work was supported by grant projects No. 66604 and No. A6066702 from the Grant Agency, Academy of Sciences of the Czech Republic and No. 526/97/0631 from the Grant Agency of the Czech Republic. Permission for the research in the Tatra National Park was kindly obtained from the Park Administration. I thank Dr J. Sutherland (Victoria, BC, Canada) for improving the English of the manuscript. I am obliged to two anonymous reviewers for critical comments.

References

Akkerhuis, G.A.J.M.J., deLey, F., Zwetsloot, H.J.C., Ponge, J.-F. and Brussaard, L. (1988) Soil microarthropods (Acari and Collembola) in two crop rotations on heavy marine clay soil. *Revue d'Écologie et de Biologie du Sol* 25, 175–202.

Alejnikova, M.M. and Martynova, E.F. (1966) Landšaftno-ekologičeskij obzor fauny počvennych nogochvostok (Collembola) srednego Povolžja. *Pedobiologia* 6, 35–64.

Edwards, C.A. and Lofty, J.R. (1975) The influence of cultivation on soil animal populations. In: Vaněk, J. (ed.) *Progress in Soil Zoology*. Academia, Praha, pp. 399–408.

Eisenbeis, G. and Wichard, W. (1985) *Atlas zur Bodenbiologie der Bodenarthropoden*. Gustav Fischer Verlag, Stuttgart.

Freckman, D.W., Blackburn, T.H., Brussaard, L., Hutchings, P., Palmer, M.A. and Snelgrove, P.V.R. (1997) Linking biodiversity and ecosystem functioning of soils and sediments. *Ambio* 26, 556–562.

Grabherr, G., Gottfried, M. and Pauli, H. (1994) Climate effects on mountain plants. *Nature* 369, 448–449.

Hill, M.O. (1979) TWINSPAN – a FORTRAN Program for Arranging Multivariate Data in an Ordered Two-way Table by Classification of the Individuals and Attributes. Ecology and Systematics. Cornell University, Ithaca.

Hoesse, B. (1981) Morphologie und Funktion des Wasserleitungssystems der terrestrischen Isopoden (Crustacea, Isopoda, Oniscoidea). *Zoomorphology* 98, 135–167.

Houšková, L. (1991) Význam půdní fauny pro bioindikaci chemického znečistění. *Autoreferát Disertační Práce k Získání Hodnosti Kandidáta Věd.* Přírodovědecká Fakulta, Masarykova Universita, Brno.

Hutson, B.R. (1978) Influence of pH, temperature and salinity on the fecundity and longevity of four species of Collembola. *Pedobiologia* 18, 163–179.

Jaeger, G. and Eisenbeis, G. (1984) pH-dependent absorption of solution by the ventral tube of *Tomocerus flavescens* (Tullberg, 1871) (Insecta, Collembola). *Revue d'Écologie et de Biologie du Sol* 21, 519–531.

Kopeszki, H. (1992) Veränderungen der Mesofauna eines Buchenwaldes bei Sauerbelastung. *Pedobiologia* 36, 295–305.

Kováč, L. (1994a) The effects of soil type on Collembolan communities in agroecosystems. *Acta Zoologica Fennica* 195, 83–93.

Kováč, L. (1994b) Spoločenstva chvostoskokov (Hexapoda, Collembola) ornej pôdy agroekosystémov Východoslovenskej nížiny a Košickej kotliny. *Autoreferát Dizertácie na Získanie Vedeckej Hodnosti Kandidáta Věd.* Entomologický Ústav ČSAV, České Budějovice.

Kováč, L. and Miklisová, D. (1995) Collembolan communities in winter wheat-clover cropping system on two different soil types. *Bulletin Entomologique Pologne* 64, 365–381.

Kubiena, W.L. (1950) *Bestimmungsbuch und Systematik der Böden Europas.* Enke-Verlag, Stuttgart.

Meyer, E. and Eisenbeis, G. (1987) The ultrastructural basis for rectal water uptake in *Trimerophorella nivicomes* (Chordeumatida, Diplopoda). In: Striganova, B.R. (ed.) *Soil Fauna and Soil Fertility.* Nauka, Moscow, pp. 28–134.

Miller, D.R. (1984) Chemicals in the environment. In: Sheehan, P.J., Miller, D.R., Butler, G.C. and Bourdeau, Ph. (eds) *Effects of Pollutants at the Ecosystem Level.* SCOPE 22, John Wiley & Sons, Chichester, pp. 7–14.

Orlóci, L. (1978) *Multivariate Analysis in Vegetation Research,* 2nd edn. Jung, The Hague.

Ruck, B. and Adams, E.W. (1988) Einfluss der Orographie auf die Schadstoffdeposition durch Feinstropfchen. In: Horsch, F., Filby, W.G., Fund, F., Gross, S., Hanisch, B., Kilz, E. and Seidel, E. (eds) *4. Statuskolloquium des PEF vom 8. bis 10. März 1988 im Kernforschungszentrum Karlsruhe.* Kernforschungszentrum, Karlsruhe, pp. 627–640.

Rusek, J. (1972) Die mitteleuropäische Agriotes- und Ectinus-Arten (Colleoptera, Elateridae) mit besonderer Berücksichtigung von A. brevis und den in Feldkulturen lebenden Arten. *Rozpravy ČSAV, Řada Matematických a Přírodních věd* 82, pp. 1–90.

Rusek, J. (1974) Ekologické aspekty v boji proti půdním škůdcům, se zvláštním zřetelem k drátovcům (Elateridae). *Sborník prací z V. Celostatni Konference o Ochraně Rostlin, Brno,* Ustav Ochrany Rostlin, Statni Zemedelske Nakladatelstvi, Praha, pp. 507–510.

Rusek, J. (1984) Zur Bodenfauna in drei Typen von Überschwemmungswiesen in Süd-

Mähren. *Rozpravy ČSAV, Řada Matematických a Přírodních věd* 94(3), pp. 1–127.

Rusek, J. (1986) Vliv herbicidů na půdní organismy a kvalitu půdy v jabloňových výsadbách. *Sborník ze Symposia Integrovaná Ochrana Ovoce, České Budějovice 1984.* Institute of Entomology, České Budějovice, pp. 69–72.

Rusek, J. (1993) Air-pollution-mediated changes in alpine ecosystems and ecotones. *Ecological Applications* 3, 409–416.

Rusek, J. (1998) Changes in mesofauna communities of arable soils. In: Pižl, V. and Tajovský, K. (eds) *Soil Zoological Problems in Central Europe.* Institute of Soil Biology, České Budějovice, pp. 173–177.

Rusek, J., Úlehlová, B. and Unar, J. (1975) Soil biological features of some alpine grasslands in Czechoslovakia. In: Vaněk, J. (ed.) *Progress in Soil Zoology.* Academia, Praha, pp. 199–215.

Streit, B. (1984) Effects of high copper concentrations on soil invertebrates (earthworms and oribatid mites): experimental results and a model. *Oecologia* 64, 381–388.

Stumpp, J. (1991) Zur Ökologie einheimischer Proturen (Arthropoda: Insecta) in Fichtenforsten. *Zoologische Beiträge*, N.F. 33, 345–432.

Ter Braak, C.J.F. (1987) CANOCO – *a* FORTRAN *Program for Canonical Community Ordination by (Partial) (Obtrended) (Canonica) Correspondence Analysis, Principal Component Analysis and Redundancy Analysis (Version 2.1).* TNO Institute of Applied Computer Science, Wageningen.

Unar, J., Unarová, M. and Šmarda, J. (1984) Vegetační poměry Tomanovy doliny a Žlebu spod Diery v Západních Tatrách. Část 1. Fytocenologické tabulky. *Folia Facultatis Scientiarum Naturalium Purkynianae Brunensis Biologia* 80, 25 (10), 1–101.

Unar, J., Unarová, M. and Šmarda, J. (1985) Vegetační poměry Tomanovy doliny a Žlebu spod Diery v Západnich Tatrách. Část 2. Charakteristika přírodních poměrů a rostlinných společenstev. *Folia Facultatis Scientiarum Naturalium Purkynianae Brunensis Biologia* 84, 26 (4), 1–77.

Webmasters in Regional and Global Contexts

13

Invertebrates and Nutrient Cycling in Coniferous Forest Ecosystems: Spatial Heterogeneity and Conditionality

T.M. Bolger[1], L.J. Heneghan[2] and P. Neville[1]

[1]Department of Zoology, University College Dublin, Belfield, Dublin 4, Ireland; [2]Environmental Sciences Program, DePaul University, Chicago, IL 60614–3298, USA

Extent and Importance of the Coniferous Forest System

The extent and importance of northern coniferous forests are generally considered in the context of the boreal system which consists of a complex of forested and partially forested ecosystems occurring in a circumpolar belt in Eurasia and North America (Apps *et al.*, 1993). The forest is primarily coniferous because of low mean temperatures and short growing seasons. The extent of the boreal coniferous forest system is approximately 1249×10^6 ha and it is estimated to contain approximately 290×10^{12} g C (Apps *et al.*, 1993) which is one of the Earth's largest terrestrial carbon pools. Within this, the soils are by far the largest reservoir, containing $247–286 \times 10^{15}$ g C which represents more than one-sixth of the Earth's soil organic matter pool (Schlesinger, 1984, 1991; Bonan and Van Cleve, 1992). In addition, the system is considered to be a significant sink for atmospheric C accounting for as much as 662×10^{12} g C year^{-1} (Apps *et al.*, 1993).

These figures alone indicate the importance of northern coniferous forests in terms of global biogeochemical cycles. However, the coniferous forest system is considerably more extensive than the boreal forests. There are mountain coniferous forests which extend further south than the boreal systems. Even under natural conditions, the boundary between boreal and temperate forests is not sharp and the situation is further complicated by the expansion of coniferous forests for commercial purposes. For example, at the beginning of the 20th century only 1.4% of Ireland's land area was forested (O'Carroll, 1984).

However, through state planting programmes and with European Union (EU) grant assistance, the cover has now reached 7% and it is expected that an additional 30,000 ha will be planted annually for the foreseeable future (Lowery, 1991). These newly planted forests are composed primarily of exotic fast-growing conifer species (95%) with only 5% broadleaved trees. Over 80% of the conifers are Sitka spruce and the remaining 20% are primarily lodgepole pine. Thus coniferous forests have a very large extent and have important roles to play in the world's biogeochemical cycles.

Determinants of Productivity and Decomposition

Productivity in boreal and cool temperate forests is largely determined by low temperature and slow rates of litter decomposition (Kimmins, 1996) which often results in low nutrient availability. Up to the middle of this century, nitrogen was the major limiting factor in most forest ecosystems of the northern hemisphere (Keeney, 1980; Tamm, 1991). However, increased use of internal combustion engines and the intensification of animal husbandry has changed this situation dramatically through the increased production of gaseous compounds of nitrogen and sulphur. In the period between 1950 and 1990 the emissions of NO_x approximately trebled in Europe and those of NH_3 doubled (Bell, 1994). Forest canopies trap these gases and aerosols very efficiently from the atmosphere, thus subjecting forest soils to acidifying pollutants (Parker, 1983). Today, the deposition of nitrogen exceeds critical loads over large parts of Europe (Aber et al., 1989; Aber, 1992) and has increased to levels that exceed vegetation demands (Encke, 1986; Ågren and Bosatta, 1988) thus mitigating the natural nutrient limitation of forest productivity.

It is generally believed that, in the climates associated with coniferous forests, processes such as decomposition are dominated by climatic conditions with the biota having little effect in determining rates of decomposition (Swift et al., 1979; Lavelle et al., 1993). Models of mass loss from litter and of the dynamics of other soil carbon pools are predicted relatively accurately and processes such as nitrification and nitrogen leaching can be modelled using temperature as a single driving variable (Berg et al., 1993, 1997). Decomposition is primarily a biological process with abiotic oxidation accounting for only a small proportion of the CO_2 evolving from soils. Generally more than 95% of carbon mineralization is associated with organismal respiration (Alexander, 1977; Lavelle et al., 1993). Thus, it is generally held that the control of decomposition is due to the regulation of microbial activity at decreasing temporal and spatial scales (Aerts, 1997).

Decomposer Fauna of Coniferous Forests

Conifer needles generally decompose slowly and this tends to lead to the formation of a mor humus (Wallwork, 1970). However, while the rate of decomposition is obviously related to the resource quality of the conifer needles, the fact that they decompose more quickly if placed on a mull humus than if placed on a mor humus (Crossley and Witkamp, 1964) indicates that factors other than climatic conditions, such as the organisms involved, have a critical role to play. The preponderance of mor humus in these forests would lead one to expect that fungal-feeding microarthropods and acid-tolerant Enchytraeidae should dominate the soil fauna. However, the fauna can be quite diverse and groups such as earthworms can represent a significant component of the faunal biomass (Petersen and Luxton, 1982). A perusal of the limited number of studies which have catalogued the soil fauna indicates that the importance of various groups can be quite variable and site dependent. For example, Bornebusch (1930) describes both mull and mor humus under spruce and shows earthworms to be dominant in some situations and virtually absent in others. Similarly, Huhta and Koskenniemi (1975) show that the dominant groups were quite different at conifer sites in Finland. While no earthworms were reported in the study of Persson *et al.* (1980), small numbers of earthworms ($1-5$ m^{-2}) were present at the site (T. Persson, personal communication).

In coniferous forests the soil fauna generally account for a relatively small proportion of the heterotrophic respiration (e.g. between 0.25 and 2.33% at two Finnish sites (Huhta and Koskenniemi, 1975)). However, they contribute significantly to the mineralization of several nutrients. This influence can be due either to direct effects such as the excretion of NH$_4$$^+$, or indirect effects such as the effects of grazing on both microbial biomass and activity (Anderson and Ineson, 1984). For example although the biomass of the soil fauna in a Scots pine forest in Sweden was only approximately 1% of that of the soil microbial biomass, it contributed between 10 and 49% of the nitrogen mineralization because it was estimated to consume 30–60% of the microbial biomass (Persson, 1983).

Earthworms

Lavelle (1983) characterized the earthworm fauna of coniferous forests as primarily epigeic. He argued that in colder regions, litter decomposes slowly thus providing a reliable food source of high energy content, while the humic materials are less digestible and nutritious for the animals and microbial activity in these horizons is generally insufficient to provide adequate nutrition for earthworms. These worms can be a major determining force in the coniferous forest system. For example, *Dendrobaena octaedra* can grow and reproduce in the organic horizons of pine forest floors to which they are introduced (McLean *et al.*, 1996b). In peaty forest soils earthworms can increase soil pH, water flux and rates of nitrogen leaching from field lysimeters and this effect is enhanced through the addition of lime (Robinson *et al.*, 1992). Haimi and Huhta (1990) showed that, in a laboratory study, *Lumbricus rubellus* and *D.*

octaedra increased the mass loss, cumulative respiration and total carbon leached from coniferous forest litter and humus, and Heungens (1969) showed that a large variety of species can accelerate the decomposition of pine litter by fragmenting it. However, earthworms do not appear to participate in the initial stages of decomposition of pine needles but have a progressively greater impact in later stages (Ponge, 1991).

Enchytraeidae

Enchytraeidae are generally abundant in coniferous forest soils with densities varying between 5700 and 134,000 m^{-2} (Didden, 1993). *Cognettia sphagnetorum* tends to be the dominant species in these circumstances as it favours the acid soils found in such forests. While enchytraeids appear capable of digesting simple organic molecules, the more complex molecules only become available to them following microbial processing and this is reflected in the patterns of resource colonization seen for conifer litter (e.g. Ponge, 1991). These worms can influence soil structure through their burrowing activity and the transportation of mineral soil particles (Didden, 1993), and several studies have shown that they can increase nitrogen mineralization. In an experiment using leaching columns, Williams and Griffiths (1989) showed that enchytraeids affected the partitioning of N and P between leachates, exchangeable forms and microbial biomass. For example, approximately 50% more mineral N was leached from Sitka spruce litter containing enchytraeids and rates of mineralization and nitrification were increased by 12% and 24%, respectively. This was apparently due to enhanced microbial activity and turnover of biomass. Similar results were obtained by Abrahamsen (1990) who showed that mineral nitrogen in the soil was increased by approximately 18% in the presence of enchytraeids. However, he also showed that this effect was reversed at 18°C and a pF (the force with which water is held in the soil) of 2.2, i.e. under warm dry conditions.

Microarthropods

The functional roles of microarthropds in decomposition processes are indirect and have been attributed mainly to their grazing on the microbial populations in the immobilization phase of decomposition (Petersen and Luxton, 1982; Seastedt, 1984). Lussenhop (1992) identified six mechanisms of interaction between soil microflora and microarthropods which are of importance: selective grazing of fungi by microarthropods; dispersal of fungal inocula by microarthropods; direct supply of mineral nutrients in urine and faeces; stimulation of bacterial activity; compensatory fungal growth due to periodic microarthropod grazing; and release of fungi from competitive stasis due to microarthropod disruption of competing mycelial networks. These processes have been well documented for forest soils (e.g. Hanlon and Anderson (1979), Parkinson *et al.* (1979), Newell (1984), Shaw (1988)). However, McLean *et al.* (1996a) have indicated that knowledge of the fungal food preferences of grazers did not allow accurate predictions of the effects of grazing on the litter community as a whole and Walsh and Bolger (1990, 1994) show that the effects on the population dynamics of the grazers are similarly unpredictable.

In examining the effects of the micro- and mesofauna on decomposition and nutrient dynamics, Setälä and Huhta (1990) found greater mass loss and respiration during the early stages of decomposition of leaf litter in the presence of communities, considered by them to have essentially the same structures as natural communities (Huhta and Setälä, 1990), than in the presence of simple communities. This, they suggested, was due to the presence of a large range of feeding categories which resulted in a more efficient use of the substrate. At later stages of decomposition the rate of mass loss was retarded in the presence of complex fauna. This suggests that the average substrate quality has been altered by the fauna either through the utilization of readily digestible components during the early stages of decomposition or to the production of faeces which are less decomposable than the parent substrate. The latter has been demonstrated for microarthropods and enchytraeids by Grossbard (1969) and Wolters (1988), respectively.

Complexity of community

The importance of micro- and mesofauna in the mineralization of N and P in laboratory experiments with coniferous forest soils was demonstrated by Huhta *et al.* (1988) and Setälä *et al.* (1990). Their experiments show that an animal community with a structure similar to that occurring naturally has a significant and positive effect on the mineralization of nitrogen and phosphorus over the 2-year duration of the incubations. This mineralization effect was reflected in a 1.6-fold increase in aboveground biomass of *Betula pendula* over a period of 45 weeks (Setälä and Huhta, 1991). They also showed that the presence of fauna increased the biomass of roots in the organic horizons but not so in the mineral soil layers and they commented that the plants appeared to invest relatively more in the roots in the absence of animals, with a root-to-shoot ratio of 0.85 in comparison with 0.75 in the faunated systems. The N content of the leaves in the faunated systems was approximately three times that of the control. A later experiment clearly showed that although a complex soil fauna had a negative impact on mycorrhiza (amounts of ergosterol on roots reduced by 80%), the same positive effect on plant growth and N uptake was still present, presumably because of the improved nutrient mobilization arising from the presence of the animals (Setälä, 1995). Bååth *et al.* (1981) studied the impact of grazers on microbial populations and the growth of Scots pine seedlings and, while they showed that the fauna had a positive effect on the release of N in microcosms, they were unable to attribute any increase in plant growth to the activity of the fauna. However, the KCl used to extract the nutrients in this study was likely to have had a significant effect on the fauna.

Conditional Results from Faunal Experiments

Recent studies have shown that many of the effects seen in the previous experiments are conditional. For example, this conditionality was seen by Hanlon (1981) where the response of fungal respiration was determined by the nutrient status of the resource on which the fungus was growing and on the grazing intensity. Similarly, Verhoef *et al.* (1989) showed that although the mobilization of ammonium and nitrate was generally enhanced in the presence of a collembolan, *Tomocerus minor*, the opposite was the case in the fermentation layer from a forest which had received high levels of N deposition from the atmosphere.

Several recent experiments have shown that environmental conditions can markedly affect the extent and nature of the impact of the total fauna on nutrient fluxes. Sulkava *et al.* (1996) showed that soil moisture and temperature exert a complex influence on soil faunal biomass and community structure. Populations of enchytraeids were affected by both temperature and moisture, while microarthropods were only affected by temperature. At low-to-medium moistures, microarthropods reduced the populations of nematodes and enchytraeids but this did not occur at higher moisture contents. The amount of KCl-extractable NH_4 was positively correlated with enchytraeid biomass but not with microarthropod biomass. Thus it would appear that the diverse microarthropod community seems to affect N mineralization indirectly by regulating the enchytraeid populations.

Setälä *et al.* (1997) also show that the abiotic environment has some control over the biotic factors affecting the relationship between plant and mycorrhizal fungus because, depending on the N status of the soil, the grazing activity by the soil fauna caused different outcomes. In N-poor soils a complex fauna reduced the amount of pine biomass, presumably through their grazing impact on the ectomycorrhizae, whereas it had no effect in N-rich soils. The complex fauna also had a negative influence on fine root biomass with the effects being clearer in N-poor soil.

Conditionality in a heterogeneous soil environment

This concept of conditionality is obviously very important in terms of whole systems which vary in their climate or nutrient status, but we believe that it should also be examined in the context of the heterogeneity within a system. The heterogeneity of the forest canopy has significant effects on the heat, water and nutrient fluxes to the soil from outside. Kleb and Wilson (1997) have shown that light and available nitrogen were significantly more variable in a forest than in a prairie. We have shown that, during July, at a Norway spruce stand in Ireland, although the average temperature in the organic horizon of the soil did not vary greatly between random locations within 400 m^2 on the forest floor, the variation in soil temperature can be as much as 22.5°C and the amplitude of the variation can be very variable. At other times of the year the differences between locations were not so great but there were still significant variations in the amplitude of the diurnal changes.

The concentrations of nutrients and volumes of throughfall were also variable. For example, there was a more than threefold difference in the amounts of water input, and the concentrations of ammonium and nitrate varied by as much as 50% and 100%, respectively, between individual collectors at a site in Ireland. The significance of this observation is added to by realizing that the locations of these collectors were not selected to maximize the effect of canopy cover but were simply random locations within a forest which received relatively low inputs of atmospheric pollution. As variations in canopy cover may be greater and dry deposition more important in areas of high deposition, we might expect that the variation in concentrations of nutrient input would be even greater at more polluted sites. A further important point is that the difference between collectors was relatively consistent on a temporal basis (Fig. 13.1).

The situation on the forest floor is very similar. The variations in the moisture content and amounts of nitrate in the soil are illustrated in Fig. 13.2. On one sampling date the water content of the soil varied from approximately 40% to more than 70% on a weight for weight basis within the stand. On the same date the amounts of nitrate varied from virtually zero to more than 5 mg 100 g^{-1} of dry soil. Thus the variety of conditions which led to different outcomes in the experiments outlined above may occur within a single stand and one can only assume that the conditionality will be evident in the biotic activity at the relatively small spatial scale discussed here.

Heneghan and Bolger (1996, 1997) have shown that comparatively minor changes in the structure of animal assemblages can affect the dynamics of several nutrients in litter. For example, when microcosms were established using animals collected from litter which had been exposed to various forms of nitrogen input, the nitrogen fluxes from a control litter were shown to be different. One might argue that such effects, measured in microcosms, have no meaning or are buffered in the field. This is illustrated in the study of Heneghan and Bolger (1992, 1999). Here undisturbed soil columns (14.4 cm diameter, 25 cm in depth) were collected from a Norway spruce stand. These were sterilized using γ-irradiation, microflora were reinoculated into all and fauna were reintroduced into half of them. The fauna clearly had an effect on decomposition in the litter layer; after 8 months there were obvious visual differences and the C:N ratios varied from 30.79 (standard error 1.98) in the presence of animals to 39.09 (SE 2.75) in the absence of animals. However, no significant differences were observed in the concentrations of either NH_4^+-N or NO_3^--N leached from the soil columns (Heneghan and Bolger, 1998). The

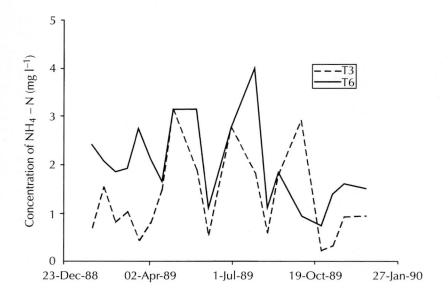

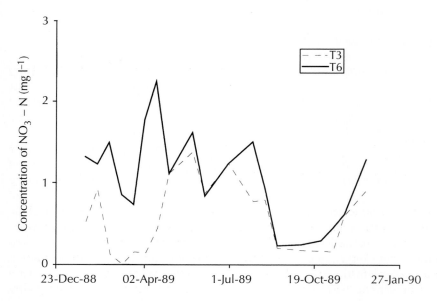

Fig. 13.1. Variation in concentration of NH$_4$-N and NO$_3$-N in two throughfall collectors within a stand of Norway spruce in south-east Ireland.

similarity of the leachates in concentrations of measured nutrients from the microcosms, with and without animals, demonstrates that the mesofauna do

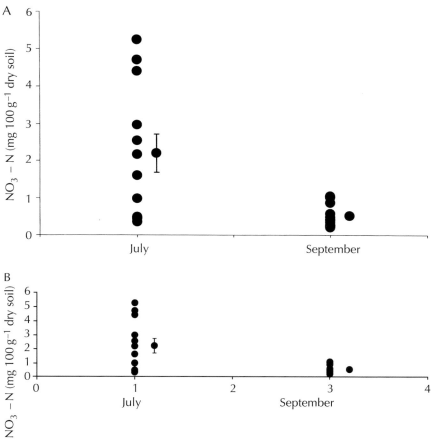

Fig. 13.2. Variation in soil moisture (A) and nitrate content (B) at random locations within a Norway spruce stand in south-east Ireland. The points represent the raw data and the associated means and standard errors.

not have a significant impact on system-level nutrient losses and that there is a substantial difference between the faunal contribution to the mobilization of nutrients depending on the scale at which the parameter is measured, i.e. the effect was seen when only the litter layer was taken into account but it was buffered out by the inclusion of the mineral soil. However, the rooting systems of many plants, both herbs and trees, are concentrated in the organic layers of the soil and thus any effect which is seen at this level may have an impact on the plant assemblage.

Until recently, plant population biologists have tended to treat soil merely as a medium for support, and the organisms living in the soil have been ignored (Watkinson, 1998). The maintenance of diversity in plant communities has traditionally been assumed to result from the partitioning of abiotic resources

(Bever *et al.*, 1997). However, plant growth and the outcome of many competitive interactions between plants may be dependent on the soil organisms present (Watkinson, 1998). Bever *et al.* (1997) found substantial negative feedback on plant growth through the soil microfloral community, suggesting that it may be involved in the maintenance of species diversity, and spatially explicit simulations indicated that plant species diversity could be maintained between locally homogeneous patches when positive feedback and dispersal occurred at local scales. This concept is based primarily on plant interactions with microflora in the rhizosphere. However, we believe that the total soil community, including the invertebrates, should be taken into account in the same way.

Conditional outcomes, faunal heterogeneity and the maintenance of plant diversity

Bearing in mind the heterogeneous distribution of microarthropods in both space and time, it is clear that the differential grazing which may occur as a consequence of this, and the conditionality which is described above, will have important implications for the functioning of the microbial community and thus for the local availability of labile nutrients including N. One of the novel results to emerge from the microcosm experiment of Heneghan and Bolger (1996, 1997) was that such variation is expressed as an altered nutrient availability in the organic horizon. It is possible to envisage, therefore, a mosaic of nutrient availability. The significance of such environmental heterogeneity for plant growth is an active area of research (Campbell and Grime, 1989; Tilman and Cowan, 1989; Campbell *et al.*, 1991). The coexistence of dominants and subdominants in species-rich herbaceous plant communities may be facilitated by the propensity of dominants to forage on a larger scale than subdominants, which can exploit nutrients with more precision (Grime, 1994). Modelling efforts which examine the response of plant competitors to differences in the way nutrients are supplied and taken up by the organisms provide an insight into how local nutrient dynamics can allow coexistence between plant species (Huston and DeAngelis, 1994). The role of microarthropods in mobilizing nutrients in circumstances of low fertility and in producing ephemeral nutrient 'hotspots' can conceivably be of importance for the maintenance of plant diversity.

Similarly, albeit in a rather unusual situation, the mosaic nature of the distribution of earthworms has been shown to have a marked effect on the structure of the vegetation in coniferous forests under certain circumstances. In general, trees acidify soil (Howard *et al.*, 1989) and the decomposition of conifer leaf litter is more acidifying than that of broadleaf trees (Howard and Howard, 1990). Soil tends to become more acidic and the humus type change to mor as coniferous trees grow. This potentially creates difficulties for tree species requiring a mull humus for seedling establishment. For example, Norway spruce (*Picea abies*), which occurs extensively in Europe, is such a species. Earthworms are of critical importance in the regeneration and sustainability of this Norway spruce forest system as has been described by Bernier and Ponge (1994) and Ponge *et al.* (1998). They studied a Norway spruce forest in the French Alps and showed that the forest consisted of a mosaic of patches representing different stages in the forest regeneration processes. They found that, as the trees grew, and as a consequence of a decline in earthworm populations, the humus form changed from mull to moder. However, as a result of a succession of earthworm species into these moder patches, the humus form later changed to a mull-like humus which allowed the establishment of the spruce. The persistence of earthworm populations and thus of the system as a whole is dependent on the spatial heterogeneity of the stand and the persistence of endogeic species in suitable patches.

Conclusion

It appears that the soil fauna can have significant impacts on the nutrient flows in coniferous forest systems. The heterogeneity in the distribution of both resources and biota have potentially important effects in the maintenance of the forest system. Because of this heterogeneity and the conditionality of many of the processes in the system, it may be difficult to predict the extent and direction of changes in systems, where the relative importance of the patch types are altered during the change. However, it is clear that the fauna, operating within the constraints of a heterogeneous environmental template, has the potential to influence the dynamics of vegetation by means of determining seedling recruitment. In this manner, operating on small local scales and by mineralizing small absolute concentrations of nutrients, soil fauna can deflect the trajectory of succession.

The importance of plant community biodiversity in influencing ecosystem productivity has been well documented (e.g. Tilman *et al.*, 1996) and the role of mycorrhizae in maintaining such diversity has recently been recognized (van der Heijden *et al.*, 1998; Read, 1998). Here we suggest that invertebrate decomposers, a third component of the system, have similarly significant specific effects on community structure and succession which are mediated through the spatial heterogeneity of the systems and the conditions under which the fauna operate.

References

Aber, J.D. (1992) Nitrogen cycling and nitrogen saturation in temperate forest ecosystems. *Trends in Ecology and Evolution* 7, 220–223.

Aber, J.D., Nadelhoffer, P., Steudler, P. and Melillo, J.M. (1989) Nitrogen saturation in Northern forest ecosystems. *BioScience* 39, 378–386.

Abrahamsen, G. (1990) Influence of *Cognettia sphagnetorum* (Oligochaeta: Enchytraeidae) on nitrogen mineralization in homogenized mor humus. *Biology and Fertility of Soils* 9, 159–162.

Aerts, R. (1997) Climate, leaf litter chemistry and leaf litter decomposition in terrestrial ecosystems: a triangular relationship. *Oikos* 79, 439–449.

Ågren, G.I. and Bosatta, E. (1988) Nitrogen saturation of terrestrial ecosystems. *Environmental Pollution* 54, 185–197.

Alexander, M. (1977) *Introduction to Soil Microbiology.* John Wiley & Sons, New York.

Anderson, J.M. and Ineson, P. (1984) Interactions between micro-organisms and soil invertebrates in nutrient flux pathways of forest ecosystems. In: Anderson, J.M., Rayner, A.D.M. and Walton, D.W.H. (eds) *Invertebrate–Microbial Interactions.* Cambridge University Press, Cambridge, pp. 59–88.

Apps, M.J., Kurz, W.A., Luxmoore, R.J., Nilsson, L.O., Sedjo, R.A., Schmidt, R., Simpson, L.G. and Vinson, T.S. (1993) Boreal forests and tundra. *Water, Air and Soil Pollution* 70, 39–53.

Bååth, E., Lohm, U., Lundgren, B., Rosswall, T., Söderström, B. and Sohlenius, B. (1981) Impact of microbial-feeding animals on total soil activity and nitrogen dynamics: a soil microcosm experiment. *Oikos* 37, 257–264.

Bell, N. (1994) *The Ecological Effects of Increased Deposition of Nitrogen.* British Ecological Society, Ecological Issues No. 5. Field Studies Council, Shrewsbury.

Berg, B., Berg, M.P., Bottner, P., Box, E., Breymeyer, A., Calvo De Anta, R., Couteaux, M.-M., Escudero, A., Gallardo, A., Kratz, W., Madeira, M., Mälkönen, E., McClaugherty, C., Meentemeyer, V., Munoz, F., Piussi, P., Remacle, J. and Virzo de Santo, A. (1993) Litter mass loss rates in pine forests of Europe and Eastern United States: some relationships with climate and litter quality. *Biogeochemistry* 20, 127–159.

Berg, M.P., Verhoef, H.A., Bolger, T., Anderson, J.M., Beese, F., Coûteaux, M.-M., Henderson, R., Ineson, P., McCarthy, F., Raubuch, M., Splatt, P. and Willison, T. (1997) Effects of air pollutant–temperature interactions on mineral-N dynamics and cation leaching in replicate forest soil transplantation experiments. *Biogeochemistry* 39, 295–326.

Bernier, N. and Ponge, J.F. (1994) Humus form dynamics during the sylvogenic cycle in a mountain spruce forest. *Soil Biology and Biochemistry* 26, 183–220.

Bever, J.D., Westover, K.M. and Antonovics, J. (1997) Incorporating the soil community into plant population dynamics the utility of the feedback approach. *Journal of Ecology* 85, 561–573.

Bonan, G.B. and Van Cleve, K. (1992) Soil temperature, N mineralization and carbon source-sink relationships in boreal forests. *Canadian Journal of Forest Research* 22, 629–639.

Bornebusch, C. (1930) The fauna of forest soil. *Det Forstl Forsoksvaesen in Danmark* 11, 1–158.

Campbell, B.D. and Grime, J.P. (1989) A comparative study of plant responsiveness to the duration of episodes of mineral nutrient enrichment. *New Phytologist* 112, 261–267.

Campbell, B.D., Grime, J.P. and Mackey, J.M.L. (1991) A trade-off between scale and precision in resource foraging. *Oecologia (Berlin)* 87, 532–538.

Crossley, D.A., Jr and Witkamp, M. (1964) Forest soil mites and mineral cycling. *Acarologia* 6, 137–146.

Didden, W.A.M. (1993) Ecology of terrestrial Enchytraeidae. *Pedobiologia* 37, 2–29.

Encke, B.G. (1986) Stickstoff und waldsterben. *Allgemeines Forstzeitschrift* 37, 922–923.

Grime, J.P. (1994) The role of plasticity in exploiting environmental heterogeneity. In: Caldwell, M.M. and Pearcy, R.W. (eds) *Exploitation of Environmental Heterogeneity by Plants: Ecophysiological Processes Above- and Belowground.* Academic Press, San Diego, pp. 1–21.

Grossbard, E. (1969) A visual record of the decomposition of ^{14}C-labelled fragments of grasses and rye added to soil. *Journal of Soil Science* 20, 38–51.

Haimi, J. and Huhta, V. (1990) Effects of endogeic earthworms on soil processes and plant growth in a coniferous forest soil. *Biology and Fertility of Soils* 10, 178–183.

Hanlon, R.D.G. (1981) Influence of grazing by Collembola on the activity of senescent fungal colonies grown on media of different nutrient concentration. *Oikos* 36, 362–367.

Hanlon, R.D.G. and Anderson, J.M. (1979) The effects of Collembola grazing on microbial activity in decomposing leaf litter. *Oecologia (Berlin)* 38, 93–99.

van der Heijden, M.G.A., Klironomos, J.N., Ursic, M., Moutoglis, P., Steitwolf-Engle, R., Boller, T., Wiemken, A. and Sanders, I.R. (1998) Mycorrhizal fungal diversity determines plant biodiversity, ecosystem variability and productivity. *Nature* 396, 69–72.

Heneghan, L. and Bolger, T. (1992) CORE Project: Studies of the role of soil invertebrates in nutrient mobilisation in soil columns. In: Teller, A., Mathy, P. and Jeffers, J.N.R. (eds) *Responses of Forest Ecosystems to Environmental Changes.* Elsevier Applied Science, London, pp. 711–712.

Heneghan, L. and Bolger, T. (1996) Effects of the components of 'acid rain' on the contribution of soil microarthropods to ecosystem function. *Journal of Applied Ecology* 33, 1329–1344.

Heneghan, L. and Bolger, T. (1997) Are soil microarthropod assemblages functionally redundant? In: Mitchell, R., Horn, D.J., Needham, G.R. and Welbourn, W.C. (eds) *Acarology IX.* Ohio Biological Survey, Columbus, pp. 561–562.

Heneghan, L. and Bolger, T. (1999) The role of microarthropod species in forest ecosystem processes: the delimitation of relevant detail. *Plant and Soil* 205, 113–124.

Heungens, A. (1969) The physical decomposition of pine litter by earthworms. *Plant and Soil* 31, 22–30.

Howard, P.J.A. and Howard, D.M. (1990) Titratable acids and bases in tree and shrub leaf litter. *Forestry* 63, 177–196.

Howard, P.J.A., Thompson, T.R.E., Hornung, M. and Beard, G.R. (eds) (1989) *An Assessment of the Principles of Soil Protection in the UK,* vol. II. Institute of Terrestrial Ecology, Grange-over-Sands.

Huhta, V. and Koskenniemi, A. (1975) Numbers, biomass and community respiration of soil invertebrates in spruce forests at two latitudes in Finland. *Annales Zoologica Fennica* 12, 164–182.

Huhta, V. and Setälä, H. (1990) Laboratory design to simulate complexity of forest floor for studying the role of fauna in the soil processes. *Biology and Fertility of Soils* 10, 155–162.

Huhta, V., Setälä, H. and Haimi, J. (1988) Leaching of N and C from birch leaf litter and raw humus with special emphasis on the influence of soil fauna. *Soil Biology and Biochemistry* 20, 875–878.

Huston, M.A. and DeAngelis, D.L. (1994) Competition and coexistence: the effects of resource transport and supply rates. *The American Naturalist* 144, 954–977.

Keeney, D.R. (1980) Prediction of soil nitrogen availability in forest ecosystems: Literature review. *Forest Science* 26, 159–171.

Kimmins, J.P. (1996) Importance of soil and role of ecosystem disturbance for sustained productivity of cool temperate and boreal forests. *Soil Science Society of America Journal* 60, 1643–1654.

Kleb, H.R. and Wilson, S.D. (1997) Vegetation effects on soil resource heterogeneity in prairie and forest. *The American Naturalist* 150, 283–298.

Lavelle, P. (1983) The structure of earthworm communities. In: Satchell, J.E. (ed.)

Earthworm Ecology: from Darwin to Vermiculture. Chapman & Hall, London, pp. 449–466.

Lavelle, P., Blanchart, E., Martin, A., Martin, S., Spain, A., Toutain, F., Barois, I. and Schaefer, R. (1993) A hierarchical model for decomposition in terrestrial ecosystems: application to soil of the humid tropics. *Biotropica* 25, 130–150.

Lowery, M. (1991) Forestry – its part in rural development. In: Mollan, C. and Maloney, M. (eds) *The Right Trees in the Right Places.* Royal Dublin Society, Dublin, pp. 35–41.

Lussenhop, J. (1992) Mechanisms of microarthropod–microbial interactions in soil. *Advances in Ecological Research* 23, 1–33.

McLean, M.A., Kaneko, N. and Parkinson, D. (1996a) Does selective grazing by mites and collembola affect litter fungal community structure? *Pedobiologia* 40, 97–105.

McLean, M.A., Kolodka, D.U. and Parkinson, D. (1996b) Survival and growth of *Dendrobaena octaedra* (Savigny) in pine forest floor material. *Pedobiologia* 40, 281–288.

Newell, K. (1984) Interactions between two decomposer basidiomycetes and a collembolan under Sitka spruce: distribution abundance and selective grazing. *Soil Biology and Biochemistry* 16, 235–239.

O'Carroll, N. (1984) *The Forests of Ireland – History, Distribution and Silviculture.* Turoe Press, Dublin.

Parker, G.G. (1983) Throughfall and stemflow in the forest nutrient cycle. *Advances in Ecological Research* 13, 57–133.

Parkinson, D., Visser, S. and Whittaker, J.B. (1979) Effects of Collembola grazing on fungal colonisation of leaf litter. *Soil Biology and Biochemistry* 11, 188–207.

Persson, T. (1983) Influence of soil animals on nitrogen mineralization. In: Lebrun, P., André, H.M., de Medts, A., Gregoire-Wibo, C. and Wauthy, G. (eds) *New Trends in Soil Biology.* Dieu-Brichart, Louvain-la-Neuve, pp. 117–126.

Persson, T., Bååth, E., Clarholm, M., Lundkvist, H., Söderström, B. and Sohlenius, B. (1980) Trophic structure, biomass dynamics and carbon metabolism in a Scots pine forest. In: Persson, T. (ed.) *Structure and Function of Northern Coniferous Forest – an Ecosystem Study. Ecological Bulletin (Stockholm)* 32, 419–459.

Petersen, H. and Luxton, M. (1982) A comparative analysis of soil fauna populations and their role in decomposition processes. *Oikos* 39(3), 287–388.

Ponge, J.F. (1991) Succession of fungi and fauna during decomposition of needles in a small area of Scots pine litter. *Plant and Soil* 138, 99–114.

Ponge, J.F., André, J., Zackrisson, O., Bernier, N., Nilsson, M.-C. and Gallet, C. (1998) The forest regeneration puzzle. *BioScience* 48, 523–530.

Read, D. (1998) Plants on the web. *Nature* 396, 22–23.

Robinson, C.H., Ineson, P., Piearce, T.G. and Rowland, A.P. (1992) Nitrogen mobilization by earthworms in limed peat soils under *Picea sitchensis. Journal of Applied Ecology* 29, 226–237.

Schlesinger, W.H. (1984) Soil organic matter: a source of atmospheric CO_2. In: Woodwell, G.M. (ed.) *The Role of Terrestrial Vegetation in the Global Carbon Cycle.* John Wiley & Sons, New York, pp. 111–127.

Schlesinger, W.H. (1991) *Biogeochemistry – an Analysis of Global Change.* Academic Press, San Diego.

Seastedt, T.R. (1984) The role of microarthropods in decomposition and mineralization processes. *Annual Review of Entomology* 29, 25–46.

Setälä, H. (1995) Growth of birch and pine seedlings in relation to grazing by soil fauna on ectomycorrhizal fungi. *Ecology* 76, 1844–1851.

Setälä, H. and Huhta, V. (1990) Evaluation of the soil fauna impact on decomposition in a simulated coniferous forest soil. *Biology and Fertility of Soils* 10, 163–169.

Setälä, H. and Huhta, V. (1991) Soil fauna increase *Betula pendula* growth – laboratory experiments with coniferous forest floor. *Ecology* 72, 665–671.

Setälä, H., Martikainen, E., Tyynismaa, M. and Huhta, V. (1990) Effects of soil fauna on leaching of nitrogen and phosphorus from experimental systems simulating coniferous forest floor. *Biology and Fertility of Soils* 10, 170–177.

Setälä, H., Rissanen, J. and Markkola, A.M. (1997) Conditional outcomes in the relationship between pine and ectomycorrhizal fungi in relation to biotic and abiotic environments. *Oikos* 80, 112–122.

Shaw, P. (1988) A consistent hierarchy in the fungal feeding preferences of the Collembola *Onychiurus armatus*. *Pedobiologia* 31, 179–187.

Sulkava, P., Huhta, V. and Laakso, J. (1996) Impact of soil faunal structure on decomposition and N-mineralisation in relation to temperature and moisture in forests soil. *Pedobiologia* 40, 505–513.

Swift, M.J., Heal, O.W. and Anderson, J.M. (1979) *Decomposition in Terrestrial Ecosystems.* Blackwell Scientific Publishers, Oxford.

Tamm, C.O. (1991) *Nitrogen in Terrestrial Ecosystems.* Ecological Studies 20, Springer-Verlag, Heidelberg.

Tilman, D. and Cowan, M.L. (1989) Growth of old field herbs on a nitrogen gradient. *Functional Ecology* 3, 425–438.

Tilman, D., Wedin, D. and Knops, J. (1996) Productivity and sustainability influenced by biodiversity in grassland ecosystems. *Nature* 379, 718–720.

Verhoef, H.A., Dorel, F.G. and Zoomer, H.R. (1989) Effects of nitrogen deposition on animal-mediated nitrogen mobilization in coniferous litter. *Biology and Fertility of Soils* 8, 255–259.

Wallwork, J.A. (1970) *Ecology of Soil Animals.* McGraw-Hill, London.

Walsh, M.I. and Bolger, T. (1990) Effects of diet on the growth and reproduction of some Collembola in laboratory cultures. *Pedobiologia* 34, 161–171.

Walsh, M.I. and Bolger, T. (1994) Effects of diet on the interactions between *Hypogastrura denticulata* Bagnall and *Onychiurus furcifer* Börner in laboratory cultures. *European Journal of Soil Biology* 29, 155–160.

Watkinson, A.R. (1998) The role of the soil community in plant population dynamics. *Trends in Ecology and Evolution* 13, 171–172.

Williams, B.L. and Griffiths, B.S. (1989) Enhanced nutrient mineralization and leaching from decomposing Sitka spruce litter by enchytraeid worms. *Soil Biology and Biochemistry* 21, 183–188.

Wolters, V. (1988) Effects of *Mesenchytraeus glandulosus* (Oligochaeta, Enchytraeidae) on decomposition processes. *Pedobiologia* 32, 387–398.

Impacts of Insects on Human-dominated and Natural Forest Landscapes

14

R.N. Coulson[1] and D.F. Wunneburger[2]

[1]*Knowledge Engineering Laboratory, Department of Entomology,* [2]*College of Architecture, Texas A&M University, College Station, TX 77843, USA*

Introduction

Insects are a ubiquitous element of forests. As a class of organisms characterized by extraordinary diversity in form, function and habit, they exert many different types of impacts. In the forestry community the subject of impact assessment has traditionally centred on valuation of negative effects caused by insect activity, for example damage or destruction of trees, incidence of insect-vectored disease or nuisance to recreationists. This view has gradually changed during the past 30 years, as a consequence of research dealing with the identification of functional roles of insects and other organisms in ecosystems (Seastedt, 1984; Seastedt and Crossley, 1984; Chapin *et al.*, 1997). Much of this book is devoted to an examination of how arthropods in particular are involved in regulating the various ecosystem processes.

All forests are influenced by the actions of humans (Noble and Dirzo, 1997). The degree of influence may vary from subtle changes in forest health associated with atmospheric pollution to drastic alteration in species composition and biodiversity resulting from harvesting. Clearly, the actions of humans and the activities of insects are closely related, but evaluating the interaction has been difficult because the common approach to impact assessment has simply involved a summary of categorical data (e.g. number of bark beetle infestations ha^{-1} of host type) without specific reference to spatial extent, resolution or forest structure. In this chapter we consider the impact of insects on forest landscapes. Our goal is to provide a sense of place to the discussion of impact assessment. We emphasize the importance of both the content and context of spatial elements forming the forest landscape, as human-caused fragmentation and natural disturbances create mosaic patterns where the specific arrangement of components can enhance or inhibit impacts of insects (Coulson *et al.*, 1999a). The specific objectives of the commentary are to: (i) define the scope (subject matter) of impact assessment; (ii) consider impact assessment in a landscape ecological context; and (iii) examine case histories that illustrate impacts of insects on landscape structure, function and rate of change.

The Scope of Impact Assessment

Herein the term *impact* is defined broadly to mean any effect resulting from the activities of insects in forests (Coulson and Witter, 1984). *Impact assessment* requires consideration of three questions: first, what are the activities of insects that cause impacts? Second, how do these activities affect forest resources? Third, how can the effects of forest insects be interpreted or valued (Fig. 14.1)? Each of these subjects is briefly examined below. Given this framework, impact assessment is represented as an anthropocentric activity.

Activities of forest insects

From a process perspective, the major impacts of insects in forests can be defined from an examination of activities associated with herbivory, carnivory and detritivory. A remarkable variety of insects is associated with each process and the following discussion is intended simply to provide a flavour of the types of activities insects are involved in that cause impacts.

In most investigations of impact, greatest emphasis is placed on evaluating the effects of herbivory. Indeed, the insect herbivores are so numerous that they are often grouped according to feeding preference, for example defoliators, sapsucking, terminal feeding, root feeding, seed and cone infesting, phloem boring, wood boring and gall makers (Coulson and Witter, 1984). Most herbivorous insects occur in the Coleoptera, Hymenoptera, Diptera, Lepidoptera, Phasmida and Orthoptera. In some instances plant modules of construction (leaves, stems, branches, cones, etc.) are the targets of herbivory and in other instances the genet is actually killed. The insect herbivores that sequester pollen and nectar for food have a particularly important impact on forests, as they also serve as pollinators of flowering plants. Most of the insect pollinators occur within the Hymenoptera, but many Lepidoptera, Diptera and Coleoptera are also involved as well as bats, other mammals and birds. Herbivory can influence the forest in many ways, by: killing selected species of trees; altering plant community species composition and age structure; weakening trees and increasing their vulnerability to plant pathogens and natural disturbances; modifying the growth form and appearance of trees; reducing food supplies used by other herbivores; limiting or enhancing plant reproduction and regeneration.

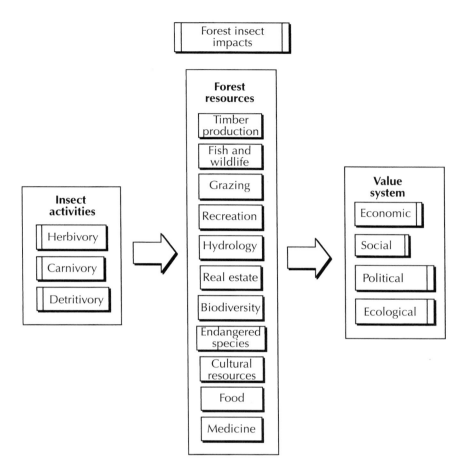

Fig. 14.1. Assessment of the impact of insects in forest landscapes involves an understanding of how the activities of insects affect forest uses in the context of a value system (Knowledge Engineering Laboratory, Texas A&M University).

Insect carnivory, as a process operating in forests, is much less understood than herbivory. There are numerous predatory species of insects in forests and the majority occur in the orders Coleoptera, Neuroptera, Hymenoptera, Hemiptera, Diptera, Mantodea and Odonata. Insect parasites (i.e. parasitoids) occur primarily in the Diptera and Hymenoptera. Most of the interest in impact of parasitoids and predators centres on their role as regulators of populations of pest herbivore species, i.e. as biological control agents. There are also a number of parasitic insects (Diptera, Hemiptera, Anoplura, Mallophaga) and other arthropods (Arachnida, ticks and mites) which serve to vector vertebrate diseases and thereby have an impact on wildlife populations, domesticated animals and humans in forests.

Detritivory is the process that results in the breakdown of dead organic matter in forests. The process takes place primarily within the litter and soil. Detritivory is addressed in detail in Coleman and Crossley (1996). The organisms involved represent a variety of taxa and the arthropods are often collectively referred to as micro- or macroarthropods based on their size. The primary role of micro- and macroarthropods is as agents that facilitate decomposition and mineralization processes (Seastedt and Crossley, 1984). The Acari dominate the microarthropods, but insects are also prominent players. Representative species occur in the orders Collembola, Protura and Diplura. The macroarthropods again include several arthropod taxa. Among the important insects are the Isoptera, Hymenoptera and Coleoptera. The impact of detritivory on forest resources centres on issues associated with nutrient dynamics and forest productivity.

Forest resources

Humans use forests in many different ways. In some instances the resources associated with forests are managed through the orchestrated activities of humans. The traditional resources addressed in forest management planning and operation include timber production, fish and wildlife, grazing, recreation, hydrology and real estate (Coulson and Witter, 1984). These resources are the basis of the multiple use forest management concept practised by the USDA Forest Service for many years. To this list we can add the following: biodiversity, endangered species, cultural resources, food and medicinal plants (Noble and Dirzo, 1997) (Fig. 14.1). Through herbivory, carnivory and detritivory, insects can have an impact on each of these resource values. In some instances, for example biodiversity and endangered species, insects are part of the resource value.

Placing value on impact

Defining the consequence or impact of herbivory, carnivory and detritivory on forest resources requires a value system. Impact can be viewed categorically from economic, social, political and ecological perspectives (Fig. 14.1). *Economic impact* is simply defined as the effect of insects (and other injurious agencies) on the monetary receipts from the production of goods and services on forest lands. *Social (axiological) impact* refers to the effects of insects on aesthetic, moral and metaphysical values associated with forests. *Political impact* refers to the actions, practices and policies employed by governments in response to insect activities on public or regulated forest properties. *Ecological impact* refers to the functional roles that insects play in forests. No judgement is made regarding positive or negative effects of the ecological roles played by insects (Coulson and Witter, 1984).

The overview presented in Fig. 14.1 suggests that assessment of insect impact is a challenging task. In some instances, the insect activity can affect multiple resources and be valued by more than one criterion. For example, herbivory by the southern pine beetle, *Dendroctonus frontalis* (Coleoptera: Scolytidae) causes mortality to mature hosts (*Pinus* spp.) which results in an economic loss to the forest owner. However, the patch created by the infestation may provide additional habitat for wildlife and thereby produce an ecological benefit. Most of the world's forests are not intensively managed for the production of goods and services. In general, concern for impact of insects on the forest resources varies directly as a function of the degree of human investment.

Impact Assessment in Landscapes

The preceding section dealt with the topical information of impact assessment. The point of departure for this discussion of landscapes is the knowledge base for the roles of insects in forest ecosystems. Therefore, our approach will be, first, to consider briefly the ecosystem perspective and, second, to examine the framework for impact assessment in forest landscapes.

Ecosystems

Many of the investigations of the functional roles of insects in forest ecosystems were conducted on sites like the Coweeta Hydrologic Laboratory, North Carolina, Hubbard Brook Experimental Forest, New Hampshire, and H.J. Andrews Experimental Forest, Oregon. The basic unit of study was the ecosystem, which Likens (1992) defined as 'A spatially explicit unit of the Earth that includes all of the organisms, along with all components of the abiotic environment within its boundaries'.[1] One of the important features of the research conducted on these experimental forests was that boundaries for the studies were delineated by watersheds, that is, the scale of the investigations was set by natural landform features. Interaction with adjacent ecosystems was thereby minimized and this circumstance facilitated measurement of nutrient input, output and internal cycling. It was also possible to manipulate the ecosystem processes and compare between different watersheds.

Much of our knowledge of the functional roles of insects in ecosystems came from investigations dealing with nutrient dynamics. Research was directed to questions of how insects were involved in regulating the rates of the basic ecosystem processes, i.e. primary production, consumption, decomposition and abiotic storage (Fig. 14.2) (Crossley *et al.*, 1984). This framework also proved to be useful in the initial studies of ecosystem management, i.e. the orchestrated modification or manipulation of the four processes. For example in forestry, primary production is modified by application of herbicides.

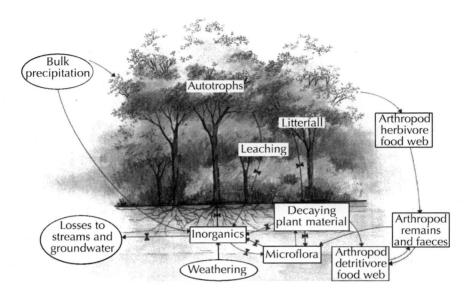

Fig. 14.2. The basic processes and pathways of nutrient cycling and energy flow within an ecosystem (modified from Crossley *et al.*, 1984).

Pesticides are used to reduce populations of consumer organisms. Harvesting and site preparation practices influence decomposition. Addition of fertilizer influences abiotic storage (Crossley *et al.*, 1984). The various ecological roles of arthropods in the nutrient dynamics of forest ecosystems were summarized by Seastedt and Crossley (1984) (Fig. 14.3) and this volume addresses and elaborates on the new discoveries.

Landscapes

Basic information associated with landscape level of ecological (or geographical) integration is summarized in Bailey (1996, 1998), Forman (1995), Forman and Godron (1986) and Zonneveld and Forman (1990). In this section we examine the spatial composition and organization of a landscape and then consider how insects interact within this framework to influence its structure, function, change and management. This discussion provides the foundation for investigating ecological impacts of insects in forest landscapes. There are numerous recent texts and reviews dealing with various specific aspects of ecology and forestry in a landscape context: Boyce (1995), Hansson *et al.* (1995), McCullough (1996), Hanski and Gilpin (1997), Polis *et al.* (1997), Tilman and Kareiva (1997), Bailey (1998), Gustafson (1998), Hof and Bevers (1998) and NRC (1998).

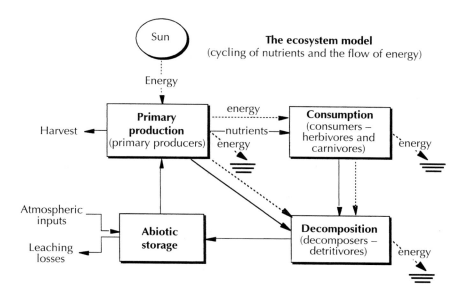

Fig. 14.3. A simplified conceptual model of elemental cycling in a terrestrial ecosystem emphasizing the presence and activities of arthropod consumers. Virtually all fluxes within ecosystems are known or believed to respond to various levels of arthropod activity (modified from Seastedt and Crossley, 1984).

Spatial composition and organization of landscapes

Art, architecture, geography and ecology use the term *landscape* in unique ways. Even within the discipline of landscape ecology there are several acceptable definitions. Examples of general definitions are 'a mosaic where a cluster of local ecosystems is repeated in similar form over a kilometers-wide area' (Forman, 1995) or 'a geographic group of site level ecosystems' (Bailey, 1998) (Fig. 14.4). The landscape environment consists of the lithosphere, landform and atmosphere. Ecologists tend to be interested in biotic attributes of the landform and perhaps the soil and atmosphere. The geographers add the lithosphere.

The actions and activities of insects in forest landscapes are guided by attributes of landscape structure. Forman (1995) summarizes in detail basic information on landscape structure. Several attributes of landscape composition and organization, pertinent to this discussion, are illustrated in Fig. 14.4 and briefly described below.

First, the fundamental unit of abstraction in the landscape is the ecosystem. This unit has discrete boundaries, which are generally visible on a map or aerial photograph, and contains both the biotic community and abiotic

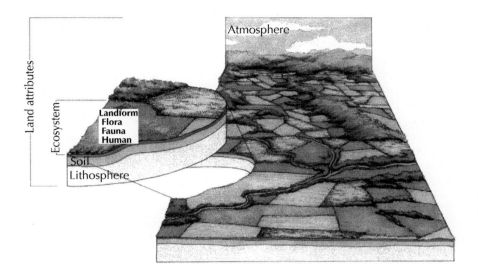

Fig. 14.4. A landscape, i.e. a mosaic where a cluster of local ecosystems is repeated in similar form over a kilometres-wide area (Forman, 1995). The fundamental unit of the landscape is the ecosystem. Landscapes have horizontal (*chorology*) and vertical (*topological*) dimension as well as specific attributes. The exploded portion of the figure illustrates an *ecosystem cluster*, where exchange of energy, materials and species is accentuated (Knowledge Engineering Laboratory, Texas A&M University).

environment. The basic ecosystem processes (primary production, consumption, decomposition and abiotic storage) operate within the boundary (Fig. 14.4). This view of the ecosystem is essentially the one expressed by Likens (1992). The fundamental unit is variously referred to in the literature as an ecosystem, ecotope, landscape element, tessera and biogeocoenose.

Second, the landscape mosaic has both horizontal and vertical dimensions. The horizontal dimension, which is referred to as the *chorology*, deals with the study of the landscape units, i.e. the kinds of landscape elements present that make up the mosaic. The vertical dimension, which is referred to as the *topology*, deals with the study of landscape attributes, i.e. the landform, lithosphere and atmosphere. A landscape typically will have a characteristic or attribute that is of particular importance and interest, such as a pine forest landscape, an agricultural landscape or a prairie landscape (Zonneveld, 1990). Each of these features (horizontal dimension, vertical dimension and characteristic) is illustrated in Fig. 14.4.

Third, another useful schema for organization of landscapes is the *ecosystem cluster*, which is defined as a group of ecosystems (landscape elements) connected by significant exchange of energy, materials and species (Forman, 1995). The exploded portion of Fig. 14.4 illustrates four closely configured ecosystems that together form a cluster. Many insects (and other animal species as well) require more than one type of ecosystem. For a particular species, dispersal strategies delineate the size of the ecosystem cluster and details of the natural history define the number needed.

Fourth, organization of the knowledge base, and hence study of landscape ecology concepts and principles, is greatly aided by the use of a simple descriptive model which recognizes three fundamental components of landscapes: patches, corridors and the matrix (Forman and Godron, 1986; Forman, 1995). This model serves as the basis for defining structure (components of the landscape and their linkages and configurations), function (quantities of flows of energy, materials and species within and among landscape elements), change (alteration in the structure and function of the ecological mosaic over time) (Forman, 1995) and management (the orchestrated modification of landscape structure, function or rate of change).

Insect activities in forest landscapes

Experimental studies of insect activities at the landscape level of ecological integration are rare. Most of the interest on insects in landscapes has centred on concerns about outbreaks. Typically, studies of insect outbreaks in forests have considered impact of herbivore populations on food and habitat resources (Barbosa and Schultz, 1987; Holling, 1988; Platt and Strong, 1989). In this context, outbreaks are considered to be autogenic disturbances that are normally observed in mesoscale landscape as levels of herbivory above an average or expected amount. Causes for change in the distribution and abundance of insects in landscape mosaics are poorly understood. In particular, activities of insects in forest landscapes have not been investigated using explicit data encompassing a broad spatial extent. A fundamental and unanswered research question in landscape ecology centres on how the spatial arrangement of ecosystems influences the distribution and abundance of organisms across complex landscape mosaics (Dunning *et al.*, 1992; Pulliam *et al.*, 1992; Turner *et al.*, 1993, 1995; Urban, 1993; Pickett and Candenasso, 1995).

One legacy of the economic importance of forest insects is a robust knowledge base on the natural history of many species. The literature is replete with examples of studies dealing with various aspects of demographics and behaviour (Coulson and Witter, 1984). For well-studied insects (e.g. the southern pine beetle (*D. frontalis*), gypsy moth (*Lymantria dispar*), Douglas-fir tussock moth (*Orgyia pseudotsugata*), the spruce budworm (*Choristoneura fumiferana*), etc.), the extant information is suitable for formulating rules to describe how the organism perceives and responds to the various elements of the landscape mosaic. The rules can be based on the results of robust empirical study or they can simply represent heuristic knowledge of human experts, i.e. knowledge based on experience. Typically, a rulebase for a specific insect consists of a blend of each type. An evaluation of the kinds, numbers and spatial arrangement of landscape elements used by the insect provides the basic information needed for an impact assessment.

The definition of habitat requirements for insects in forest landscapes is a straightforward task amenable to direct observation or experimentation. How insects link habitat elements together in time and space is a considerably more difficult question. The answer to this question centres on an understanding of how species move within landscape mosaics. The subject of species movement in landscapes has been summarized by Forman (1995). Conclusions pertinent to this discussion include the following: (i) the landscape mosaic is the template across which the search for food and habitat takes place; (ii) spatial elements (ecosystems) of the landscape mosaic vary widely in their suitability; (iii) insects have targets which generally are suitable patches containing resources and conditions necessary for survival, growth and reproduction; (iv) species movement is a function of spatial attributes of each mosaic, and within a specific landscape both the context and content of the elements are important variables influencing the activity; (v) the arrangement of suitable and unsuitable elements has an important effect on the spread of species; and (vi) the spread of a species is especially dependent on the arrangement of corridors which can act as conduits, filters, barriers, habitat, sources and sinks (Forman, 1995). Therefore, in addition to knowledge of species response to the various elements of landscape heterogeneity, we must also have an understanding of how the insect moves within the mosaic. Both types of information are rarely available for insects in general and forest species in particular.

Case Histories of Insect Impact in Forest Landscapes

At the onset of this discussion, we indicated that most efforts at insect impact assessment in forest landscapes have involved summary of categorical data without specific reference to spatial extent and mosaic structure. This approach provides a useful means for accounting social, economic, ecological and political impacts of insects in forest landscapes. However, it does little to explain the mechanism of impact or the reasons for variation in the amount observed under different conditions. These two issues are at the forefront of the research agenda for impact assessment. The solidification of general principles of landscape ecology and the development of new tools and technologies (GIS, GPS, remote sensing, digital map and image processing, spatial statistics and modelling, etc.) are helping to advance understanding in this research frontier.

Nevertheless, there are good examples of the different kinds of impacts insects can have on landscapes. We have identified three case histories that serve to illustrate how insects can influence landscape structure, function and change. These three topics are obviously related, but in our discussion we emphasize (isolate) each subject. The examples include the southern pine beetle, *Dendroctonus frontalis* (impact on landscape structure); the honey bee,

Apis mellifera (Hymenoptera: Apidae) (impact on landscape function); and the balsam woolly adelgid, *Adelges piceae* (impact on landscape change).

Southern pine beetle impact on forest landscape structure

The southern pine beetle (Fig. 14.5) is a cryptic insect that spends most of its life history in the inner bark of host trees. It reproduces within the host and brood development is completed there. The host tree provides a protected habitat as well as food resources. Upon completion of a life cycle, the brood adults disperse and colonize a new host. Herbivory by *D. frontalis* results in mortality to the host tree.

There are three types of habitat targets required by the insect in a forest landscape mosaic: acceptable host species, susceptible habitat patches and lightning-struck hosts. *Acceptable host species* include the commercially important species of southern yellow pines: loblolly (*Pinus taeda*), shortleaf (*Pinus echinata*) and slash (*Pinus elliotti*), and occasionally longleaf (*Pinus palustris*). *Susceptible habitat patches* include stands containing mature loblolly, shortleaf pine and slash pine with high basal area and stagnate radial growth. Such stands are considered to be high hazard. These stands are important in *D. frontalis* epidemiology as they represent habitat patches and they are suitable for growth (enlargement) of infestations. In some instances the infestations can occupy several hectares. A pine forest landscape mosaic generally consists of a collection of patches (stands) that range from low to high hazard.

Fig. 14.5. An adult southern pine beetle (from the Southern Forest Insect Work Conference photographic slide series).

Lightning-struck hosts represent a special instance of an acceptable host species. The insect can locate these hosts and easily colonize them. Presumably, *D. frontalis* identifies the hosts from resin volatiles produced as a consequence of lightning striking the tree. The strike also diminishes the effectiveness of the resin system as a defence mechanism. Lightning-struck hosts function as epi-centres for the initiation of multiple-tree infestations, refuges for dispersing bee-tles and stepping stones that link *D. frontalis* populations in habitat patches (Coulson *et al.*, 1999a,b).

Schowalter *et al.* (1981) considered the role of *D. frontalis* and fire in maintenance of structure and function of the south-eastern coniferous forest. They suggested that herbivory by the insect served to truncate ecosystem development at a time when the forest had become stagnant or 'overconnected'. In the Holling (1992) scheme of ecosystem succession, *D. frontalis* serves as the agent of creative destruction and its actions result in the release of tightly bound biomass and nutrients associated with the large trees of old-growth forests.

How this scenario plays out in an actual pine forest is very much a function of the structure of the landscape. Figure 14.6 represents a specific instance of herbivory that occurred over several years in the Little Lake Creek Wilderness Area on the Sam Houston National Forest in south-east Texas. This 1495 ha landscape was vegetated primarily with uniform old-growth loblolly pine (Fig. 14.6a). Several infestations of *D. frontalis* occurred in the forest and created isolated patches (Fig. 14.6b). Regeneration of pines within the patches followed and this circumstance introduced a new age structure to the forest landscape. Later, a massive infestation of *D. frontalis* resulted in mortality to the remaining old-growth pines (Fig. 14.6c). Again, regeneration of pines followed. However, the structure of the pine forest landscape was fundamentally altered by the sequence of herbivory. Growth of pines in the initial infestations had progressed and these patches now contained the oldest age class present in the landscape (Fig. 14.6d). Herbivory by *D. frontalis* introduced a new age structure and resulted in fragmentation of the forest landscape. Infestations of *D. frontalis* will occur again as the forest matures, but the impact will be ameliorated by the fact that large contiguous areas of old-growth pine no longer exist.

This example illustrates how herbivory by *D. frontalis* has an impact on forest landscape structure (i.e. components of the landscape and their linkages and configurations). As wilderness areas within National Forests are not managed for commercial purposes, the forest uses (Fig. 14.1) affected are primarily fish and wildlife, recreation, hydrology, biodiversity, endangered species and cultural resources. The impacts are mainly ecological and social.

The impact of honey bees on forest landscape function

When European colonists arrived in the 17th century, there were no honey bees, *Apis mellifera* (Fig. 14.7), in North America. Although more than 3500 native species of bees occurred, none produced useful amounts of wax or honey.

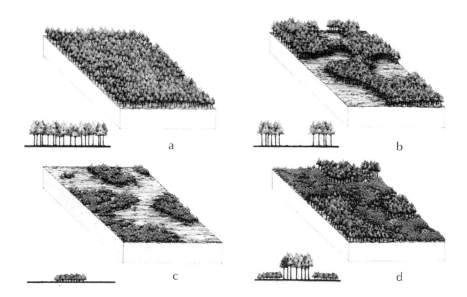

Fig. 14.6. Change in landscape structure brought about by herbivory of the southern pine beetle (Knowledge Engineering Laboratory, Texas A&M University). (a) Pine forest landscape consisting of contiguous old-growth (60 years +) yellow pine, which provides suitable habitat and food for the southern pine beetle. (b) Several infestations of the southern pine beetle occurring during the same season created isolated patches. Rates of decomposition of infested host trees proceeds rapidly in the southern USA. Regeneration of pines within the patches commonly occurs. (c) A large infestation of the southern pine beetle followed and the remaining old-growth pines were killed. Again, regeneration of pines occurred. (d) The structure of the forest landscape was fundamentally altered by the sequence of herbivory. Herbivory by the southern pine beetle introduced a new age structure and resulted in fragmentation of the forest landscape.

A. mellifera was introduced into the USA sometime between 1620 and 1650. By 1850 the insect, which became known as the 'European bee', was established on both the east and west coasts and during the next 50 years the insect would spread throughout North America. Some were carried overland by settlers and others dispersed from feral colonies that had escaped from their keepers (Batra, 1996).

Feral honey bees are European bees that have become established as colonies in the natural landscape (Fig. 14.8). Feral colonies occur as a result of a behaviour known as swarming, in which managed bees abandon the hive and search out new habitat sites. This behaviour is one way in which social insects increase the number of colonies. Feral *A. mellifera* serve to pollinate native plants, which provide food and habitat for wildlife and also vegetation cover for millions of hectares of forests and uncultivated lands.

Fig. 14.7. Honey bees are the dominant pollinators of flowering plants in forest landscapes (courtesy of the Texas Agricultural Extension Service).

The commercial beekeepers recognize the importance of maintaining viable feral *A. mellifera* colonies. Throughout the history of beekeeping in North America, new strains of bees have been introduced into the commercial market. These bees are usually associated with specific regions of Europe (e.g. German bees, Italian bees, etc.) and all were imported because they possessed attributes valued by beekeepers (e.g. good temperament, disease resistance, adaptability to specific climates, honey production, etc.). Recently, with the immigration of African bees into Texas, a new strain has entered the amalgam. Today, the feral colonies represent the aggregate gene pool of all the imported strains, whereas the commercial hives reflect the latest preference of the beekeepers. The genetic diversity associated with the feral populations of *A. mellifera* is looked to as a resource for investigation of subjects ranging from disease and parasite resistance to acclimatization.

Two species of parasitic mites have been recently introduced into the USA: the honey bee tracheal mite, *Acarapis woodi*, and the *Varroa* mite, *Varroa jacobsoni*. These parasites cause mortality to *A. mellifera* colonies and are considered to be the most serious pests affecting commercial beekeeping. Once established in managed hives, the tracheal mite and *Varroa* mite were soon discovered in feral bee colonies. The degree of impact on feral populations is the subject of considerable speculation and loss estimates range from 25 to 90% of colonies.

Fig. 14.8. Feral honey bees are European bees that have become established as colonies in natural landscapes. Parasitization by introduced mites has greatly reduced the presence of honey bees in forests (Knowledge Engineering Laboratory, Texas A&M University).

Since its introduction into North American landscapes, *A. mellifera* has become the dominant pollinator of native and introduced flowering plants. This insect forages broad areas and many different plant species. However, before the introduction of *A. mellifera*, flowering plants were pollinated primarily by native species of insects. Insect pollinators occur principally in the orders Hymenoptera, Lepidoptera, Diptera and Coleoptera. The prominent members of the Hymenoptera occur in the superfamily Apoidea (bees) which includes the Apidae (the honey bee, bumble bees), Anthophoridae (digger bees, carpenter bees), Halictidae (sweat bees) and Megachilidae (leaf-cutter bees).

The constituency and extent of the native fauna of pollinators was not of much concern until the feral *A. mellifera* colonies declined (Buchmann and Nabhan, 1996). As feral *A. mellifera* have become the dominant pollinator species in our natural and managed landscapes, it is not clear whether the native species (i.e. those pollinator species present before the introduction of the honey bee) still persist today. These species may have been displaced and without the feral honey bees, suitable pollinators of flowering plants may not exist in forest landscapes (Allen-Wardell *et al.*, 1998).

Most plants move twice during their life cycle: once as a pollen grain and once as a seed. Therefore the pollination mutualists and seed dispersal mutualists play a major role in landscape function (Bronstein, 1995). In the Holling (1992) scheme of ecosystem succession, the pollinators (*A. mellifera* and other species) facilitate the transition from exploitation to conservation. Although many variables are involved in floristics, plant species diversity in forested landscapes is in part dependent on the presence of insect pollinators.

This example was chosen to illustrate the impact of *A. mellifera* on landscape function, i.e. how the role of the insect as a pollinator can influence quantities and flows of energy, materials and species within and among landscape elements. In this example the impact occurs because the insect has been reduced in number within the forest landscape as a consequence of the activities of parasitic mites. As a result, the critical process of pollination may not occur. Given this circumstance, all of the forest uses (Fig. 14.1) are affected. The impact is economic, social, political and ecological.

The impact of the balsam woolly adelgid on landscape change

The balsam woolly adelgid, *Adelges piceae* (Fig. 14.9), is an exotic species that was introduced into the USA *c.* 1900 on infested nursery stock imported from Europe (Fig. 14.9). The insect is broadly distributed in North America and Canada and infests a variety of fir (*Abies*) species. Herbivory often results in mortality to the tree (Coulson and Witter, 1984).

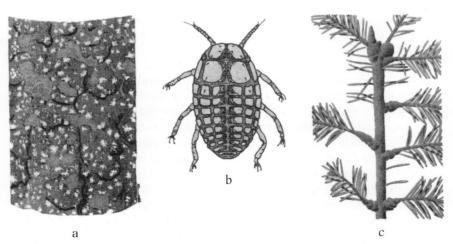

a   c

Fig. 14.9. Balsam woolly adelgid (a) sessile stage, (b) crawler stage and (c) typical damage to branches of fir (modified from Novak *et al.*, 1976).

Fig. 14.10. Fraser fir in the southern Appalachian mountains (USA) infested and killed by the balsam woolly adelgid (Knowledge Engineering Laboratory, Texas A&M University).

In the mid-1950s, *A. piceae* was discovered in the spruce–fir forests of the southern Appalachian mountains, USA. Infestations occurred in Fraser fir (*A. fraseri*) (Fig. 14.10). The spruce–fir forests in the south have little commercial timber value but they are highly prized for their scenic beauty. The forests occur in prominent tourist destinations and recreational sites, such as The Black Mountains (Mount Mitchell), The Great Smoky Mountains National Park, the Balsam Mountains and the Mount Rodgers National Recreational Area.

Activity of *A. piceae* in the southern Appalachian mountains has been monitored closely and effects on the spruce–fir forests noted (Dull *et al.*, 1988). The spatial pattern of mortality in Fraser fir has proved to be difficult to explain. This species occurs at high elevations (1314–2037 m) and is therefore subject to extreme fluxes in weather conditions and direct exposure to atmospheric deposition. *Adelges piceae* is often identified as the cause for mortality, but this is probably the result of the interaction of herbivory with abiotic environmental conditions.

Regeneration of the Fraser fir after herbivory by *A. piceae* does not necessarily follow. In some instances natural regeneration does occur (Fig. 14.11a)
but in other cases dense thickets of blackberry (*Rubus* spp.) vegetate the sites
formerly occupied by the tree (Fig. 14.11b). In the Holling (1992) scheme of
ecosystem succession, the transition from reorganization to exploitation can
produce two fundamentally different results: (i) natural regeneration of the
original spruce–fir forest; or (ii) replacement with blackberry thickets. As few
other woody plant species can survive at the high elevations where the
spruce–fir forests occur, the structure of the landscape is significantly changed
in comparison with the original condition.

This example illustrates how herbivory by *A. piceae* has an impact on
landscape change (alteration in the structure and function of the ecological mosaic
over time). The forest uses affected are primarily recreation, hydrology, biodiversity
and perhaps endangered species. The impacts are mainly social and ecological.

Conclusions

1. The degree of human imprint on the Earth's ecosystems in general (Vitousek
et al., 1997) and forests in particular (Noble and Dirzo, 1997) is a subject of

a   b

Fig. 14.11. (a) In some instances natural regeneration of fir occurs following
infestation by the balsam woolly adelgid. (b) In other instances the sites are covered
with dense thickets of blackberry (Knowledge Engineering Laboratory, Texas A&M
University).

concern, debate and speculation. This subject is also the focus of considerable scientific enquiry and one aspect of the research agenda centres on developing an understanding of the roles of insects in forest landscapes.

2. Assessment of impact of insects in forest landscapes involves understanding how the activities of insects affect forest uses in the context of a value system.

3. Much of the research on insects in ecosystems has focused on evaluating their roles in regulating the rates of the basic ecosystem processes: primary production, consumption, decomposition and abiotic storage.

4. The spatial framework needed to investigate impacts of insects in landscapes has been examined and general principles of species movement within the mosaic considered in the published literature.

5. In forest landscapes, insects have direct impacts on structure, function and rate of change. Probable impacts of insects in forest landscapes can be inferred through an examination of how the natural history of a particular species operates at broad spatial scales (spatial extent). As few rigorous studies of insects in landscapes have been conducted, this research arena offers many challenges and the opportunity to contribute new knowledge to ecological science as well as the applied disciplines of entomology and forestry.

6. Examples were chosen to illustrate how insects can impact landscape structure (*D. frontalis*), function (*A. mellifera*) and change (*A. piceae*). The case histories were used to demonstrate how information on natural history could be used in context with the developing corpus of knowledge on landscape ecology. Tools (GIS, GPS, remote sensing, digital map and image processing, spatial statistics and modelling, etc.) needed to conduct rigorous studies of roles of insects in forest landscapes are now available.

Acknowledgements

We acknowledge and thank A.M. Bunting for technical assistance in the preparation of the text and figures for this chapter, and Drs F.P. Hain, T.D. Schowalter, F.M. Stephen and two anonymous reviewers, for their critique of the manuscript.

Note

1. In the applied literature dealing with ecosystem management, the definition of exactly what constitutes an ecosystem is vague with regard to scale, i.e. range and resolution. The term ecosystem is often used to describe the environment around a special organism of interest, for example the forest ecosystem, the cotton ecosystem or the apple orchard ecosystem. Further, the term is also used to describe land areas of interest such as the Yellowstone Ecosystem or the Everglades Ecosystem (Coulson *et al.*, 1999b).

References

Allen-Wardell, G. *et al.* (1998) The potential consequence of pollinator declines on the conservation of biodiversity and stability of food crop yields. *Conservation Biology* 12, 8–17.

Bailey, R.G. (1996) *Ecosystem Geography.* Springer, New York.

Bailey, R.G. (1998) *Ecoregions.* Springer, New York.

Barbosa, P. and Schultz, J.C. (eds) (1987) *Insect Outbreaks.* Academic Press, San Diego.

Batra, S. (1996) Unsung heroes of pollination. *Natural History* 106, 42–43.

Boyce, S.G. (1995) *Landscape Forestry.* John Wiley & Sons, New York.

Bronstein, J.L. (1995) The plant-pollinator landscape. In: Hansson, L., Fahrig, L. and Merriam, G. (eds) *Mosaic Landscapes and Ecological Processes.* Chapman & Hall, New York.

Buchmann, S.L. and Nabhan, G.P. (1996) The *Forgotten Pollinators.* Island Press, Washington, DC.

Chapin, F.S., III, Walker, B.H., Hobbs, R.J., Hopper, D.U., Lawton, H.H., Sala, O.E. and Tilman, D. (1997) Biotic control of the function of ecosystems. *Science* 277, 500–504.

Coleman, D.C. and Crossley, D.A., Jr (1996) *Fundamentals of Soil Ecology.* Academic Press, San Diego.

Coulson, R.N. and Witter, J.A. (1984) *Forest Entomology.* John Wiley & Sons, New York.

Coulson, R.N., McFadden, B.A., Pulley, P.E., Lovelady, C.N., Fitzgerald, J.W. and Jack, S.B. (1999a) Heterogeneity of forest landscapes and the distribution and abundance of the southern pine beetle. *Forest Ecology and Management* 114, 471–485.

Coulson, R.N., Saarenmaa, H., Daugherity, W.C., Rykiel, E.J., Jr, Saunders, M.C. and Fitzgerald, J.W. (1999b) A knowledge system environment for ecosystem management. In: Klopatek, J. and Gardner, R. (eds) *Landscape Ecological Analysis.* Springer-Verlag, New York, pp. 57–79.

Crossley, D.A., Jr, House, G.J., Snider, R.M., Snider, R.J. and Stinner, B.R. (1984) The positive interactions in agroecosystems. In: Lowrance, R., Stinner, B.R. and House, G.J. (eds) *Agricultural Ecosystems.* John Wiley & Sons, New York.

Dull, C.W., Ward, J.D., Brown, H.D., Ryan, G.W., Clerke, W.H. and Uhler, R.J. (1988) Evaluation of spruce and fir mortality in the southern Appalachian mountains. USDA Forest Service Protection Report R8-PR 13.

Dunning, J.B., Danielson, B.J. and Pulliam, H.R. (1992) Ecological processes that affect populations in complex landscapes. *Oikos* 65, 169–175.

Forman, R.T.T. (1995) *Land Mosaics.* Cambridge University Press, New York.

Forman, R.T.T. and Godron, M. (1986) *Landscape Ecology.* John Wiley & Sons, New York.

Gustafson, E.J. (1998) Quantifying landscape spatial pattern: what is the state of the art? *Ecosystems* 1, 143–156.

Hanski, I. and Gilpin, M.E. (eds) (1997) *Metapopulation Biology: Ecology, Genetics and Evolution.* Academic Press, San Diego.

Hansson, L., Fahrig, L. and Merriam, G. (eds) (1995) *Mosaic Landscapes and Ecological Processes.* Chapman & Hall, New York.

Hof, J. and Bevers, M. (1998) *Spatial Optimization for Managed Ecosystems.* Columbia University Press, New York.

Holling, C.S. (1988) The role of insects in structuring the boreal landscape. In: Shugart, H.H., Leemans, R. and Bonan, G.B. (eds) *A Systems Analysis of the Global Boreal Forest.* Cambridge University Press, New York.

Holling, C.S. (1992) Cross-scale morphology, geometry, and dynamics of ecosystems. *Ecological Monographs* 62, 447–502.

Likens, G.E. (1992) Reflections on ecosystem science in relation to environmental policymaking. In: *The Ecosystem Approach: Its Use and Abuse*. Ecology Institute, Olendorf/Luhe, Germany.

McCullough, D.R. (1996) *Metapopulations and Wildlife Conservation*. Island Press, Washington, DC.

Noble, I.R. and Dirzo, R. (1997) Forests as human-dominated ecosystems. *Science* 277, 522–525.

Novak, V., Hrozinka, F. and Stary, B. (1976) *Atlas of Insects Harmful to Forest Trees*, Vol. 1. Elsevier Scientific, New York.

NRC (1998) *Forested Landscapes in Perspective*. National Academy Press, Washington, DC.

Pickett, S.T.A. and Candenasso, M.L. (1995) Landscape ecology: spatial heterogeneity in ecological systems. *Science* 269, 331–334.

Platt, W.J. and Strong, D.R. (eds) (1989) Gaps in forest ecology feature. *Ecology* 70, 535–576.

Polis, G.A., Anderson, W.B. and Holt, R.D. (1997) Toward an integration of landscape and web ecology: the dynamics of spatially subsidized food webs. *Annual Review of Ecology and Systematics* 28, 289–316.

Pulliam, H.R., Dunning, J.B., Jr and Liu, J. (1992) Population dynamics in complex landscapes: a case study. *Ecological Applications* 2, 165–177.

Seastedt, T.R. (1984) The role of microarthropods in decomposition and mineralization processes. *Annual Review of Entomology* 29, 25–46.

Seastedt, T.R. and Crossley, D.A., Jr (1984) The influence of arthropods on ecosystems. *BioScience* 34, 157–161.

Schowalter, T.D., Coulson, R.N. and Crossley, D.A., Jr (1981) The role of southern pine beetle and fire in the maintenance of structure and function of the southeastern coniferous forest. *Environmental Entomology* 10, 821–825.

Tilman, D. and Kareiva, P. (eds) (1997) *Spatial Ecology*. Princeton University Press, Princeton, New Jersey.

Turner, M.G., Romme, W.H., Gardner, R.H., O'Neill, R.V. and Kratz, T.K. (1993) A revised concept of landscape equilibrium: disturbance and stability on scaled landscapes. *Landscape Ecology* 8, 213–227.

Turner, M.G., Arthaud, G.J., Engstrom, R.T., Hejl, S., Liu, J., Loeb, S. and McKelvey, K. (1995) Usefulness of spatially explicit population models in land management. *Ecological Applications* 5, 12–16.

Urban, D.L. (1993) Landscape ecology and ecosystem management. In: Covington, W.W. and DeBano, J.F. (eds) *Sustainable Ecological Systems: Implementing an Ecological Approach to Land Management*. USDA Forest Service General Technical Report RM-247.

Vitousek, P.M., Mooney, H.A., Lubchenco, J. and Melillo, J.M. (1997) Human domination of Earth's ecosystems. *Science* 277, 494–499.

Zonneveld, I.S. (1990) Scope and concepts of landscape ecology as an emerging science. In: Zonneveld, I.S. and Forman, R.T.T. (eds) *Changing Landscapes: an Ecological Perspective*. Springer-Verlag, New York.

Zonneveld, I.S. and Forman, R.T.T. (eds) (1990) *Changing Landscapes: an Ecological Perspective*. Springer-Verlag, New York.

Soil Fauna and Controls of Carbon Dynamics: Comparisons of Rangelands and Forests across Latitudinal Gradients

15

T.R. Seastedt

Department of EPO Biology and Institute of Arctic and Alpine Research, University of Colorado, Boulder, CO 80309–0450, USA

Introduction

The contention that soil arthropods are 'webmasters of ecosystems' is based on two well-established hypotheses: (i) 'the litter fauna has a pronounced influence on the rate of breakdown and release of nutrient minerals from forest litter' (Crossley and Witkamp, 1964, p. 891); and (ii) 'soil arthropods have the capacity to influence, perhaps to regulate, nutrient uptake by plants' (Crossley, 1977, p. 53). Given that nutrient limitation controls plant productivity and species composition in most terrestrial ecosystems (e.g. Chapin, 1980; Vitousek and Howarth, 1991), the direct and indirect effects of soil arthropods can significantly alter ecosystem characteristics in most terrestrial communities.

This conclusion that soil arthropods function as important controls on ecosystem processes appears to be largely opaque to the general scientific community. This may be because the attributes of the fauna are often adequately captured and incorporated into abiotic, proxy variables that are much easier to measure and summarize. This simplification has advantages; however, it is important to recognize the situations and conditions where such relationships will not be valid.

The scientific community has accepted the general model of controls on decomposition and mineralization processes as articulated by Swift *et al.* (1979) (Fig. 15.1). Heal *et al.* (1997) indicate that over 1000 papers on litter decomposition have been published since the appearance of their benchmark text on soil decomposition processes, but their general model remains valid. The

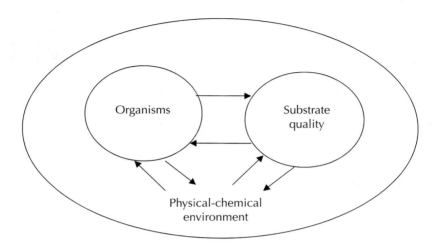

Fig. 15.1. The Swift *et al.* (1979) model of factors affecting decomposition. All components exhibit the capability to modify the others, albeit at different temporal and spatial scales. Controls by climate and soil texture (physical and chemical factors) are often acknowledged as the regional controls on soil carbon storage.

understanding of decomposition and mineralization requires adequate documentation of abiotic and biotic structure of the decomposer community, as influenced by the physical and chemical environment. The importance of each control can be experimentally established by holding two of the three components constant and then manipulating the magnitude or variability of the third component. The extent to which a value can deviate due to manipulation of the experimental variable then demonstrates the strength or relative importance of that component. The importance of soil arthropods in the decomposition and mineralization process was established with this procedure (Crossley and Witkamp, 1964; Swift *et al.*, 1979).

    While this experimental approach to understanding the relative importance of climate, substrate quality and biota in affecting decomposition processes of a given ecosystem is appropriate, there are some conceptual pitfalls that should be explicit. If one experimentally removes all sources of variation except for the variable of interest, that variable receives credit for the observed pattern and therefore often assumes substantial importance. Hierarchy theory has demonstrated that, depending upon scale considerations, 'important' variables become invisible or insignificant when evaluated at higher or lower levels of resolution (e.g. Allen and Starr, 1982). Lavelle *et al.* (1993) and Beare *et al.* (1995) provide examples of how the climate or the soil environment can constrain activities of fauna. This approach is also inadequate if many of the patterns we see in descriptive studies are generated by interactions among the various components. These interactions cannot be understood without more complex experimentation.

In comparing decomposer activities across sites, the activities of the biota are often strongly constrained by substrate quality and climate variables. Decay is fast for simple molecular compounds in warm, moist environments, and slow for complex substrates in cold, dry deserts or in cold, wet sites. Thus, a significant part of the variance in these processes can be accounted for using only substrate quality and climate, and assuming the biota will function as some sort of constant, constrained by the other variables. This leads investigators to a well-known conclusion: 'When climate and site conditions are constant on a temporal and spatial scale, decomposition rates are regulated primarily by the chemical composition and physical structure of the organic matter' (Berg *et al.*, 1998, p. 170). Even if we implicitly acknowledge biotic factors as causal mechanisms, why invest time measuring details about the composition and activity of the biota? This is especially true if soil food webs are moderately non-linear systems. Decomposer effects are generated by the specific trophic dynamics of the soil food web, which is under strong biotic (top-down control) as well as abiotic (bottom-up control) constraints (e.g. Moore *et al.*, 1988; Wardle and Lavelle, 1997). Thus, strong seasonal and year-to-year variability based on non-linear interactions should be anticipated in these systems, particularly at the subsystem or population level. Food web structure may produce either amplification or attenuation of specific processes which would lead to additional variation or uncertainty about functional relationships.

Meentemeyer (1978) presented a simple empirical model of decomposition that explained pattern at regional and global scales. The model used lignin and actual evapotranspiration (AET) as independent variables to estimate decomposition (Fig. 15.2). Subsequent ecosystem models that have been used in global change analyses have almost uniformly used these measurements as proxy variables for decomposer biotic activity. 'This ability to predict the rate of a central ecosystem process when only a few important variables are known is important to the understanding and prediction of ecosystem function at the global level' (Aber and Melillo, 1991, p. 188).

If biota are important, then are they uniformly important across ecosystem types so that their effects can be adequately contained within the proxy variables of the Meentemeyer model? In this chaper, I compared ecosystem characteristics of two different terrestrial ecosystems occurring in similar climates and asked: (i) how decomposition in these systems differed; (ii) how the composition of soil fauna differed; and (iii) if changes in the abundance or composition of the decomposer biota influenced ecosystem patterns. Second, I extended these comparisons across a latitudinal gradient and repeated this set of questions. While this comparative approach has obvious limitations, the exercise provides an overview of the prerequisite complexity required for experimental studies, if such studies are to provide generality and adequately portray the role of fauna in ecosystem processes at regional and global scales.

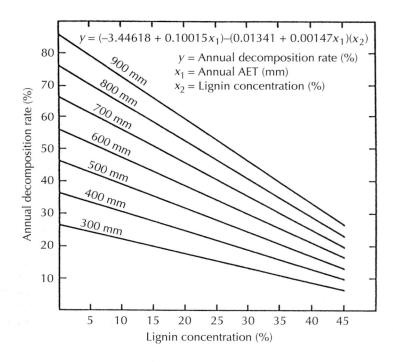

Fig. 15.2. The Meentemeyer (1978) model of climate (actual evapotranspiration) and substrate quality (lignin) controls on decomposition (reprinted with permission from Aber and Melillo, 1991).

Comparison of Rangeland and Forest Ecosystems

Grasslands of the world occupy somewhat over 20% of the terrestrial surface (Curry, 1987a); inclusion of more arid regions as well as alpine and tundra increases this estimate to almost 50% of the Earth's surface (Chapin *et al.*, 1996). The natural woodland–grassland boundaries are generally maintained by fire. Rangelands are bordered by hot deserts at one extreme and by glaciers and bogs at the other, spanning areas from the semitropics to the arctic regions. Rangelands show wide variation in temperatures but a more limited range of precipitation (Fig. 15.3). These lands therefore exhibit a somewhat restricted set of AET values, ranging from about 200 to 1000 mm, generated by water limitation at high annual temperatures and energy limitation at cold sites. Temperate zone rangelands are associated with three soil orders, Mollisols, Histosols and Aridisols, as well as including areas containing newer soils, the Entisols and Inceptisols. Tropical savannas may also be composed of Oxisols and Ultisols. Soil acidity can range from very basic in the arid grasslands to moderately acidic in the colder, wetter grasslands and tundra. Thus, the unifying

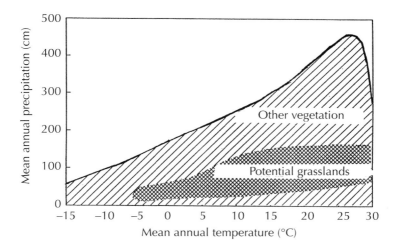

Fig. 15.3. The relationship between grasslands and other terrestrial ecosystems found across temperature and moisture gradients. The grasslands exhibit large ranges in temperatures, but relatively small ranges in precipitation. Tundra areas would extend the grasslands to the coldest of ecosystem types (redrawn from Whittaker, 1975).

features of grasslands would seem few except for the fact that graminoid plant species often compose the dominant vegetation.

Comparison of decomposition and mineralization of grasslands and forests

The temperate rangelands and tundra that are dominated by grasses and sedges do appear to possess a unifying characteristic. The soils of these systems store a relatively larger amount of carbon and associated limiting nutrients such as N and P than do woody ecosystems occupying similar climates (Jenny, 1930) (Table 15.1). While these estimates lack the carbon reserves of deeper soils, which can be substantial (Stone *et al.*, 1993), this pattern probably holds. Given that these grasslands have reduced plant productivity or are at most equal in productivity to corresponding forests (e.g. Whittaker, 1975), this pattern in soil organic matter (SOM) indicates that decomposition processes of grassland soils are, on average, slower than those observed in forests occupying the same climatic zone. This pattern, therefore, represents an ecosystem characteristic that cannot be explained by climate, even when observed across a large temperature gradient. Thus, substrate quality and biota remain as possible controlling factors explaining this phenomenon.

Before attempting to interpret this pattern, however, a second complication is noted. The trend for higher carbon storage in grasslands holds for most

Table 15.1. Global distribution of organic matter in terrestrial ecosystems (from Schlesinger, 1977).

Ecosystem type	Soil organic matter (kg C m^{-2})
Tropical forest	10.4
Temperate forest	11.8
Boreal forest	14.9
Woodland/shrubland	6.9
Tropical savanna	3.7
Temperate grassland	19.2
Tundra	26.1

temperate and cold region comparisons, but data in Table 15.1 indicate that tropical savanna soils do not store more carbon than their forest counterparts. Schlesinger (1997), while acknowledging a small database, did not believe that the pattern which Jenny (1930) observed continued into tropical areas. Lavelle (1988a) cites a number of studies that suggest that grassland-savannas have reduced carbon storage compared with forests. Soil carbon storage is higher beneath trees in grassland savanna areas (Scholes and Archer, 1997). The loss of soil carbon with conversion of tropical forests to grasslands has been documented by Johnson and Wedin (1997). Moreover, lower SOM levels can also be found in comparisons of selected temperate grasslands with forest counterparts (e.g. Zak *et al.*, 1994).

In contrast to those studies, however, are the works of Cerri *et al.* (1991), Chone *et al.* (1991), Bonde *et al.* (1992) and Feigl *et al.* (1995) which indicate that SOM increases in areas where tropical forests have been converted to grasslands. Moreover, this increase in C storage may be enhanced with introduced grasses (Fisher *et al.*, 1994). This issue therefore remains unresolved in the tropics. And, in the desert regions of the south-west USA, grassland sites also appear to have about 10% more carbon in soils than do adjacent or replacement shrublands (Schlesinger and Pilmanis, 1998). Thus, inconsistencies in soil carbon storage patterns between grasslands and woodlands of tropical and semitropical regions exist which are not explained by climatological variables or by the dominant vegetation present at the site.

Parton *et al.* (1987) stressed that soil texture, in addition to temperature and moisture concerns, was a very important variable influencing soil carbon storage. SOM is less likely to be stabilized in coarser soils, and these sites have reduced carbon storage. In areas where grasslands and forests co-occur, forests are more likely to be found on coarser-textured soils. Thus, the patterns correlated with vegetation could be generated by texture differences, alone. Some, but certainly not all of the inconsistencies presented above may result from soil texture differences in the paired comparisons. Texture may constrain the efficiency of the biota in the consumption of SOM. Hassink *et al.* (1993a,b) demonstrated that greater mineralization of C and N in sandy soils was related to enhanced protozoan grazing of bacteria on these sites. The absence of protected sites explained the more rapid turnover of microbes and enhanced N flux observed in coarser soils.

However, causal relationships among soil texture, vegetation and carbon storage are complex. Vegetation types within the temperate zone, if given sufficient time, appear capable of converting Mollisol-type soils to Alfisol- or Spodosol-type soils and vice versa (Holtmeier and Broll, 1992; Willis *et al.*, 1997). Thus, the vegetation–climate interaction strongly influences the soil type and probably affects soil texture. Moreover, there are examples where carbon storage differences occur when soil texture is not different between vegetation types (Pauker and Seastedt, 1996, and unpublished results). Thus, while recognizing texture as an important variable influencing soil carbon storage, texture alone cannot entirely explain the differences in soil carbon storage observed between rangelands and forests.

Schlesinger (1977) pointed out that a major portion of the grassland detritus is deposited *in situ* by roots, whereas a higher percentage of detritus in forests is from canopy litter. However, in terms of actual carbon inputs, the data suggest that grasslands rarely, if ever, match the productivity of adjacent forests. While root detritus is more intimately associated with older carbon and mineral soil, exactly why this carbon should be less vulnerable to mineralization is not apparent. While the soil microclimate may, on average, be cooler than the surface, the enhanced moisture content of the soil may compensate and decomposition of litter in soil is equal or faster than surface litter decomposition in grasslands (Seastedt *et al.*, 1992). Also, new data summaries (Casper and Jackson, 1997, p. 552) challenge the dogma that grasslands actually do exhibit greater root abundance at greater depths. Collectively, then, changes related to climate, soil texture and rooting distributions of plants all appear to be inadequate to explain the grassland–forest patterns in carbon sequestration.

How do the decomposer invertebrate fauna differ in grasslands and forests in similar climates?

The grasslands include a huge amount of diversity in soil invertebrate composition. The warmer, dry grasslands are dominated by termites and other arthropods, while more temperate, semihumid grasslands are dominated by earthworms and secondarily by macroarthropods such as millipedes. Further north or higher in elevation, the colder grasslands and tundra are again dominated by arthropods, but in these regions the microarthropods rather than macroarthropods or termites dominate the fauna (Douce and Crossley, 1977; Swift *et al.*, 1979; Petersen and Luxton, 1982).

While the biomass of soil invertebrates in both tropical forests and tropical grasslands is very large (Lavelle, 1988b), the general decline in invertebrate biomass from the tropics to the poles described by Swift *et al.* (1979) has some interesting deviations in the temperate zone. The arid regions around 30° latitude produce a soil invertebrate fauna dominated by termites and microarthropods (Whitford *et al.*, 1988). Macroinvertebrate abundance and biomass, generated largely by earthworms, is potentially greatest further north in the temperate, semihumid grasslands (e.g. Seastedt *et al.*, 1987, 1988). Nematodes are very abundant, approaching densities of 10^7 organisms m^{-2} in the humid grasslands (Todd, 1996). While data on other microfauna are sparse, the biomass and activities of the microfauna in the temperate semihumid grasslands are suggested to match or exceed numbers of similar organisms found in more northerly ecosystems. Thus, a temperate zone maximum in soil invertebrate fauna appears to occur in the semihumid temperate grasslands, with lower densities and biomass of fauna in both colder and hotter grasslands found in temperate zone regions. This pattern deviates significantly from the general global pattern presented in Swift *et al.* (1979).

Work from humid and semihumid grasslands (e.g. Seastedt, 1984a; O'Lear, 1996), dry grasslands (Leetham and Milchunas, 1985; Walter *et al.*, 1987), desert (Santos *et al.*, 1984; Whitford *et al.*, 1988), alpine tundra (O'Lear and Seastedt, 1994) and arctic tundra (Douce and Crossley, 1977) indicates that prostigmatid mites are the most numerous of the microarthropods in these ecosystems. This observation was also noted for European grasslands (Curry, 1987b). These patterns of microarthropod abundance are very different from those reported for forests. Prostigmatids include many trophic groups but may be dominated by nematode-feeders and other invertebrate predators (Walter, 1987; Whitford *et al.*, 1988). This is not surprising given the abundance of nematodes in these systems, but the inference is that the trophic function of the Oribatida, which is probably a combination of detritivory and fungivory, is of secondary importance. The overall trophic classification of the group may be closer to predators than detritus-feeders; i.e. the fauna occupy a higher trophic level on average. It is intriguing to suggest a causal relationship between the trophic composition of the fauna and the differences in SOM storage in grasslands and forests. 'Arthropods tend to have the greatest effect on decomposition and nutrient cycling in systems dominated by fungi. Typically, these are forest systems' (Moore *et al.*, 1988, p. 434).

We know that at least portions of the soil food web in grasslands are very sensitive to changes in resource quality and quantity. Frequent fires in grasslands appear to have a positive effect on macroarthropods (Seastedt, 1984b) and earthworms (James, 1991), presumably due to enhanced detritus inputs from unburned portions of the plant (i.e. roots and often surface litter). Termites in tropical regions do not respond positively to either fire or grazing (Decaens *et al.*, 1994); this pattern may be reversed in more temperate regions (e.g. O'Lear *et al.*, 1996). Todd (1996) reported a substantial amount of sensitivity of nematode trophic groups to grassland management practices. Thus, resource quality is clearly an issue in structuring grassland food webs, as it is in forests (Cheng and Wise, 1997).

While the composition and abundance of the decomposer fauna do appear to be sensitive to nitrogen availability, the effects of this nitrogen availability on soil respiration activities of both grasslands and forests are less evident. Neither Pastor *et al.* (1987) nor Bryant *et al.* (1998) were able to demonstrate that exogenous nitrogen availability had any effect on the early stages of litter decay. Hunt *et al.* (1988) did, however, show that nitrogen availability had a modest, positive effect on decomposition rate. Seastedt and Crossley (1983) and Prescott (1995) could not demonstrate a statistically significant response of forest litter treated with enhanced nitrogen. Using a different approach, Wardle *et al.* (1997) demonstrated that mixed litter containing highly decomposable material (high nitrogen, low lignin) did occasionally stimulate litter decay of more recalcitrant litter such that the average rate of litter decay was increased. It is very possible that the presence of high quality substrates containing labile carbon and nutrients stimulates microbes and associated fauna, and the presence and activity of this biota then influences the decay of other associated materials. However, when nitrogen alone is added, decomposition is not stimulated. This is consistent with the notion that microbes are often energy rather than nutrient limited and that the response of this group is critical to carbon flux of litter. Microarthropods, however, do exhibit numerical responses to the addition of inorganic nitrogen to litter (Seastedt and Crossley, 1983). Perhaps high nitrogen–low labile carbon could create systems where over-grazing of the microflora is more likely, thereby constraining carbon flux.

Does the decomposer system function less efficiently in grasslands?

The soil organic matter found in the surface (A) horizon of rangelands is largely composed of processed detritus, i.e. invertebrate faeces and associated microbial debris. The key question is whether this processing by invertebrates is positive, neutral or negative in terms of plant matter decay and SOM stability. Jastrow and Miller (1999) indicate that the stabilization of SOM is much faster than the processes involved in humification. SOM is stabilized as a result of biochemical recalcitrance, as influenced by chemical stabilization and physical protection. Jastrow and Miller state that the recalcitrance may be due to chemical characteristics of the substrate itself (i.e. lignin content), by materials produced by soil fungi and other soil organisms, or from transformations during decomposition, including incorporation into the excrement of soil meso- and microfauna (Kooistra and van Noordwijk, 1996). Work done or cited in Jastrow (1996), Jastrow *et al.* (1996) and Jastrow and Miller (1998) suggests that the combination of fibrous rooting and associated mycorrhizae of these roots provides the macrostructure (a fibrous bag of sorts) that holds macro- and microaggregates.

Grazing and fragmentation activities of microarthropods and larger invertebrates in the soil would, superficially at least, appear only to disrupt the stabilization process. Faecal production by arthropods as well as casting by earthworms is known to influence subsequent decay dynamics of this material in the short term (e.g. Blair *et al.*, 1995). However, long-term effects remain poorly documented. Anderson (1988) was unable to demonstrate that fauna could influence the humus content of microcosm soils. In semihumid to semi-arid zones, the burrowing activities of larger arthropods and earthworms may significantly enhance the hydraulic conductivity of soils, but since water export is minor, effects on dissolved or particulate organic matter export would be minor as well.

The conversion of decaying plant materials and microbes into faecal material may alter subsequent decomposition rates (e.g. Webb, 1977; Blair *et al.*, 1995). Faecal material, like unconsumed, decaying plant litter, exhibits a long-term decomposition rate that may not be strongly correlated with its short-term decay dynamics. While the short-term effect of conversion of litter to faecal material appears to generate an increase in substrate decay, the longer-term dynamics may be neutral or even negative. This fact is perhaps best known for earthworms. Heal *et al.* (1997) cite work reported by Zech (1991) indicating a very significant decline in the long-term decomposition of substrates converted to faecal material by earthworms. Martin (1991) provided similar results for earthworms in a tropical site. Both studies suggest that the physical mixing of mineral clays with recalcitrant organic matter in the guts of inverte-brates produces compact soil casts that may contribute to subsequent aggregate stability and reduced decay rates of this material. Similar studies are unknown for arthropods, although Webb (1977) provided a conceptual model for how arthropod faecal material could affect surface area–volume relationships of substrates and decomposition rates. Most arthropods do not ingest mineral soil to the extent observed in earthworms or tropical termites, so the physical association of mineral clays with organics in faecal material may not be a common function of this group, at least not in ecosystems found in temperate climates.

A more compelling argument for increased carbon storage can be made by considering the effects of the decomposer fauna on plant productivity (Fig. 15.4). If, via stimulation of limiting nutrient mineralization, plant detritus inputs are enhanced by the activities of fauna feeding on labile components of litter and associated microflora, then the fauna effects on long-term decom-position may not strongly influence soil carbon storage. Indeed, if the argument that the mineralization effects of fauna outweigh their effects on carbon flux (via direct and indirect effects on soil respiration) holds, then the net result would be enhanced carbon storage. This argument has potential merit for three reasons. As mentioned in the introduction, nitrogen is intensely limiting in most terrestrial ecosystems and is particularly limiting in non-tropical regions. The effects on nitrogen mineralization by fauna are well established, even under soil $C:N$ ratios that favour nitrogen immobilization (Anderson, 1988). Second, this same process that makes nitrogen available to plants may also influence the nitrogen environment of recalcitrant substrates. Berg (1986), Fog (1988) and Berg *et al.* (1998) report that enhanced availability of nitrogen appears to suppress the subsequent rate of decomposition of older litter substrates. Berg *et al.* (1998) stated that increased nitrogen may reduce microbial species diversity, change the dominant microbial decomposers from fungi to bacteria and

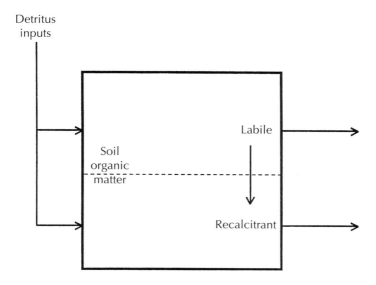

Fig. 15.4. The soil organic matter component of soils is controlled by the amount of inputs and rate of outputs. Fauna could enhance soil organic matter storage by stimulating increased inputs via enhanced mineralization of labile substrates while concurrently having a neutral or negative effect on the decomposition of recalcitrant materials. Mechanisms that may facilitate sequestration include the changes caused by fauna to the physical and chemical environment of the recalcitrant materials.

potentially repress the formation of enzymes with lignolytic capabilities. Thus, one can make the argument that the mineralization effects of fauna enhance both plant productivity and may favour enhanced sequestration of SOM in soils by reducing rates of decay of more recalcitrant substrates. Finally, there is the argument that, given the combined top-down and bottom-up controls on the decomposer food web, occasional overgrazing of microbes by fauna, a mechanism that would reduce carbon flux but enhance mineralization, is a potential common outcome of soil food web interactions (Wardle and Lavelle, 1997). Arthropods function at several trophic levels in this process. To state that overgrazing is more common among the trophic food webs of the grasslands is pure speculation, but this potential mineralization mechanism deserves further empirical and modelling efforts.

Global Patterns of Ecosystem Productivity and Decomposition

Observations by Jenny *et al.* (1949) and Olson (1963) on the patterns of detritus inputs and decomposition across latitudinal gradients and the consequences these fluxes have on carbon storage are now an important topic of global change (Schlesinger, 1977). Decomposition rates diminish faster in more northerly ecosystems than do rates of plant productivity (NPP) (Fig. 15.5). The inference is that decomposition is generally more temperature limited while plant NPP is more responsive to other variables (e.g. Schlesinger, 1977; Schimel, 1995). Just as grassland systems appear to have decomposers unable to match the decomposition rates of their forest counterparts, the decomposers of more northerly ecosystems appear to be unable to consume the annual productivity of plants, a feat easily accomplished by tropical counterparts. Why the asymmetry?

The exponential (Q_{10}) relationship between temperature and respiration is well known. However, to argue that cold temperatures are the reason for enhanced SOM storage in northerly environments is at most a proximate explanation. If resources are available and other constraints are not limiting, we should expect to observe biotic strategies to obtain those resources. Lower per capita respiration rates could be compensated by higher densities and biomass of decomposer flora and fauna, thereby keeping SOM levels reduced. Clearly, additional environmental constraints must exist to preclude this response.

Whitford *et al.* (1981) and Santos *et al.* (1984) demonstrated that when soil biotic activity is not adequately captured by actual evapotranspiration

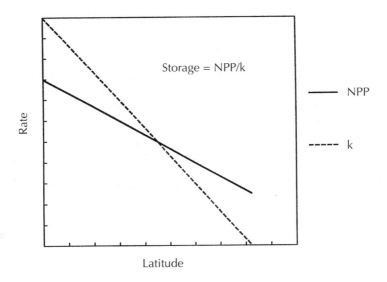

Fig. 15.5. The global pattern of plant productivity (NPP) and decay rate of soil organic matter (k). Decomposition rates decline relatively faster than NPP along an increasing latitudinal gradient, creating the largest soil organic matter reservoirs in more northerly environments (Beck, 1971).

measurements, the decomposition model based on AET will not predict decay. In those studies fauna were able to circumvent water restrictions. O'Lear *et al.* (1996) showed that decomposition of a common substrate on sites with identical microclimates was different if land management histories (and, presumably, the soil decomposer food web) were different between sites. Recently, Heneghan *et al.* (1999) provided data suggesting that site-specific contributions of fauna in tropical forests could independently influence decomposition in spite of similarities in climate and substrate quality. In their study, site variations in biota superceded abiotic sources of variation. Thus, whenever the biotic activity is not linearly coupled to climatic constraints, abiotic proxy variables obviously cannot be expected to predict decomposition processes. This coupling between biotic and abiotic factors is clearly different both between grasslands and forests and across latitudinal gradients. The diversity of biotic functions is sufficiently significant to impose deviations in patterns of global carbon storage and flux.

Unlike the rangeland–forest comparison discussed earlier, one cannot easily distinguish between the relative importance of the three controlling variables discussed in the Swift *et al.* (1979) model. Climate, substrate quality and biota all covary across this gradient. As indicated by the Meentemeyer (1978) model, substrate quality appears to be amplified in more favourable environments, while climate appears to dominate in cold regions (Couteaux *et al.*, 1995). The reasons for the climate–substrate quality interaction are not obvious, but two observations are offered. The problems associated with freezing have yet to be solved in an evolutionary sense by most macroinvertebrates. The absence of this group is not compensated for, completely, by meso- and microinvertebrates, and those groups are also limited in their ability to feed at temperatures favouring ice nucleation. Second, when climate limitation is no longer expressed on the composition of the biota, substrate quality becomes relatively more important (Couteaux *et al.*, 1995). Competition by the decomposer flora and fauna for high-quality resources appears to be intense. With a full spectrum of soil fauna present, selectivity may be enhanced, thereby amplifying the importance of substrate quality in the Meentemeyer (1978) model. This observation does not conflict with the contention that fauna are relatively more important in the decay of low quality substrates (Seastedt, 1984c; Couteaux *et al.*, 1991; Tian *et al.*, 1997; Wardle *et al.*, 1997). Rather, it emphasizes that, when given a choice of resources, the fauna will select high-quality materials first, and thereby enhance the difference in decay rates between high- and low-quality materials available at the same site at the same time.

Swift *et al.* (1979) pointed out the absence of decomposer macrofauna and termites in more northerly environments. Certainly the absence of these groups might seem to be a logical covariate explaining the inability of decomposition to keep up with production. Decomposition is disproportionately faster in the tropics than in colder regions, and the reason is a relatively larger influence of fauna – mostly macroinvertebrates – in tropical regions. Also, in functional terms, arthropods in general and termites in particular have found evolutionary mechanisms that allow them to function very efficiently in warm temperatures and arid environments. Thus, organic matter is rapidly consumed in spite of a potential water limitation on NPP. However, in colder regions, the fauna has yet to solve the problem of the absence of oxygen in wet sites that compose a large percentage of northern terrestrial environments. Nor have the fauna developed the ability to maintain feeding activities under conditions that involve potential freezing of tissues. This last activity is worth emphasizing. Arthropods can remain active below freezing temperatures, and can survive extreme cold via supercooling mechanisms (Cannon and Block, 1988). However, feeding in general and especially particulate feeding cannot occur, as ice formation from this activity would be lethal to otherwise well-adapted organisms. Entire trophic groups must terminate feeding activities during periods when freezing is likely. Hence, while aerobic environments of the tropics are accessible to active feeding by arthropods throughout the year, the same is not true of the colder regions. This argues for a threshold-type model rather than a simple exponential relationship for decay. Increased densities of meso- and microfauna in cold regions do compensate, somewhat, for the lack of year-round activity, but unprocessed detritus and partially decomposed litter becomes a significant fraction of carbon storage in these systems (Schlesinger, 1977). Permafrost ultimately terminates any decomposer activity in the lower soil horizons of most northerly systems, while fairly frequent fire becomes a significant mechanism for carbon removal in boreal systems. These physical processes both exclude and diminish the significance of soil fauna in the carbon flux of these systems.

Decades ago, the scientific community spent a considerable amount of time pondering why the terrestrial world is green, and argued that top-down controls on herbivory were significant. Do we now suggest that northerly climates and temperate grasslands are 'brown' (i.e. contain large standing crops of organic matter) because of a similar mechanism? A more modern interpretation suggests that both predation and resource quality are significant in herbivore regulation, and detritivore regulation probably exhibits similar controls (Wardle and Lavelle, 1997). However, it is difficult to suggest that the quality of detritus of alpine or arctic tundra is, on average, of lower quality than that of a tropical forest. When data are finally available, the opposite may be true. However, microbial colonization of detritus may ultimately determine 'quality' from the decomposer food web's perspective. If the microbial growth is ice- and temperature-limited in the northern environments, then the quality of detritus in these regions may, from the perspective of the invertebrate food web, be inferior.

Conclusions

Comparisons of rangelands and forests across latitudinal gradients result in patterns in soil carbon storage that are not explained by simple climate–substrate quality properties. Two areas of research are needed to test the significance of the soil invertebrate fauna in these patterns. First, the interaction between effects of fauna on decomposition and mineralization and the quality of decomposing substrates needs to be explored across regional gradients. The tentative hypothesis is that macroinvertebrates and termites in warmer ecosystems selectively feed on high-quality substrates, thereby enhancing the difference in the rate of decay between high- and low-quality substrates. Second, the net effects of fauna on the decay and sequestration of older organic matter needs detailed experimentation. The role of the soil fauna in carbon storage is largely a net effect between stimulation of decomposition and indirect effects of fauna on physical and chemical protection of these substrates. Fauna–soil texture interactions may also be critically important in this process. The data suggest that the net effect of fauna on stabilization will be greater in colder regions. Whether this is due to the presence of newer soils (i.e. a variety of clays, a dominant cation exchange component, etc.), to temperature-related chemical and structural changes in organic residues, or to as yet undescribed differences, remains a significant research challenge.

Acknowledgements

Mentoring by D.A. Crossley, Jr., an activity that has continued for over 20 years, must be assigned at least partial blame for some of this. The Konza Prairie and Niwot Ridge Long-Term Ecological Research (LTER) programmes have funded a number of research efforts contributing to data and research ideas used in this chapter. Grizelle Gonzalez, Patrick Lavelle, David Coleman and an anonymous reviewer graciously commented upon an earlier draft of this manuscript, and I thank them for many positive suggestions and improvements. I apologize for attempting such an overwhelming topic, but I assume that readers would rather be frustrated than bored.

References

Aber, J.D. and Melillo, J.M. (1991) *Terrestrial Ecosystems*. Saunders College Publishing, Philadelphia, Pennsylvania.

Allen, T.F.H. and Starr, T.B. (1982) *Hierarchy: Perspectives for Ecological Complexity*. University of Chicago Press, Chicago, Illinois.

Anderson, J.M. (1988) Invertebrate-mediated transport processes in soils. *Agriculture, Ecosystems and Environment* 24, 5–19.

Beare, M.H., Coleman, D.C., Crossley, D.A., Hendrix, P.F. and Odum, E.P. (1995) A hierarchical approach to evaluating the significance of soil biodiversity to biogeochemical cycling. *Plant and Soil* 31, 1–18.

Beck, L. (1971) Bodenzoologische gliederung und charakterisierung des amazonischen Regenwaldes. *Amazoniania* 3, 69–132.

Berg, B. (1986) Nutrient release from litter and humus in coniferous forest soils – a mini review. *Scandinavian Journal of Forest Research* 1, 359–369.

Berg, B., Kniese, P.P., Roomer, R. and Verhoef, H.A. (1998) Long-term decomposition of successive organic substrate in a nitrogen-saturated Scots pine forest soil. *Forest Ecology and Management* 107, 159–172.

Blair, J.M., Parmelee, R.W. and Lavelle, P. (1995) Influences of earthworms on biogeochemistry. In: Hendrix, P.F. (ed.) *Earthworm Ecology and Biogeography in North America*. Lewis Press, Chelsea, Michigan, pp. 128–158.

Bonde, T.A., Christensen, B.T. and Cerri, C.C. (1992) Dynamics of soil organic matter as reflected by natural ^{13}C abundance in particle size fractions of forested and cultivated Oxisols. *Soil Biology and Biochemistry* 24, 275–277.

Bryant, D.M., Holland, E.A., Seastedt, T.R. and Walker, M.D. (1998) Analysis of litter decomposition in alpine tundra. *Canadian Journal of Botany* 76, 1295–1304.

Cannon, R.J. and Block, W. (1988) Cold tolerance of microarthropods. *Biological Reviews of the Cambridge Philosophical Society* 63, 23–77.

Casper, B.B. and Jackson, R.B. (1997) Plant competition belowground. *Annual Review of Ecology and Systematics* 28, 545–570.

Cerri, C.C., Vokoff, B. and Andreaux, F. (1991) Nature and behavior of organic matter in soils under natural forest, and after deforestation, burning and cultivation, near Manaus. *Forest Ecology and Management* 38, 247–257.

Chapin, F.S. (1980) The mineral nutrition of wild plants. *Annual Review of Ecology and Systematics* 11, 233–260.

Chapin, F.S. *et al.* (1996) Rangelands in a Changing Climate: Impacts, adaptations and mitigation. In: Watson, R.T., Zinyowera, M.C. and Moss, R.H. (eds) *Climate Change 1995: Impacts, Adaptations and Mitigation of Climate Change: Scientific-Technical Analyses*. Contribution of Working Group II to the Second Assessment Report of the Intergovernmental Panel on Climate Change. Cambridge University Press, Cambridge, pp. 133–158.

Cheng, B. and Wise, D.H. (1997) Responses of forest-floor fungivores to experimental food enhancement. *Pedobiologia* 41, 316–326.

Chone, T., Andreux, F., Correa, J.C., Volkoff, B. and Cerri, C.C. (1991) Changes in organic matter in an oxisol from the Central Amazonian forest during eight years as pasture, determined by ^{13}C isotopic composition. In: Berthelin, J. (ed.) *Diversity of Environmental Biogeochemistry*. Elsevier, New York, pp. 397–405.

Couteaux, M.M., Mousseau, M., Celerier, M.L. and Bottner, P. (1991) Increased atmospheric CO_2 and litter quality: decomposition of sweet chestnut leaf litter with animal food webs of different complexities. *Oikos* 61, 54–64.

Couteaux, M.M., Bottner, P. and Berg, B. (1995) Litter decomposition, climate and litter quality. *Trends in Ecology and Evolution* 10, 63–66.

Crossley, D.A., Jr (1977) The roles of terrestrial saprophagous arthropods in forest soils: Current status of concepts. In: Mattson, W.J. (ed.) *The Role of Arthropods in Forest Ecosystems*. Springer-Verlag, New York, pp. 49–56.

Crossley, D.A., Jr and Witkamp, M. (1964) Forest soil mites and mineral cycling. *Acarologia* (Supplement, International Congress of Acarology 1963), 137–143.

Curry, J.P. (1987a) The invertebrate fauna of grassland and its influence on productivity. I. The composition of the fauna. *Grass and Forage Science* 42, 103–120.

Curry, J.P. (1987b) The invertebrate fauna of grassland and its influence on productivity. III. Effects on soil fertility and plant growth. *Grass and Forage Science* 42, 325–341.

Decaens, T., Lavelle, P., Jimenez Jaen, J.J., Escobar, G. and Rippstein, G. (1994) Impact of land management on soil macrofauna in the Oriental Llanos of Colombia. *European Journal Soil Biology* 30, 157–168.

Douce, K.G. and Crossley, D.A., Jr (1977) Acarina abundance and community structure in an arctic coastal tundra. *Pedobiologia* 17, 32–42.

Feigl, B.J., Melillo, J. and Cerri, C.C. (1995) Changes in the origin and quality of soil organic matter after pasture introduction in Rondonia (Brazil). *Plant and Soil* 175, 21–29.

Fisher, M.J., Rao, I.M., Ayarza, M.A., Lascano, C.E., Sanz, J.I., Thomas, R.J. and Vera, R.R. (1994) Carbon storage by introduced deep-rooted grasses in the South American savannas. *Nature* 371, 236–238.

Fog, K. (1988) The effect of added nitrogen on the rate of decomposition of organic matter. *Biological Reviews* 63, 433–462.

Hassink, J., Bouwman, L.A., Zwart, K.B. and Brussaard, L. (1993a) Relationships between habitable pore space, soil biota and mineralization rates in grassland soils. *Soil Biology and Biochemistry* 25, 47–55.

Hassink, J., Bouwman, L.A., Zwart, K.B., Bloem, J. and Brussaard, L. (1993b) Relationships between soil texture, physical protection of organic matter, soil biota and C and N mineralization in grassland soils. *Geoderma* 57, 105–128.

Heal, O.W., Anderson, J.M. and Swift, M.J. (1997) Plant litter quality and decomposition: an historical overview. In: Cadisch, G. and Giller, K.E. (eds) *Driven by Nature: Plant Litter Quality and Decomposition.* CAB International, Wallingford, pp. 3–30.

Heneghan, L., Coleman, D.C., Zou, X., Crossley, D.A., Jr and Haines, B.L. (1999) Soil microarthropod contributions to decomposition dynamics: tropical and temperate comparisons of a single substrate (*Quercus prinus* L.). *Ecology* 80, 1873–1882.

Holtmeier, F.K. and Broll, G. (1992) The influence of tree islands and microtopography on pedoecological conditions in the forest–alpine tundra ecotone on Niwot Ridge, Colorado Front Range. *Arctic and Alpine Research* 24, 216–228.

Hunt, H.W., Ingham, E.R., Coleman, D.C., Elliott, E.T. and Reid, C.P.P. (1988) Nitrogen limitation of production and decomposition in prairie, mountain meadow, and pine forest. *Ecology* 69, 1009–1016.

James, S.W. (1991) Soil, nitrogen, phosphorus and organic matter processing by earthworms in tallgrass prairie. *Ecology* 72, 2101–2109.

Jastrow, J.D. (1996) Soil aggregate formation and the accrual of particulate and mineral-associated organic matter. *Soil Biology and Biochemistry* 28, 665–676.

Jastrow, J.D. and Miller, R.M. (in press) Soil aggregate stabilization and carbon sequestration: feedbacks through organomineral associations. In: Lal, R., Kimble, J.M., Follett, R.F. and Stewart, B.A. (eds) *Soil Processes and the Carbon Cycle.* CRC Press, New York.

Jastrow, J.D., Boutton, T.W. and Miller, R.M. (1996) Carbon dynamics of aggregate-associated organic matter estimated by carbon-13 natural abundance. *Soil Science Society of America Journal* 60, 801–807.

Jenny, H. (1930) A study on the influence of climate upon the nitrogen and organic matter content of the soil. University of Missouri College of Agriculture, Agricultural Experiment Station Research Bulletin 152, Columbia, Missouri.

Jenny, H., Gessel, S.P. and Bingham, F.T. (1949) Comparative study of decomposition rates of organic matter in temperate and tropical regions. *Soil Science* 68, 419–432.

Johnson, N.C. and Wedin, D.A. (1997) Soil carbon, nutrients and mycorrhizae during conversion of dry tropical forest to grassland. *Ecological Applications* 7, 171–182.

Kooistra, M.J. and van Noordwijk, M. (1996) Soil architecture and distribution of organic matter. In: Carter, M.R. and Steward, B.A. (eds) *Structure and Organic Matter Storage in Agricultural Soils*. CRC Press, Boca Raton, Florida, pp. 15–56.

Lavelle, P. (1988a) Biological processes and productivity of soils in the humid tropics. In: Dickinson, R.E. (ed.) *The Geophysiology of Amazonia: Vegetation and Climate Interactions*. John Wiley & Sons, New York, pp. 175–214.

Lavelle, P. (1988b) Assessing the abundance and role of invertebrate communities in tropical soils: aims and methods. *Journal of African Zoology* 102, 275–283.

Lavelle, P., Blanchart, E., Martin, A., Martin, S., Spaine, A., Toultain, F., Barios, I. and Schaefer, R. (1993) A hierarchical model for decomposition in terrestrial ecosystems – application to soils of the humid tropics. *Biotropica* 25, 130–150.

Leetham, J.W. and Milchunas, D.G. (1985) The composition and distribution of soil microarthropods in the shortgrass steppe in relation to soil water, root biomass, and grazing by cattle. *Pedobiologia* 28, 321–325.

Martin, A. (1991) Short and long-term effects of the endogenic earthworm *Millsonia anomala* of tropical savannas on soil organic matter. *Biology and Fertility of Soils* 11, 234–238.

Meentemeyer, V. (1978) Macroclimate and lignin control of litter decomposition rates. *Ecology* 59, 465–472.

Moore, J.C., Walter, D.E. and Hunt, H.W. (1988) Arthropod regulation of micro- and mesobiota in belowground detrital food webs. *Annual Review of Entomology* 33, 419–439.

O'Lear, H.A. (1996) Effects of altered soil water on litter decomposition and micro-arthropod composition. MS thesis, Kansas State University Manhattan, Kansas, USA.

O'Lear, H.A. and Seastedt, T.R. (1994) Landscape patterns of litter decomposition in alpine tundra. *Oecologia* 99, 95–101.

O'Lear, H.A., Seastedt, T.R., Briggs, J., Blair, J.M. and Ramundo, R.A. (1996) Fire and topographic effects on decomposition rates and nitrogen dynamics of buried wood in tallgrass prairie. *Soil Biology and Biochemistry* 28, 323–330.

Olson, J.S. (1963) Energy storage and the balance of producers and decomposers in eco-logical systems. *Ecology* 44, 322–331.

Petersen, H. and Luxton, M. (1982) A comparative analysis of soil fauna populations and their role in decomposition processes. *Oikos* 39, 287–388.

Parton, W.J., Schimel, D.S., Cole, C.V. and Ojima, D.S. (1987) Analysis of factors controlling soil organic matter levels in great plains grasslands. *Soil Science Society of America Journal* 51, 1173–1179.

Pastor, J., Stillwell, M.A. and Tilman, D. (1987) Little bluestem litter dynamics in Minnesota old fields. *Oecologia* 72, 327–330.

Pauker, S.J. and Seastedt, T.R. (1996) Effects of mobile tree islands on soil carbon storage in tundra ecosystems. *Ecology* 77, 2563–2567.

Prescott, C.E. (1995) Does nitrogen availability control rates of litter decomposition in forests? *Plant and Soil* 168, 83–88.

Santos, P.F., Elkins, N.Z., Steinberger, Y. and Whitford, W.G. (1984) A comparison of surface and buried *Larrea tridentata* leaf litter decomposition in North American hot deserts. *Ecology* 65, 278–284.

Schimel, D.S. (1995) Terrestrial ecosystems and the carbon cycle. *Global Change Biology* 1, 77–91.

Schlesinger, W.H. (1977) Carbon balance in terrestrial detritus. *Annual Review of Ecology and Systematics* 8, 51–81.

Schlesinger, W.H. and Pilmanis, A.M. (1998) Plant–soil interactions in deserts. *Biogeochemistry* 42, 169–187.

Scholes, R.J. and Archer, S.R. (1997) Tree–grass interactions in savannas. *Annual Review of Ecology and Systematics* 28, 517–544.

Seastedt, T.R. (1984a) Microarthropods of burned and unburned tallgrass prairie. *Journal of the Kansas Entomological Society* 57, 468–476.

Seastedt, T.R. (1984b) Belowground macroarthropods of annually burned and unburned tallgrass prairie. *American Midland Naturalist* 111, 405–408.

Seastedt, T.R. (1984c) The role of microarthropods in decomposition and mineralization processes. *Annual Review of Entomology* 29, 25–46.

Seastedt, T.R. and Crossley, D.A., Jr (1983) Naphthalene and artificial throughfall effects on forest floor nutrient dynamics: a field microcosm study. *Soil Biology and Biochemistry* 15, 159–165.

Seastedt, T.R., Todd, T.C. and James, S.W. (1987) Experimental manipulations of arthropod, nematode and earthworm communities in a North American tallgrass prairie. *Pedobiologia* 30, 9–18.

Seastedt, T.R., James, S.W. and Todd, T.C. (1988) Interactions among soil invertebrates, microbes and plant growth in tallgrass prairie. *Agriculture, Ecosystems and Environment* 24, 219–228.

Seastedt, T.R., Parton, W.J. and Ojima, D.S. (1992) Mass loss and nitrogen dynamics of decaying litter of grasslands: the apparent low nitrogen immobilization potential of root detritus. *Canadian Journal of Botany* 70, 384–391.

Stone, E.L., Harris, W.G., Brown, R.B. and Kuehl, R.J. (1993) Carbon storage in Florida Spodosols. *Soil Science Society of America Journal* 57, 179–182.

Swift, M.J., Heal, O.W. and Anderson, J.M. (1979) *Decomposition in Terrestrial Ecosystems.* Blackwell Scientific Publications, Oxford.

Tian, G., Brussaard, L., Kang, B.T. and Swift, M.J. (1997) Soil fauna-mediated decomposition of plant residues under constrained environmental and residue quality conditions. In: Cadisch, G. and Giller, K.E. (eds) *Driven by Nature: Plant Litter Quality and Decomposition.* CAB International, Wallingford, pp. 125–134.

Todd, T.C. (1996) Effects of management practices on nematode community structure in tallgrass prairie. *Applied Soil Ecology* 3, 235–246.

Vitousek, P.M. and Howarth, R.W. (1991) Nitrogen limitation on land and in the sea: How can it occur? *Biogeochemistry* 13, 87–115.

Walter, D.E. (1987) Belowground arthropods of semiarid grasslands. In: Capinera, J.L. (ed.) *Rangeland Pest Management: a Shortgrass Prairie Perspective.* Westview Press, Boulder, Colorado, pp. 271–290.

Walter, D.E., Kethley, J. and Moore, J.C. (1987) A heptane flotation method for recovering microarthropods from semiarid soils, with comparison to the Merchant-Crossley high gradient extraction method and estimates of microarthropod biomass. *Pedobiologia* 30, 221–232.

Wardle, D.A. and Lavelle, P. (1997) Linkages between soil biota, plant litter quality and decomposition. In: Cadisch, G. and Giller, K.E. (eds) *Driven by Nature: Plant Litter Quality and Decomposition.* CAB International, Wallingford, pp. 107–124.

Wardle, D.A., Bonner, K.I. and Nicholson, K.S. (1997) Biodiversity and plant litter: experimental evidence which does not support the view that enhanced species richness improves ecosystem function. *Oikos* 79, 247–258.

Webb, D.P. (1977) Regulation of deciduous forest litter decomposition by soil arthropod feces. In: Matson, W.J. (ed.) *The Role of Arthropods in Forest Ecosystems*. Springer-Verlag, New York, pp. 57–69.

Whitford, W.G., Meentemeyer, V., Seastedt, T.R., Cromack, K., Jr, Crossley, D.A., Jr, Santos, P., Todd, R.L. and Waide, J.B. (1981) Tests of the AET model of litter decomposition in deserts and a clear-cut forest. *Ecology* 62, 275–277.

Whitford, W.G., Stinnet, K. and Steinberger, Y. (1988) Effects of rainfall supplementation on microarthropods on decomposing roots in the Chihuahuan Desert. *Pedobiologia* 31, 147–155.

Whittaker, R.H. (1975) *Communities and Ecosystems*, 2nd edn. MacMillan, New York.

Willis, J.J., Braun, M., Surnelgi, P. and Toth, A. (1997) Does soil change cause vegetation change or vice versa? A temporal perspective from Hungary. *Ecology* 78, 740–750.

Zak, D.R., Tilman, D., Parmenter, R., Rice, C.W., Fisher, F.M., Vose, J., Milchunas, D. and Martin, C.M. (1994) Plant production and soil microorganisms in late successional ecosystems: a continental-scale study. *Ecology* 75, 2333–2347.

Zech, W. (1991) Litter decomposition and humification in forest soils. In: Van Breeman, N. (ed.) *Decomposition and Accumulation of Organic Matter in Terrestrial Ecosystems: Research Priorities and Approaches*. Ecosystems Research Report 1. Commission of the European Communities, Brussels, pp. 46–51.

Soil Processes and Global Change: Will Invertebrates Make a Difference?

16

P.M. Groffman and C.G. Jones

Institute of Ecosystem Studies, Box AB, Millbrook, NY 12545, USA

Introduction

Predicting the effects of global change on terrestrial ecosystems is one of the great challenges facing ecologists and soil scientists (Ojima *et al.*, 1991; Coleman *et al.*, 1992; Field *et al.*, 1992; Melillo *et al.*, 1993). Evaluating which components of global change (e.g. temperature, precipitation, atmospheric chemistry, land use change) will have significant effects on soil processes important to ecosystem function (e.g. nitrogen (N) mineralization, trace gas fluxes, soil carbon (C) storage, hydrologic N loss) is a current major focus (Walker and Steffen, 1996). These evaluations require identification of the key components mediating ecosystem response at the large (e.g. ecosystem, landscape, regional) scales relevant to global change issues.

In this paper, we ask two questions. First, do we need to consider soil invertebrates in an analysis of global change? Second, if so, what conceptual and practical tools can we use to facilitate this consideration? These questions arise from a dominant paradigm in soil ecology: soil invertebrates have large effects at small scales (mm–cm), but they do not 'matter' at larger scales (Anderson, 1995) (Fig. 16.1). This paradigm challenges us to 'make a case' for the importance of invertebrates in an arena where they are often neglected. In the sections that follow, we first present the case 'against' invertebrates, giving an example of how landscape and regional scale analysis of soil responses to global change can be accomplished without considering invertebrates. We then present two cases 'for' invertebrates, examples of global change questions that cannot be answered without explicitly considering invertebrates. We end by considering some general principles that can be used to help predict where and when invertebrates might play key roles in global change.

© CAB *International* 2000. *Invertebrates as Webmasters in Ecosystems*
(eds D.C. Coleman and P.F. Hendrix)

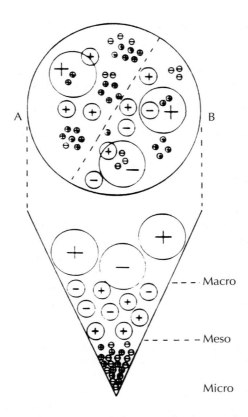

Fig. 16.1. Interactions of the functional domains of soil invertebrates. Micro-, meso- and macrofauna have functional domains of different sizes (represented by circle diameter). The fauna can have positive or negative effects on soil processes, e.g. N mineralization. If these effects are all in the same direction at one scale, they will have an observable effect at a larger scale (e.g. semicircle A). Alternatively, different effects can cancel each other out, resulting in no net effect at a larger scale (e.g. semicircle B) (see Chapter 1). (From Anderson, 1995.)

A dominant theme that emerges from this analysis is that the reason why there is uncertainty about the importance of invertebrates at large scales is that there have been too few ecosystem-scale experiments on the role of invertebrates. We suggest that if importance can be demonstrated at the ecosystem scale, then importance at landscape, regional and global scales is likely (Matson and Vitousek, 1990; Groffman, 1991). The importance of ecosystem-scale work is particularly appropriate for discussion in a volume highlighting the work of D.A. Crossley, Jr. Although his research has always focused on invertebrates, his determined dedication to ecosystem-scale questions and experiments is, in our view, what distinguishes him from the vast majority of past and present soil and invertebrate ecologists.

The Case 'Against' Invertebrates: Using the CENTURY Model to Study Soil Processes at a Regional Scale

The CENTURY model (Parton *et al.*, 1988) is perhaps the most widely used tool for landscape and regional-scale analysis of soil processes relevant to global change. The model was originally designed to simulate soil C levels in different agricultural management systems and does not include soil invertebrates. CENTURY has been modified to simulate a range of ecosystem processes (e.g. net primary production, N mineralization) and to run over large scales given spatially explicit input data on key driving variables (e.g. soil texture, land use). For example, Burke *et al.* (1994) simulated long-term changes in soil carbon levels, N mineralization and aboveground net primary productivity for the entire Great Plains region of the USA, driving the CENTURY model with input data on soil texture and land use organized in a geographic information system.

The widespread use of CENTURY and other models (e.g. Jenkinson and Rayner, 1977; Rastetter *et al.*, 1992) that do not include soil invertebrates, to address soil process questions relevant to global change, reflects the dominant paradigm that invertebrates are not important at large scales. Conceptually, we can justify this paradigm by assuming that faunal effects are subsumed within larger-scale controls such as soil texture, or that both positive and negative effects of invertebrates balance out at smaller scales, yielding no net effect at larger scales (e.g. Fig. 16.1). Because it is difficult to run complex models at large scales, model simplification is a practical necessity (i.e. avoid adding components unless absolutely necessary). To challenge this paradigm and approach, we must demonstrate specific cases where the failure to include invertebrates in these models critically limits our ability to depict real changes in the environment. Below we present examples of two global change questions that cannot be addressed without explicit consideration of soil invertebrates.

Case 1: Carbon Additions to Soils

The effect of adding C to soil is an old topic that has received renewed attention in recent years. There are studies dating back several decades which analysed the response of N mineralization and soil C storage to additions of different C-containing materials (Paul and Clark, 1996). However, recent concerns about increases in atmospheric CO_2 levels (Zak *et al.*, 1993; Curtis *et al.*, 1994) and waste disposal (Cole *et al.*, 1985) have created new interest in soil processing of added C. Many of these studies have found puzzling results. Despite strong expectations that C additions would result in a decrease in N availability, several studies have found increases in N availability (Zak *et al.*, 1993; Hungate *et al.*, 1997; Jones *et al.*, 1998).

We have conducted studies of soil responses to added C at the Hubbard Brook Experimental Forest in the White Mountains of New Hampshire. We added sodium acetate (32 g C m^{-2}) to forest plots dominated by American beech (*Fagus grandifolia*), sugar maple (*Acer saccharum*) and yellow birch (*Betula lutea*). Experiments with Na-bicarbonate showed that the levels of Na added with the acetate were not high enough to affect microbial biomass and activity (Groffman, 1999). Soils on the plots were acidic (pH 3.9) Typic Haplorthods. We then measured soil respiration, microbial biomass C and N content, soil inorganic N levels and potential net N mineralization in these plots 2, 10 and 58 days after addition. Additions were made in early August, and were compared with water-addition controls.

We expected to observe an increase in microbial respiration, microbial biomass C and N, and a decrease in potential net N mineralization and soil inorganic N levels in response to acetate addition. These expectations were based on the assumption that soil microbes are C-limited and that the acetate additions would stimulate activity (respiration), growth and N immobilization to support this growth (Paul and Clark, 1996). Complete results from the study are reported in Groffman (1999). As expected, we observed a significant increase in soil respiration in the acetate-addition plots. The increase was large enough to suggest that all the C that we added was respired away within 30 days. However, in contrast to expectations, we observed no increase in microbial biomass C and a significant increase in soil inorganic N levels (Table 16.1).

There are two possible explanations for our puzzling results. First, it is possible that the activity of soil microbes in these soils was C-limited but growth was limited by some other factor. While we expect that added substrates will stimulate both growth and respiration, it is possible for substrates to be metabolized without growth (Huntjens, 1979; Persson *et al.*, 1990; Scheu, 1993; Christensen *et al.*, 1996). While this scenario could explain the respiration and biomass results, it would not account for the observed increase in inorganic N.

Table 16.1. Soil inorganic N (ammonium only) and microbial biomass C, 10 and 58 days after application of acetate to soils at the Hubbard Brook Experimental Forest. Inorganic N was extracted with 2 M KCl and was analysed colorimetrically. Microbial biomass C was measured with the chloroform fumigation incubation method, and was calculated as the flush of CO_2 released following fumigation and incubation divided by 0.41. Respirable C was calculated as the CO_2 produced in a 10 day incubation of unfumigated soil. Soil nitrate was undetectable. Data from Groffman (1999).

Treatment	Inorganic N (mg N kg^{-1})		Microbial biomass C (mg C kg^{-1})		Respirable C (mg C kg^{-1})	
	10 days	58 days	10 days	58 days	10 days	58 days
Acetate	158 (22)[a]	8 (2)[a]	1903 (130)[a]	2191 (253)[a]	925 (31)[a]	m
Control	64 (17)[b]	12 (2)[a]	1793 (135)[a]	2200 (204)[a]	769 (79)[a]	662 (63)

Values are mean (standard error) of three replicate plots. Values within a column followed by different superscripts are significantly ($P < 0.05$) different in a one-way analysis of variance with a Fisher's protected least significant difference test.
m, missing data.

An alternative explanation is that the C addition stimulated both microbial growth and activity but the new biomass was grazed by soil invertebrates, resulting in no net increase in microbial biomass and an increase in N mineralization associated with invertebrate predation. Increases in faunal grazing of microbial biomass in response to C inputs are not uncommon (Christensen *et al.*, 1992; Rønn *et al.*, 1996) and grazing is often linked to N mineralization (Rosswall and Paustian, 1984; Hunt *et al.*, 1987; Kuikman *et al.*, 1990). The transient nature of the increase in inorganic N observed (it was gone by 58 days) is consistent with the idea of a 'pulse' of growth and predation stimulated by the C addition (Christensen *et al.*, 1992).

Hungate *et al.* (1997) observed similarly surprising responses to C additions (via exposure of whole ecosystems to elevated CO_2) in annual grasslands. They observed increases in N mineralization, plant N uptake and microbial biomass C in grasslands exposed to twice ambient CO_2 for a full growing season. They suggested that the increased CO_2 had led to increased soil water (via decreases in plant transpiration) and that the increase in soil water had stimulated microbial growth as well as protozoan grazing and associated N mineralization

These examples represent clear cases where a lack of consideration of the importance of soil invertebrates results in erroneous predictions for soil ecosystem responses to a common type of 'global change', i.e. C additions to soils. The dominant paradigm assumes that increases in C availability will increase microbial growth and that available N will be immobilized to support this growth. However, these examples, which are consistent with several studies that have observed increases in N availability in response to increases in soil C supply induced by increased levels of atmospheric CO_2 (Zak *et al.*, 1993; Jones *et al.*, 1998), suggest that food web dynamics can alter these expected responses. Clearly, regional and global scale analyses of the effects of increases in atmospheric CO_2 on soil N availability will need to consider soil invertebrate dynamics and food web structure.

Case 2: Earthworms and N Saturation

Earthworms have received considerable attention in soil ecology (Edwards and Bohlen, 1996). However, there have been surprisingly few studies that have addressed the importance of earthworms to fundamental ecosystem functions. We ask: does the presence of worms fundamentally affect the ability of ecosystems to retain atmospheric N, produce trace gases of interest to atmospheric chemistry and physics, and to sequester C?

We ask our question about the fundamental effects of earthworms on ecosystem N retention in the context of some puzzling patterns in watershed ecosystem N cycling, which have been observed in forests of the north-eastern USA. Rates of atmospheric deposition in this region have increased in recent decades (Lovett, 1994) just as rates of storage of N in aggrading forest biomass have decreased as forests throughout the region have matured. These changes have led to concern about N saturation, a condition of N excess where eco-systems are unable to absorb atmospheric N inputs (Aber *et al.*, 1989). In the advanced stages of N saturation, rates of hydrologic and gaseous N losses increase greatly.

Despite these concerns about regional N saturation, forests in the north-eastern USA have shown a surprising lack of N saturation. Several studies have found near 100% retention of large doses (> 500 kg N ha^{-1}) of N added to north-eastern forests over multiple years (Aber *et al.*, 1993; Norton *et al.*, 1994; Christ *et al.*, 1995; Magill *et al.*, 1996, 1997). At the Hubbard Brook Experimental Forest, where there is a more than 30-year record of streamwater chemistry, nitrate (NO_3^-) concentrations in the stream draining the biogeo-chemical reference watershed are at their lowest levels ever, despite the fact that the forests in this watershed have not been aggrading for at least 10 years (Driscoll *et al.*, unpublished observations). Decreases in NO_3^- concentrations over the past 20 years have also been observed at several other locations in the north-eastern USA (Goodale and Aber, 1998). Forests in the north-eastern USA appear to be much less susceptible to N saturation than forests in other regions which have shown multiple symptoms of N saturation, in many locations (Stams *et al.*, 1991; Johnson, 1992; Hedin *et al.*, 1995; Lepisto *et al.*, 1995; Wright *et al.*, 1995; Peterjohn *et al.*, 1996; Mitchell *et al.*, 1997). Interestingly, models of N saturation (e.g. Aber *et al.*, 1989) do not consider invertebrates.

We suggest that the surprising lack of N saturation in north-eastern USA forest soils may be linked to low levels of soil invertebrate, especially earthworm, activity in these soils. Many of these soils do not contain earthworms. Native species of worms are thought to be rare in this region due to their elimination by Pleistocene glaciation (Gates, 1976; Reynolds, 1995). Soils in this region are characterized by the presence of thick organic horizons and low pH (Zak *et al.*, 1994). Several studies have shown that colonization of these soils by earth-worms can result in rapid consumption of these organic horizons (Langmaid, 1964; Alban and Berry, 1994). Removal of these horizons, which have a high potential to absorb N, may account for observations where earthworms have stimulated N turnover and loss from forest soils (Haimi and Huhta, 1990; Haimi and Boucelham, 1991; Robinson *et al.*, 1992; Scheu and Parkinson, 1994).

Our own work suggests that ecosystems colonized by earthworms have fundamentally different abilities to retain and cycle N than ecosystems without worms. We have been studying oak-dominated forests along an urban to rural gradient in the New York City metropolitan area for the past 10 years. The gradient includes 27 oak-dominated forest stands in urban, suburban and rural areas which are located on the same geologic and soil substrates and are of similar age, composition and disturbance history (McDonnell and Pickett, 1990). Earthworms are much more abundant in the urban and suburban stands than in the rural stands, probably due to more inadvertent introduction by humans in the urban and suburban areas (Steinberg *et al.*, 1997). Worm-colonized stands are very rare in rural areas, but worm-free stands can be found in the urban and suburban areas.

By comparing rural and urban stands with no history of earthworms with worm-colonized urban stands, it is possible to draw inferences about the ecosystem-scale effects of earthworms on N cycling. In a laboratory microcosm study, soils from worm-colonized ecosystems had much higher rates of hydrologic (Fig. 16.2) and gaseous (Fig. 16.3) N loss than soils from ecosystems not colonized by earthworms. These results suggest that earthworm colonization may produce an ecosystem with a much lower capacity to retain atmospheric N. It is interesting to note that Fig. 16.2 suggests that the long-term effect of earthworm colonization is much greater than the short-term effect. While microcosms from worm-colonized soils had much higher N leaching than microcosms from worm-free soils, adding worms to microcosms did not increase leaching.

These studies suggest that the lack of earthworms may at least partially explain the lack of N saturation in the north-east USA relative to other regions where earthworms are more common. We suggest that colonization of northeastern USA forest soils by exotic species of earthworms may be a major factor affecting N dynamics in this region over the next 100 years or so (Groffman and Bohlen, 1999).

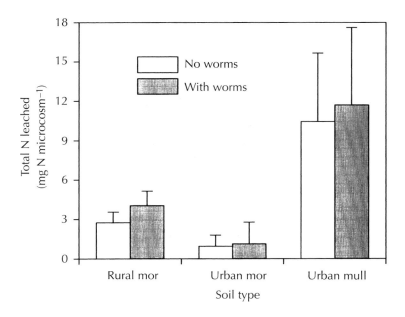

Fig. 16.2. Effects of earthworms on nitrogen leaching in an 8 week laboratory incubation of oak forest soil microcosms with (urban mull) and without (rural mor, urban mor) a previous history of earthworms. Over 80% of the nitrogen leached was as nitrate. Values are mean with standard errors. From Pouyat *et al.* (unpublished observations).

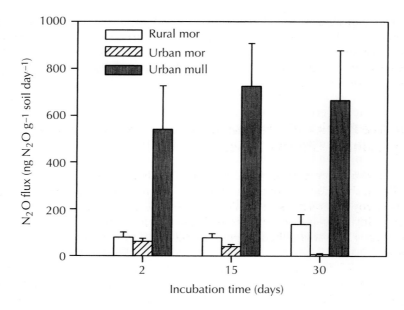

Fig. 16.3. Effects of earthworms on N_2O flux in a laboratory incubation of oak forest soil microcosms with (urban mull) and without (rural mor, urban mor) a previous history of earthworms. Values are means with standard errors. From Bohlen *et al.* (unpublished observations).

The Verdict?

Our examples of 'C additions' and 'earthworms and N saturation' suggest that there are indeed global change questions that require consideration of soil invertebrates. The analyses suggest that neglect of invertebrates in current global change efforts (e.g. with the CENTURY model) may arise more from practical concerns about model size and lack of data than from hard data suggesting that invertebrates are not important.

Indeed, some current critical uncertainties in regional C and N modelling might benefit from a consideration of invertebrates. For example, the question of whether conversion of ploughed agroecosystems to no-tillage agroecosystems results in an increase in soil C storage is a major uncertainty in several regions (Paustian *et al.*, 1997). Long-term studies (led by D.A. Crossley, Jr) in Georgia have shown that this conversion results in major changes in soil invertebrate food web structure (Beare *et al.*, 1992). Variation in the nature and extent of these changes may explain the confusing and variable soil C storage results that have been observed at different sites around the world.

Despite the evidence in support of the importance of invertebrates to global change, making a case for their inclusion in the global change research arena will continue to be a 'hard sell' (Coleman *et al.*, 1992). There is a critical need for more experiments that demonstrate the importance of invertebrates at the ecosystem scale. There is also a strong need to develop approaches to evaluating invertebrate dynamics at ecosystem, landscape and regional scales. Developing these approaches will greatly aid inclusion of invertebrates in large-scale models and analyses.

Can We Predict Where and When Invertebrates Will Be Important to Global Change?

The short answer is no. Given that there have been so few ecosystem-scale studies of the effects of invertebrates on ecosystem function, we cannot yet propose any general rules for predicting where and/or when it will be important to explicitly address invertebrates in global change assessments. However, we can suggest how we might develop a database for formulation of these general rules.

Figure 16.4 illustrates the challenge we face. Soil processes are driven by several key variables including parent material, time, climate, land use and C inputs (Amundson and Jenny, 1997). In global change studies, we need to address changes in climate and C inputs for a given land use type. In most cases, we hope to be able to assess the effect of these changes on soil processes without explicitly considering invertebrates, i.e. treat them as 'subsumed biology'. It is worth noting that this subsumed biology is susceptible to change, for example via invasion by exotics, or via changes in the distribution, abundance and activities of indigenous local biota. The challenge is to determine just where and when we need to consider this subsumed biology and where we can ignore it.

One way to develop a database on 'where/when invertebrates might matter' is to conduct detailed comparisons of model predictions with observed data. Figure 16.5 shows the general relationship between some observed values for a soil process, and that expected from driving variables for a given land use type. There are two ways that observations may differ from predictions. The first type of deviation is 'local' deviation of a point from a generally robust regression line. This type of residual deviation is suggestive of the importance of some local biota, perhaps an exotic, which must be considered. An example of this type of deviation would be north-eastern forests that are not responding to N saturation as expected because of a lack of earthworms. The second type of deviation is a more general deviation from a predicted pattern, i.e. the slope or intercept of the regression line is different. An example of this type of deviation would be the observation that N availability increased in response to C inputs when decreases were predicted (via altered trophic interactions in soil food webs). We suggest that continual evaluation of the way that observed values differ from predictions will eventually lead to the emergence of general principles for predicting just where and when soil invertebrates must be considered in global change analyses.

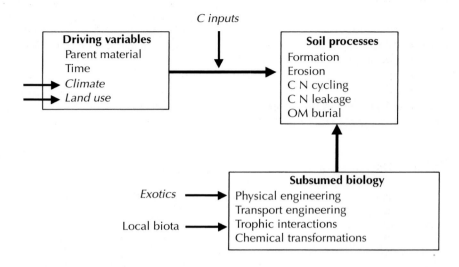

Fig. 16.4. Soil processes and global change as affected by driving variables and subsumed biology. Subsumed biology refers to activities of the biota, including ecosystem engineering activities (Jones *et al.*, 1994, 1997). Italicized font indicates variables affected by global change.

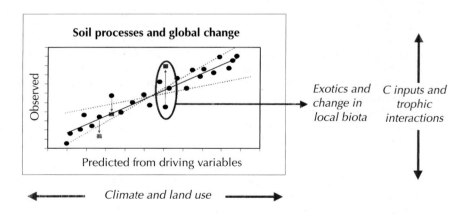

Fig. 16.5. General relationships between observed and predicted soil processes with global change for a given land use type. Italicized font indicates variables affected by global change. Two types of deviations from predictions are shown: local (residual deviations of a point from the regression line); and general (altered slope or intercept of the regression line). See text for detailed explanation.

Acknowledgements

It was a great honour to be asked to participate in the symposium and book honouring the work of D.A. Crossley, Jr. We thank Paul Hendrix and Dave Coleman for organizing this effort and we greatly value the interactions at the symposium and the comments of two anonymous reviewers which greatly increased the quality of this paper. We thank the Mary Flagler Cary Charitable Trust and the Andrew W. Mellon Foundation for financial support. This is a contribution to the programme of the Institute of Ecosystem Studies.

References

Aber, J.D., Nadelhoffer, K.J., Steudler, P. and Melillo, J.M. (1989) Nitrogen saturation in northern forest ecosystems. *BioScience* 39, 378–386.

Aber, J.D., Magill, A., Boone, R., Melillo, J.M., Steudler, P. and Bowden, R. (1993) Plant and soil responses to chronic nitrogen additions at the Harvard Forest, Massachusetts. *Ecological Applications* 3, 156–166.

Alban, D.H. and Berry, E. (1994) Effects of earthworm invasion on morphology, carbon, and nitrogen of a forest soil. *Applied Soil Ecology* 1, 243–249.

Amundson, R. and Jenny, H. (1997) On a state factor model of ecosystems. *BioScience* 47, 536–543.

Anderson, J.M. (1995) Soil organisms as engineers: Microsite modulation of macroscale processes. In: Jones, C.G. and Lawton, J.H. (eds) *Linking Species and Ecosystems.* Chapman & Hall, New York, pp. 94–106.

Beare, M.H., Parmelee, R.W., Hendrix, P.F., Cheng, W., Coleman, D.C. and Crossley, D.A., Jr (1992) Microbial and faunal interactions and effects on litter nitrogen and decomposition in agro-ecosystems. *Ecological Monographs* 62, 569–591.

Burke, I.C., Lauenroth, W.K., Parton, W.J. and Cole, C.V. (1994) Interactions of land use and ecosystem structure and function: a case study in the central Great Plains. In: Groffman, P.M. and Likens, G.E. (eds) *Integrated Regional Models: Interactions Between Humans and Their Environment.* Chapman & Hall, New York, pp. 79–95.

Christ, M., Zhang, Y., Likens, G.E. and Driscoll, C.T. (1995) Nitrogen retention capacity of a northern hardwood forest soil under ammonium sulfate additions. *Ecological Applications* 5, 802–812.

Christensen, S., Griffiths, B.S., Ekelund, F. and Rønn, R. (1992) Huge increases in bacteriovores on freshly killed barley roots. *FEMS Microbiology Ecology* 86, 303–310.

Christensen, S., Rønn, R., Ekelund, F., Andersen, B., Damgaard, J., Friberg-Jensen, U., Jensen, L., Kiil, H., Larsen, B., Larsen, J., Riis, C., Thingsgaard, K., Thirup, C., Tom-Petersen, A. and Vesterdal, L. (1996) Soil respiration profiles and protozoan enumeration agree as microbial growth indicators. *Soil Biology and Biochemistry* 11, 865–868.

Cole, D.W., Henry, C.L. and Nutter, W.L. (1985) *The Forest Alternative for Treatment and Utilization of Municipal and Industrial Wastes.* University of Washington Press, Seattle, Washington.

Coleman, D.C., Odum, E.P. and Crossley, D.A., Jr (1992) Soil biology, soil ecology, and global change. *Biology and Fertility of Soils* 14, 104–111.

Curtis, P.S., O'Neill, E.G., Teeri, J.A., Zak, D.R. and Pregitzer, K.S. (1994) Belowground responses to rising atmospheric CO_2: implications for plants, soil biota and ecosystem processes. *Plant and Soil* 165, 1–6.

Edwards, C.A. and Bohlen, P.J. (1996) *Biology and Ecology of Earthworms*. Chapman & Hall, London.

Field, C.B., Chapin, F.S., III, Matson, P.A. and Mooney, H.A. (1992) Responses of terrestrial ecosystems to the changing atmosphere: a resource-based approach. *Annual Review of Ecology and Systematics* 23, 201–235.

Gates, G.E. (1976) More on earthworm distribution in North America. *Proceedings of the Biological Society of Washington* 89, 467–476.

Goodale, C.I. and Aber, J.D. (1998) The long-term effects of disturbance on nitrogen cycling in northern hardwood forests. *Abstracts of the 83rd Annual Meeting of Ecological Society of America*, Baltimore, Maryland, p. 170.

Groffman, P.M. (1991) Ecology of nitrification and denitrification in soil evaluated at scales relevant to atmospheric chemistry. In: Whitman, W.B. and Rogers, J. (eds) *Microbial Production and Consumption of Greenhouse Gases: Methane, Nitrogen Oxides and Halomethanes*. American Society of Microbiology, Washington, DC, pp. 201–217.

Groffman, P.M. (1999) Carbon additions increase nitrogen availability in northern hardwood forest soils. *Biology and Fertility of Soils* 29, 430–433.

Groffman, P.M. and Bohlen, P.J. (1999) Soil and sediment biodiversity: cross-system comparisons and large scale effects. *BioScience* 49, 139–148.

Haimi, J. and Boucelham, M. (1991) Influence of a litter feeding earthworm, *Lumbricus rubellus*, on soil processes in a simulated forest floor. *Pedobiologia* 35, 247–256.

Haimi, J. and Huhta, V. (1990) Effects of earthworms on decomposition processes in raw humus forest soil: a microcosm study. *Biology and Fertility of Soils* 10, 178–183.

Hedin, L.O., Armesto, J.J. and Johnson, A.H. (1995) Patterns of nutrient loss from unpolluted, old-growth temperate forests: evaluation of biogeochemical theory. *Ecology* 76, 493–509.

Hungate, B.A., Chapin, F.S., III, Zhong, H., Holland, E.A. and Field, C. (1997) Stimulation of grassland nitrogen cycling under carbon dioxide enrichment. *Oecologia* 109, 149–153.

Hunt, H.W., Coleman, D.C., Ingham, R.E., Elliott, E.T., Moore, J.C., Rose, S.L., Reid, C.P.P. and Morley, C.R. (1987) The detrital food web in a shortgrass prairie. *Biology and Fertility of Soils* 3, 17–68.

Huntjens, J.L. (1979) A sensitive method for continuous measurement of the carbon dioxide evolution rate of soil samples. *Plant and Soil* 53, 529–534.

Jenkinson, D.S. and Rayner, J.H. (1977) The turnover of soil organic matter in some of the Rothamsted Classical Experiments. *Soil Science* 123, 298–305.

Johnson, D.W. (1992) Nitrogen retention in forest soils. *Journal of Environmental Quality* 21, 1–12.

Jones, C.G., Lawton, J.H. and Shachak, M. (1994) Organisms as ecosystem engineers. *Oikos* 69, 373–386.

Jones, C.G., Lawton, J.H. and Shachak, M. (1997) Positive and negative effects of organisms as physical ecosystem engineers. *Ecology* 78, 1946–1957.

Jones, T.H., Thompson, L.J., Lawton, J.H., Bezemer, T.M., Bardgett, R.D., Blackburn, T.M., Bruce, K.D., Cannon, P.F., Hall, G.S., Howson, G., Jones, C.G., Kampichler, C., Kandeler, E. and Ritchie, D.A. (1998) Impacts of rising atmospheric carbon dioxide on model terrestrial ecosystems. *Science* 280, 441–443.

Kuikman, P.J., Jansen, A.G. and Van Veen, J.A. (1990) Protozoan grazing and the turnover of soil organic carbon and nitrogen in the presence of plants. *Biology and Fertility of Soils* 10, 22–28.

Langmaid, K.K. (1964) Some effects of earthworm invasion in virgin podzols. *Canadian Journal of Soil Science* 44, 34–37.

Lepisto, A., Andersson, L., Arheimer, B. and Sundblad, K. (1995) Influence of catchment characteristics, forestry activities and deposition on nitrogen export from small forested catchments. *Water, Air and Soil Pollution* 84, 81–102.

Lovett, G.M. (1994) Atmospheric deposition of nutrients and pollutants in North America: an ecological perspective. *Ecological Applications* 4, 629–650.

Magill, A.H., Downs, M.R., Nadelhoffer, K.J., Hallett, R.A. and Aber, J.D. (1996) Forest ecosystem response to four years of chronic nitrate and sulfate additions at Bear Brooks Watershed, Maine, USA. *Forest Ecology and Management* 84, 1–3.

Magill, A.H., Aber, J.D., Hendricks, J.J., Bowden, R.D., Melillo, J.M. and Steudler, P.A. (1997) Biogeochemical response of forest ecosystems to simulated chronic nitrogen deposition. *Ecological Applications* 7, 402–415.

Matson, P.A. and Vitousek, P.M. (1990) Ecosystem approach to a global nitrous oxide budget. *BioScience* 40, 667–672.

McDonnell, M.J. and Pickett, S.T.A. (1990) The study of ecosystem structure and function along urban–rural gradients: an unexploited opportunity for ecology. *Ecology* 71, 1231–1237.

Melillo, J.M., McGuire, A.D., Kicklighter, D.W., Moore, B., III, Vorosmarty, C.J. and Schloss, A.L. (1993) Global climate change and terrestrial net primary production. *Nature* 363, 234–240.

Mitchell, M.J., Iwatsubo, G., Ohrui, K. and Nakagawa, Y. (1997) Nitrogen saturation in Japanese forests: an evaluation. *Forest Ecology and Management* 97, 39–51.

Norton, S.A., Kahl, J.S., Fernandez, I.J., Rustad, L.E., Scofield, J.P. and Haines, T.A. (1994) Response of the West Bear Brook Watershed, Maine, USA, to the addition of $(NH_4)_2SO_4$: 3-year results. *Forest Ecology and Management* 68, 61–73.

Ojima, D.S., Kittel, T.G.F., Rosswall, R. and Walker, B.H. (1991) Critical issues for understanding global change effects on terrestrial ecosystems. *Ecological Applications* 1, 316–325.

Parton, W.J., Stewart, J.W.B. and Cole, C.V. (1988) Dynamics of carbon, nitrogen, phosphorus and sulfur in cultivated soils: a model. *Biogeochemistry* 5, 109–131.

Paul, E.A. and Clark, F.E. (1996) *Soil Microbiology and Biochemistry*, 2nd edn. Academic Press, San Diego.

Paustian, K., Collins, H.P. and Paul, E.A. (1997) Management controls on soil carbon. In: Paul, E.A., Elliott, E.T., Paustian, K. and Cole, C.V. (eds) *Soil Organic Matter in Temperate Agroecosystems: Long-Term Experiments in North America*. CRC Press, Boca Raton, Florida, pp. 15–49.

Persson, A., Molin, G. and Weibull, C. (1990) Physiological and morphological changes induced by nutrient limitation of *Pseudomonas fluorescens* 378 in continuous culture. *Applied and Environmental Microbiology* 56, 686–692.

Peterjohn, W.T., Adams, M.B. and Gilliam, F.S. (1996) Symptoms of nitrogen saturation in two Appalachian hardwood forest ecosystems. *Biogeochemistry* 35, 507–522.

Rastetter, E.B., McKane, R.B., Shaver, G.R. and Melillo, J.M. (1992) Changes in C storage by terrestrial ecosystems: how C–N interactions restrict responses to CO_2 and temperature. *Water, Air and Soil Pollution* 64, 327–344.

Reynolds, J.W. (1995) The distribution of earthworms (Annelida, Oligochaeta) in North

America. In: Mishra, P.C., Behera, N., Senapati, B.K. and Guru, B.C. (eds) *Advances in Ecology and Environmental Sciences.* Ashish Publishing House, New Delhi, pp. 133–153.

Robinson, C.H., Ineson, P., Piearce, T.G. and Rowland, A.P. (1992) Nitrogen mobilization by earthworms in limed peat soils under *Picea sitchensis. Journal of Applied Ecology* 29, 226–237.

Rønn, R., Griffiths, B.S., Ekelund, F. and Christensen, S. (1996) Spatial distribution and successional pattern of microbial activity and micro-faunal populations on decomposing barley roots. *Journal of Applied Ecology* 33, 662–672.

Rosswall, T. and Paustian, K. (1984) Cycling of nitrogen in modern agricultural systems. *Plant and Soil* 76, 3–21.

Scheu, S. (1993) Analysis of the microbial nutrient status in soil microcompartments; earthworm feces from a basalt–limestone gradient. *Geoderma* 56, 55–58.

Scheu, S. and Parkinson, D. (1994) Effects of earthworms on nutrient dynamics, carbon turnover and microorganisms in soils from cool temperate forests of the Canadian Rocky Mountains – laboratory studies. *Applied Soil Ecology* 1, 113–125.

Stams, A.J.M., Booltink, H.W.G., Lutke-Schipholt, I.J., Beemsterboer, B., Woittiez, J.R.W. and van Breemen, N. (1991) A field study on the fate of ^{15}N-ammonium to demonstrate nitrification of atmospheric ammonium in an acid forest soil. *Biogeochemistry* 13, 241–255.

Steinberg, D.A., Pouyat, R.V., Parmelee, R.W. and Groffman, P.M. (1997) Earthworm abundance and nitrogen mineralization rates along an urban–rural land use gradient. *Soil Biology and Biochemistry* 29, 427–430.

Walker, B.H. and Steffen, W. (eds) (1996) *Global Change and Terrestrial Ecosystems.* Cambridge University Press, Cambridge.

Wright, R.F., Roelofs, J.G.M., Bredemeier, M., Blanck, K., Boxman, A.W., Emmett, B.A., Gundersen, P., Hultberg, H., Kjønaas, O.J., Moldan, F., Tietema, A., van Breemen, N. and van Dijk, H.F.G. (1995) NITREX: responses of coniferous forest ecosystems to experimentally changed deposition of nitrogen. *Forest Ecology and Management* 71, 163–169.

Zak, D.R., Pregitzer, K.S., Curtis, P.S., Teeri, J.A., Fogel, R. and Randlett, D.L. (1993) Elevated atmospheric CO_2 and feedback between carbon and nitrogen cycles. *Plant and Soil* 151, 105–117.

Zak, D.R., Tilman, D., Parmenter, R.R., Rice, C.W., Fisher, F.M., Vose, J., Milchunas, D. and Martin, C.W. (1994) Plant production and soil microorganisms in late-successional ecosystems: a continental-scale study. *Ecology* 75, 2333–2347.

Index